Challenges and New Trends in Power Electronic Devices Reliability

Challenges and New Trends in Power Electronic Devices Reliability

Editors

Elio Chiodo

Pasquale De Falco

Luigi Pio Di Noia

MDPI • Basel • Beijing • Wuhan • Barcelona • Belgrade • Manchester • Tokyo • Cluj • Tianjin

Contents

Preface to "Challenges and New Trends in Power Electronic Devices Reliability"

This book is a collection of papers regarding recent approaches to the evaluation of reliability in power electronic devices.

The book is addressed to all academic and industrial researchers involved in the evaluation of risk, reliability, and availability in power electronic converters for energy and transportation applications.

The authors involved in the book are as follows:

Alberto Belli, Università Politecnica delle Marche, Italy

Alessandro Ciocia, Politecnico di Torino, Italy

Angela Amato, Politecnico di Torino, Italy

Ariya Sangwongwanich, Aalborg University, Denmark

Baoming Li, Nanjing University of Science and Technology, China

Bassel Diban, University of Bologna, Italy

Carmelo Barbagallo, University of Catania, Italy

Chenyuan Wang, Wuhan University, China

Chuankun Wang, Wuhan University, China

Chun-Te Lee, Cheng-Shiu University, Taiwan

Elio Chiodo, University of Napoli Federico II, Italy

Federica Pinti, Università Politecnica delle Marche, Italy

Filippo Spertino, Politecnico di Torino, Italy

Francesco Iannuzzo, Aalborg University, Denmark

Frede Blaabjerg, Aalborg University, Denmark

Gabriele Casali, Politecnico di Torino, Italy

Gabriele Malgaroli, Politecnico di Torino, Italy

Giacomo Scelba, University of Catania, Italy

Giovanni Mazzanti, University of Bologna, Italy

Giovanni Susinni, University of Catania, Italy

Giuseppe Scarcella, University of Catania, Italy

Han Yang, National Space Science Center, Chinese Academy of Sciences, China

Jianwei Han, Institute of Astronomy and Space, University of Chinese Academy of Sciences, China

Jiufu Wei, Nanjing University of Science and Technology, China

Lie Li, Wuhan University, China

Lorenzo Palma, Università Politecnica delle Marche, Italy

Luigi Pio Di Noia, University of Napoli Federico II, Italy

Mario Cacciato, University of Catania, Italy

Massimo Gattari, iGuzzini Illuminazione S.p.A, Italy

Monika Sandelic, Aalborg University, Denmark

Paola Pierleoni, Università Politecnica delle Marche, Italy

Pasquale De Falco, University of Napoli Parthenope, Italy

Ping-Tsan Ho, Cheng-Shiu University, Taiwan

Qian Chen, National Space Science Center, Chinese Academy of Sciences, China

Rui Chen, Institute of Astronomy and Space, University of Chinese Academy of Sciences, China

Santi Agatino Rizzo, University of Catania, Italy

Xiaoxin Wu, Wuhan University, China

Xuan Wang, National Space Science Center, Chinese Academy of Sciences, China

Yanan Liang, National Space Science Center, Chinese Academy of Sciences, China

Yigang He, Wuhan University, China

Zhenxiao Li, Nanjing University of Science and Technology, China

The guest editors Prof. Elio Chiodo, Dr. Pasquale De Falco, and Dr. Luigi Pio Di Noia would like to thank Ms. Stephanie Wang, Assistant Managing Editor of MDPI, for her precious collaboration in the success of this Special Issue.

Elio Chiodo, Pasquale De Falco, Luigi Pio Di Noia
Editors

electronics

MDPI

Editorial

Challenges and New Trends in Power Electronic Devices Reliability

Elio Chiodo [1,*], Pasquale De Falco [2,*] and Luigi Pio Di Noia [3,*]

[1] Department of Industrial Engineering, University of Naples Federico II, 80125 Naples, Italy
[2] Department of Engineering, University of Naples Parthenope, 80143 Naples, Italy
[3] Department of Electrical Engineering and Information Technology, University of Naples Federico II, 80125 Naples, Italy
* Correspondence: chiodo@unina.it (E.C.); pasquale.defalco@uniparthenope.it (P.D.F.); luigipio.dinoia@unina.it (L.P.D.N.)

check for updates

Citation: Chiodo, E.; De Falco, P.; Di Noia, L.P. Challenges and New Trends in Power Electronic Devices Reliability. *Electronics* **2021**, *10*, 925. https://doi.org/10.3390/electronics10080925

Received: 29 March 2021
Accepted: 7 April 2021
Published: 13 April 2021

1. Introduction

Power electronic devices are expected to play an ever more fundamental role in unlocking the potentialities of smart power systems and in developing more electric ground and air transportation systems. The reliability of power electronic devices at different hierarchical levels (single component, single device, installation and system) becomes a crucial point in this framework, as failures may determine technical, economical and safety issues that should be carefully addressed at the design and maintenance stages.

Power electronic devices are subject to thermal, electrical and mechanical stresses, which can be assessed through consolidated, traditional techniques [1–4]. However, today these devices are expected to operate under challenging environmental conditions (e.g., high altitudes in more electric aircrafts or high temperatures on photovoltaic (PV) installations), undermining the effectiveness of traditional approaches that are typically based on historical failure data, fault rates or past observed scenarios. In fact, the rapid evolution of power electronic technologies and the ever more challenging operating frameworks pose severe limitations on the trustworthiness of available reliability data, as they are typically related to incoherent operating conditions [1–4].

2. The Special Issue

The *Electronics* Special Issue on Challenges and New Trends in Power Electronic Devices Reliability collected contributions on methodologies and approaches for the analysis of reliability in power electronic devices at different hierarchical levels, covering multiple fields of applications ranging from PVs to motor drives to LED lamps. The editorial resume of contributions is listed below:

- The reliability of PV AC/DC converters was analyzed in [5], with a finalization related to maintenance and repair activities. A 46 MW large-scale grid-connected PV plant was considered in the reliability analysis, with a detailed comparison to the maintenance reports collected for a three-year span, also leading to an estimation of the overall losses due to unavailability of the system. The reliability of a hybrid PV–battery installation was considered in [6], with a comparison of the DC- and AC-coupled configurations. Therefore, DC/DC and DC/AC converter units were considered, and the reliability was evaluated at different hierarchical levels;
- The gallium nitride high-electron mobility transistor (GaN HEMT) technology was assessed in [7] with reference to its typical cascode structure. Single-event effects on the cascode were studied using a technical computer-aided design for heavy ion experiments. An interesting outcome of the research was that the enhancement of the ionization mechanism at the gate edge may increase the performance in terms of leakage;

- The lifetime estimation of discrete SiC power MOSFETs for motor drives was performed in [8]. The thermal stress was modeled through a tool that exploited the Coffin–Manson theory, rainflow counting and Miner's rule for the estimation of the lifetime under variable mission profiles, allowing for the generalization of the procedure to different final purposes;
- A multi-chip IGBT module failure monitoring method was presented in [9] to track solder layer fatigue or bond wire fall-off. The method was based on the module transconductance, which was able to represent the failure mechanism of the IGBT and the relationship between chip failure, bond wire failure and the transmission characteristic curve of the IGBT module. Thermal stress on the IGBT operating in microgrids was evaluated in [10]. The aging process of the IGBT was considered in an online evaluation through a fusion algorithm that combined condition monitoring and reliability evaluation. An electrothermal coupling model obtained the junction temperature data, considering the microgrid inverter topology and the IGBT features, and a segmented long short-term memory algorithm was exploited to predict the aging process;
- The work in [11] quantified the impact of overvoltages on high-power thyristors, investigating the effects of the initial voltage of the energy storage capacitor, the discharge time intervals and the load resistance on the reverse recovery currents. On the basis of the outcomes, an improved topology was then developed to damp the surge energy and to mitigate the reverse recovery currents;
- A reliability forecasting model that targeted the insulation of power components subjected to varying harmonics was presented in [12]. The model focused on the role played by low percentiles of time to failure, typically selected as the rated life in the framework of the modern probabilistic design of components, and it treated all the odd voltage harmonics from the fifth to the twenty-fifth;
- LED lamps were investigated in [13,14]. The design of a new LED lamp concept, with attention to energy-saving, comfort and reliability purposes was presented in [13]. The lifetime of LED lamps was considered in [14], with a Norris–Landberg model that analyzed the probability of failure and the lifetime with reference to solder joint cracks. Over 1800 events were considered in the experimental validation process;
- A review of the state of the art of the condition monitoring of power electronics devices is in [15].

3. Discussion and Future Contributions

The potentialities unlocked by high-performance power electronic devices in smart power systems and more electric transportation are well recognized, although studies on the reliability of these components in challenging and severe conditions are far from being established. The integration of power electronics in different topologies and hybrid systems, together with the analysis of still unexplored fields such as full electric aircrafts and decentralized smart grids, will likely be the object of extensive research in the next decade. Future contributions on the reliability of power system components cannot neglect these aspects and should try to overcome the common bottleneck of the lack of field data in real-world challenging contexts.

Author Contributions: E.C., P.D.F. and L.P.D.N. were equally involved in the conceptualization of the Special Issue and in its entire editorial process. Writing—original draft preparation, P.D.F.; writing—review and editing, E.C., P.D.F. and L.P.D.N. All authors have read and agreed to the published version of the manuscript.

Funding: This research received no external funding.

Acknowledgments: E.C., P.D.F. and L.P.D.N. gratefully thank the authors for providing their contributions to the Special Issue and all the reviewers involved at any stage of the peer review processes. E.C., P.D.F. and L.P.D.N. also acknowledge the *Electronics* Editorial Office for maintaining the rigorous revision process and for ensuring the timely publication of the contents.

Conflicts of Interest: The authors declare no conflict of interest.

References

1. Yang, S.; Xiang, D.; Bryant, A.; Mawby, P.; Ran, L.; Tavner, P. Condition monitoring for device reliability in power electronic converters: A review. *IEEE Trans. Power Electron.* **2010**, *25*, 2734–2752. [CrossRef]
2. Song, Y.; Wang, B. Survey on reliability of power electronic systems. *IEEE Trans. Power Electron.* **2012**, *28*, 591–604. [CrossRef]
3. Wang, H.; Liserre, M.; Blaabjerg, F. Toward reliable power electronics: Challenges, design tools, and opportunities. *IEEE Ind. Electron. Mag.* **2013**, *7*, 17–26. [CrossRef]
4. Falck, J.; Felgemacher, C.; Rojko, A.; Liserre, M.; Zacharias, P. Reliability of power electronic systems: An industry perspective. *IEEE Ind. Electron. Mag.* **2018**, *12*, 24–35. [CrossRef]
5. Spertino, F.; Amato, A.; Casali, G.; Ciocia, A.; Malgaroli, G. Reliability Analysis and Repair Activity for the Components of 350 kW Inverters in a Large Scale Grid-Connected Photovoltaic System. *Electronics* **2021**, *10*, 564. [CrossRef]
6. Sandelic, M.; Sangwongwanich, A.; Blaabjerg, F. Reliability Evaluation of PV Systems with Integrated Battery Energy Storage Systems: DC-Coupled and AC-Coupled Configurations. *Electronics* **2019**, *8*, 1059. [CrossRef]
7. Liang, Y.; Chen, R.; Han, J.; Wang, X.; Chen, Q.; Yang, H. The Study of the Single Event Effect in AlGaN/GaN HEMT Based on a Cascode Structure. *Electronics* **2021**, *10*, 440. [CrossRef]
8. Barbagallo, C.; Rizzo, S.A.; Scelba, G.; Scarcella, G.; Cacciato, M. On the Lifetime Estimation of SiC Power MOSFETs for Motor Drive Applications. *Electronics* **2021**, *10*, 324. [CrossRef]
9. Wang, C.; He, Y.; Wang, C.; Li, L.; Wu, X. Multi-Chip IGBT Module Failure Monitoring Based on Module Transconductance with Temperature Calibration. *Electronics* **2020**, *9*, 1559. [CrossRef]
10. Wang, C.; He, Y.; Wang, C.; Wu, X.; Li, L. A Fusion Algorithm for Online Reliability Evaluation of Microgrid Inverter IGBT. *Electronics* **2020**, *9*, 1294. [CrossRef]
11. Wei, J.; Li, Z.; Li, B. Investigation of Reverse Recovery Current of High-Power Thyristor in Pulsed Power Supply. *Electronics* **2020**, *9*, 1292. [CrossRef]
12. Mazzanti, G.; Diban, B.; Chiodo, E.; De Falco, P.; Di Noia, L.P. Forecasting the Reliability of Components Subjected to Harmonics Generated by Power Electronic Converters. *Electronics* **2020**, *9*, 1266. [CrossRef]
13. Lee, C.-T.; Ho, P.-T. Energy-Saving Research on New Type of LED Sensor Lamp with Low-Light Mode. *Electronics* **2020**, *9*, 1649. [CrossRef]
14. Pinti, F.; Belli, A.; Palma, L.; Gattari, M.; Pierleoni, P. Validation of Forward Voltage Method to Estimate Cracks of the Solder Joints in High Power LED. *Electronics* **2020**, *9*, 920. [CrossRef]
15. Susinni, G.; Rizzo, S.A.; Iannuzzo, F. Two Decades of Condition Monitoring Methods for Power Devices. *Electronics* **2021**, *10*, 683. [CrossRef]

electronics

MDPI

Article

Reliability Evaluation of PV Systems with Integrated Battery Energy Storage Systems: DC-Coupled and AC-Coupled Configurations

Monika Sandelic *, **Ariya Sangwongwanich** and **Frede Blaabjerg**

Department of Energy Technology, Aalborg University, 9220 Aalborg, Denmark; ars@et.aau.dk (A.S.); fbl@et.aau.dk (F.B.)

* Correspondence: mon@et.aau.dk; Tel.: +45-9181-2347

Received: 23 August 2019; Accepted: 17 September 2019; Published: 19 September 2019

Abstract: Deployment of a battery energy storage system for the photovoltaic (PV) application has been increasing at a fast rate. Depending on the number of power conversion units and their type of connection, the PV-battery system can be classified into DC- and AC-coupled configurations. The number of the components and their electrical loading directly affects the reliability of each of the configurations. Hence, in order to assure high efficiency and lifetime of the PV-battery system, reliability assessment of power conversion units (representing the most reliability-critical system components) is necessary. With respect to that, in this paper, a reliability assessment of the PV-battery system is performed and a comparison of the DC- and AC-coupled configuration reliability is conducted. In the analysis, all parts of the power conversion system, i.e., DC/DC and DC/AC converter units, are taken into consideration and component-, converter- and system-level reliability is assessed. A case study of 6 kW PV system with integrated 3 kW/7.5 kWh battery system has shown that higher reliability is achieved for DC-coupled configuration. The obtained results indicate that the probability of failure for the 15% of the population for DC-coupled configuration occurs 7 years later than that is a case for AC-coupled configuration. Finally, the presented analysis can serve as a benchmark for lifetime and reliability assessment of power conversion units in PV-battery systems for both configuration types. It provides information about differences in electrical and thermal loading of the power conversion units and resulting reliability of the two configurations.

Keywords: photovoltaic system; battery; DC-coupled configuration; AC-coupled configuration; mission profile; reliability

1. Introduction

In recent years, Photovoltaic (PV) power capacity has increased more than any other type of generation technology. In 2018, the addition of PV power installed capacity of 100 GW was higher than all other technology types combined, and now accounts for 505 GW globally [1]. PV power generation is heavily dependable on the environmental conditions—solar irradiation and temperature. In order to increase PV system flexibility and to provide more dispatchable energy, integration of battery systems has been considered as a viable solution. Historically, the high cost of this storage technology has been the main barrier for its deployment. However, the declining cost of battery systems in recent years has enabled its commercialization. Lithium-ion is predominated technology type due to its merits suitable for a PV application—a fast response, scalability and low self-discharge. Its cost has declined for an average of 23% per year from 2010 to 2015, as reported in [2]. It is expected that continuous reduction in cost will further continue, which is then reflected in the expected increase of installed PV-battery systems. Precisely, 55% of annual energy storage deployments are expected to be coupled with PV systems by 2023 [3]. In such a case, system architecture and its impact on overall system performance are becoming an important topic.

In general, PV-battery power processing units consists of the three main components. Those are (1) PV panels representing power generation unit; (2) battery representing a storage unit and; (3) power electronic interface representing power conversion unit. Depending on the number of power electronic components and their type of connection, two main system configurations are available DC- and AC-coupled configurations. The main difference between two lies in the point of connection (POC) for the battery unit which can either be connected in the DC-link or at the point of common coupling (PCC).

System configuration does not only directly influence the operation, but also the cost, efficiency, lifetime and, consequently, reliability of the PV-battery system. In terms of operation, DC-coupled configuration has lower operational flexibility, as the total power delivered to the load is limited by the inverter capacity [4]. In cases of high load demand, PV and battery unit power capacity may be sufficient, but inverter capacity will limit the amount of the delivered power. In AC-coupled configuration, a greater amount of power can be delivered to supply the load, where both the PV and the battery inverter can supply the load at the same time. Cost of AC-coupled configuration is higher due to the additional DC/AC conversion unit which makes higher complexity of the system design and its balance. As reported in [5], DC-coupled configurations yield on average 1% lower total cost than AC-coupled configurations. However, in the case in which battery is integrated into the already existing PV system, the cost of the DC-coupled configuration could be higher. Already existing DC/AC conversion unit may need to be modified to accommodate multiple DC connections as well as a bi-directional DC/AC inverter. This would then additionally increase the installation cost of such a system. On the other hand, the AC-coupled battery system can be easily added to the existing PV system at the PCC, as shown in Figure 1. As a general rule, the efficiency of a system decreases as the number of power conversion units increases. Hence, the AC-coupled configuration has lower efficiency due to the additional DC/AC conversion unit. Nonetheless, as stated in [5], higher efficiency of power electronic units nowadays has caused the difference in system efficiency to become smaller compared to the early stage of development of the PV-battery systems.

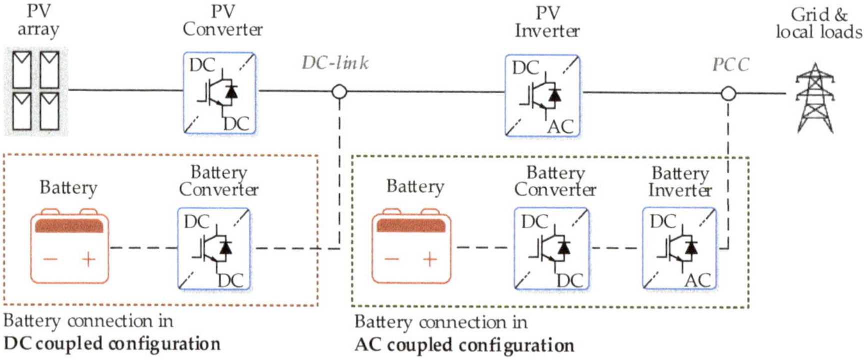

Figure 1. System diagram of the photovoltaic (PV) system with integrated battery energy storage system. Point of connection (POC) for the battery can either be at the DC-link for DC-coupled configuration or point of common coupling (PCC) for AC-coupled configuration.

In [6], the residential PV-battery systems were studied from a control point of view. A techno-economic analysis of a PV-battery system is investigated for the installation site in Greece [7] with DC-coupled configuration. Similarly, the benefits of connecting battery to a PV system in a DC-coupled configuration are investigated in [8], while AC-coupled configuration is investigated in [9]. Two configurations are compared in [10], where the performance of the PV-system connected in each of the configuration is further analyzed. However, most of the research on comparison of the two configurations is done from the economic point of view. In [11], the installed cost benchmark is proposed and the authors focused on researching potential cost-reduction opportunities of the PV-battery system. Similar research is conducted in [5] where the main focus is also put on the cost-effectiveness of a such system.

However, an important aspect that has not yet been researched is related to the lifetime and reliability of the PV-battery system connected in the two aforementioned configurations. This evaluation is necessary as it gives information on differences in the reliability of the two configurations resulting from the different number of components and their electrical loading. In general, information about the reliability of the system and its components is critical for adequate system operation and related economic profitability. In [12], reliability of the inverter unit for the DC-coupled configuration is analyzed. However, the reliability of the remaining system components, such as the DC/DC converters, have not been investigated. Thus, it is necessary to investigate each of the reliability-critical system components. If such an approach is used, it will yield information on which part of the power electronic interface is prone to failure the most. Moreover, by comparing the reliability of the two systems, additional information on the choice of the adequate configuration type for PV-battery system is obtained. Considering that, a reliability benchmark for PV-battery system connected in DC- and AC-coupled configuration is here presented.

With respect to that, an overview of the PV-battery system configurations is provided in Section 2 along with the implemented energy management strategy. The power converters electrical and thermal loading in the DC- and AC-coupled configurations is investigated in Section 3. The procedure for reliability evaluation is presented in Section 4. Reliability analysis is carried out with a case study of the real on-site measurement data in Section 5 where the reliability of the components and systems in DC- and AC-coupled configuration are compared. The conclusion of the carried work is presented in Section 5.

2. PV-Battery System

2.1. Self-Consumption Control Strategy

A self-consumption control strategy is the system's energy management strategy that is defining the power flow among the units. It is also directly influencing the loading of the power conversion units. The key target of this control strategy is to maximize the internal power consumption of the battery and PV power before purchasing power from the grid. The main reasoning for increased implementation of this control strategy is seen in its cost-effectiveness. In the last 8 years, residential electricity prices in the United States and Europe have grown for 17% and 43% respectively [13]. On the other hand, the average cost of electricity produced from PV has shown significant decrease [14]. With respect to that, maximization of the internal consumption is seen as the most favourable option that has been widely adopted in residential level PV-battery systems. Examples of commercially available products can be found in [15,16].

In Figure 2, a typical daily profile of a residential unit with indicated PV generation and load demand during winter month is shown. The profile is characterized by the periods of high PV generation and low load demand in the midday and early afternoon hours. Contrary, low PV power generation and high load demand occur in the evening hours. Due to the misalignment between the PV generation and the load demand, a direct use of self-consumption is limited in the residential PV systems. Thus, the battery storage unit is employed to absorb the excess power during the day and then supply the load in the evening hours. Accordingly, the requested power to be absorbed or delivered by the battery can be defined as:

$$P_{bat\_req} = P_{PV} - P_{load} \tag{1}$$

where P_{PV} is power generated by the PV unit and P_{load} is power demanded by the residential load at the AC side.

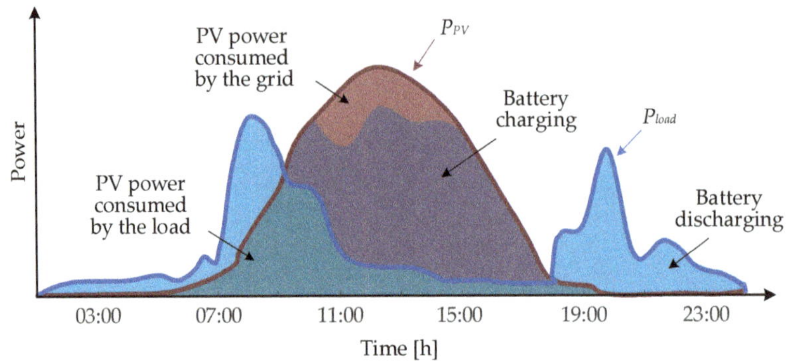

Figure 2. A typical daily profile of the PV power generation and load demand and battery (dis)charge periods defined based on the self-consumption strategy.

The requested power can be absorbed/delivered by the battery only when the battery is not fully charged/discharged. Hence, the actual power absorbed/delivered by the battery is P_{bat}. It represents a certain amount of the requested power P_{bat_req} that can actually be absorbed/delivered based on the battery availability. If excess power cannot be absorbed by the battery, it is delivered to the grid. This situation occurs between 11:00 and 15:00, where the brown area in Figure 2 represents the power that is being sent to the grid. Similarly, in cases in which the combined power of PV and battery units is not sufficient to cover load demand, the power is delivered from the grid. It is worth mentioning that the difference between the requested and available battery power is directly related to the battery capacity and power rating.

2.2. Characterization of Configuration Types

2.2.1. DC-Coupled Configuration

The DC/DC interface in the DC-coupled configuration consists of the PV and battery connected converters. DC/AC interface consists of a single inverter unit. The system diagram is shown in Figure 3.

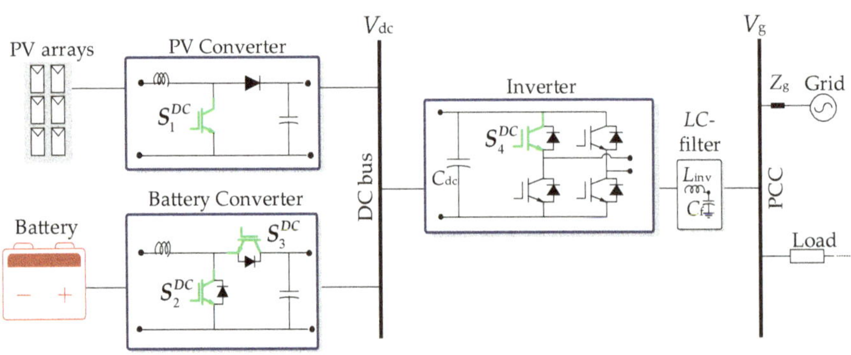

Figure 3. A system diagram of the PV-battery connected in the DC-coupled configuration with the detailed converter topology.

- PV converter: A uni-directional DC/DC converter connected to the PV panel and is used to step up the voltage at the PV panel terminals to one of the DC bus. Furthermore, by implementing the maximum power point tracking control, the PV panel generation is maximized. Different topologies can be employed to assist this requirement. However, for simplification, a boost converter is implemented in the system under study.
- Battery converter: A bi-directional DC/DC converter connected to the battery. When the battery is discharging, the DC/DC converter is operating as a step-up converter, e.g., boost converter. In the case in which battery is charging and power is being absorbed by the battery, the DC/DC converter is operating as a step-down converter, e.g., buck converter.
- Inverter: A DC/AC inverter which main purpose is to transfer power to the AC side, as well as synchronize with the AC grid. Its loading is based on the power being delivered from both the PV and the battery units to the load. In this paper, a full-bridge single-phase inverter with four power devices is used in the system under study.

2.2.2. AC-Coupled Configuration

A system diagram of the AC-coupled configuration is shown in Figure 4. Its DC/DC interface configuration is the same as in the case of the DC-coupled configuration. DC/AC interface consists of two units—the PV and the battery inverters.

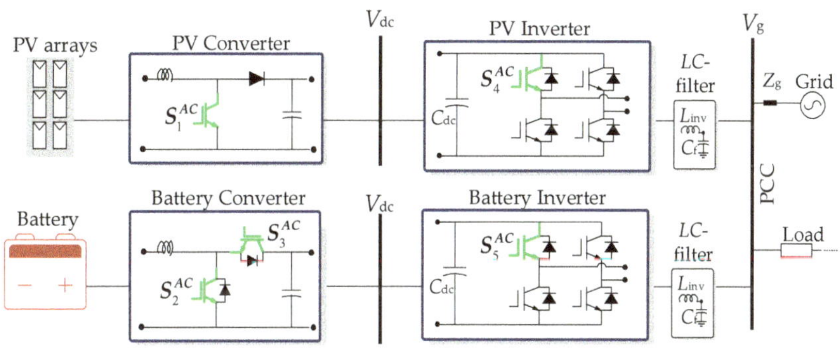

Figure 4. A system diagram of the PV-battery system connected in the AC-coupled configuration with the detailed converter topology.

- PV inverter: A DC/AC inverter whose main purpose is to transfer the power generated by the PV panel to the AC side, similar to that in the DC-coupled configuration.
- Battery inverter: A bi-directional DC/AC converter connected to the battery. When the battery is discharging, the battery inverter is operating in the inverter mode. When the battery is charging, the battery inverter is operating in the rectifier mode. Its loading is only determined based on the excess PV power that is being absorbed and delivered by the battery.

The same topology as for the DC-coupled configuration is employed for the AC-coupled configuration for the comparison purposes in the further part of the paper. This refers that DC/DC interface is implemented through the boost convert and bi-directional buck-boost converter for PV and battery converter respectively. Similarly, the DC/AC interface is implemented by means of a full-bridge single-phase inverter with four power devices. In the next section, power converters loading in each configuration is examined more into detail.

3. Power Converters Loading

In order to analyze the operational impact on the component reliability, a power converter loading during a typical one-day operation of the PV-battery system is studied. The key target is to investigate the difference in the electrical and thermal loading of the components in the two configurations. In general, power converter loading is directly related to the thermo-mechanical stress of its components such as power devices, and it is the cause of the component's failure, reflecting the reliability of the system [17].

3.1. Electrical Loading of Power Converters

Electrical loading of the power converters is investigated by the means of their input power. For DC/DC interface, input power to the PV converter equals the power generated by the PV array, P_{PV}. The electrical loading of the PV converter is only influenced by the environmental conditions at the installation site (solar irradiance and ambient temperature). Contrary, the electrical loading of the battery converter is the result of the input PV power, load demand, implemented energy management strategy and available battery capacity. In the case of the DC/AC interface, input power is defined differently for the two configurations.

3.1.1. DC-Coupled Configuration

In the DC-coupled configuration, the DC/AC interface is represented through the inverter (Figure 3). The input power to this unit is defined as:

$$P_{inv} = P_{PV}^{conv} - P_{bat}^{conv} \tag{2}$$

where P_{PV}^{conv} and P_{bat}^{conv} are output power of the PV and battery converter respectively, and are defined as:

$$P_{PV}^{conv} = P_{PV} - P_{PV}^{loss} \tag{3}$$

$$P_{bat}^{conv} = P_{bat} - P_{bat}^{loss} \tag{4}$$

where P_{PV}^{loss} and P_{bat}^{loss} represent converter losses defined based on the their respective efficiency curves. P_{PV} is power generated by the PV panel and P_{bat} is power absorbed (positive power) or delivered (negative power) to the battery.

Electrical loading of the DC/AC inverter differs for battery discharging and charging. This is graphically represented in Figure 5, where the inverter input power for two cases is highlighted.

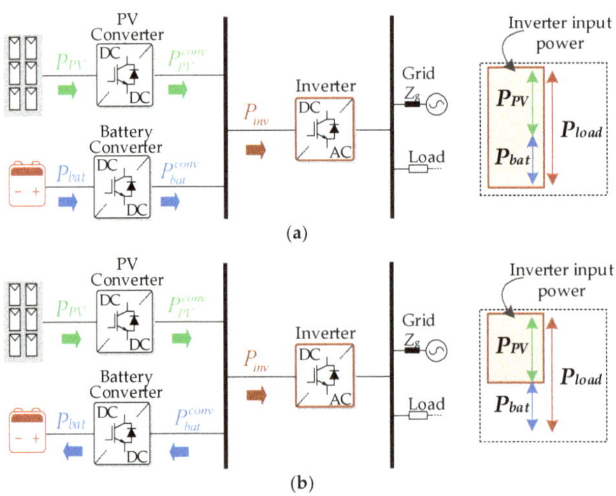

Figure 5. Electrical loading of the power electronic converters for the DC-coupled configuration with the indicated power flow: (**a**) battery discharging; and (**b**) battery charging.

As observed, the electrical loading in case of the battery discharging will be higher, as it will account for the power generated from PV and delivered by the battery.

3.1.2. AC-Coupled Configuration

The DC/AC interface for the AC-coupled configuration is represented through the PV and the battery inverters (Figure 4). The input power for each unit is defined as:

$$P_{PV}^{inv} = P_{PV} - P_{PV}^{loss} \tag{5}$$

$$P_{bat}^{inv} = P_{bat} - P_{bat}^{loss} \tag{6}$$

where P_{PV}^{loss} and P_{bat}^{loss} represent converter losses defined based on the their respective efficiency curves.

The graphical representation of the electrical loading in the case of the battery charging and discharging is shown in Figure 6 with the highlighted PV inverter input power.

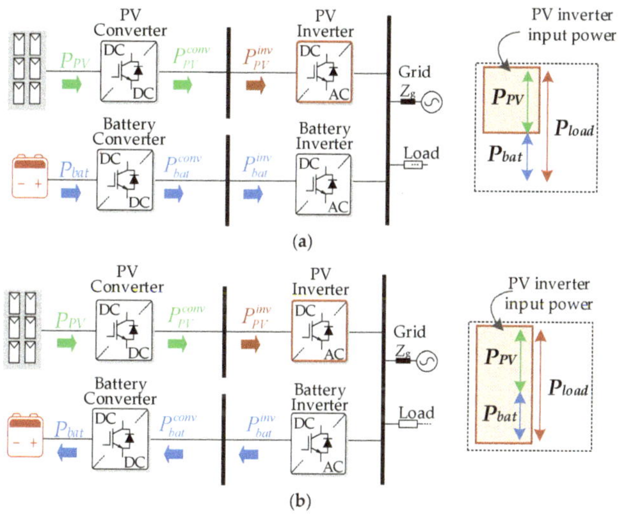

Figure 6. Electrical loading of the power electronic converters for the AC-coupled configuration with the indicated power flow: (**a**) battery discharging; and (**b**) battery charging.

As in the case of the DC-coupled configuration, the electrical loading of the PV inverter is examined. In case of the battery charging, the electrical loading will be higher. This is because the excess power generated by the PV is sent to the battery at the AC side.

3.1.3. One-Day Operation

An example of the electrical loading of the PV and battery converter during one-day operation is shown in Figure 7, where P_{PV} and P_{bat} are input to the PV and battery converter respectively. Two time intervals are distinguished. (1) battery charging period marked with the green area under the curve and (2) battery discharging period marked with the red area under the curve. On the example of the two time intervals, the electrical loading of the DC/AC interface is examined and compared for the two configurations in order to exemplify the power flow shown in Figures 5 and 6.

In the case of battery discharge, the input power for the inverter in the DC-coupled configuration equals to the sum of PV and battery power. For the same amount of power that has to be delivered to the load in the AC-coupled configuration, the electrical loading of the PV inverter is the PV power, while power being delivered from the battery is transferred through the battery inverter. During the battery discharge, the electrical loading of the DC-coupled configuration inverter is higher than the one of the PV inverter in AC-coupled configuration.

In case of battery charging, the input power for the inverter in the DC-coupled configuration equals to the amount of PV power that is needed to supply the load demand. The rest of the PV power is transferred to the battery at the DC bus (i.e., the excess PV power is input to the battery converter). In the case of the AC-coupled configuration, the electrical loading of the PV inverter is equal to the total power generated from the PV panels. At the AC side, this power is then divided to supply the load and the excess that is input to the battery inverter. In such a case, the electrical loading of the AC-coupled configuration PV inverter is higher than the one of the inverter in DC-coupled configuration.

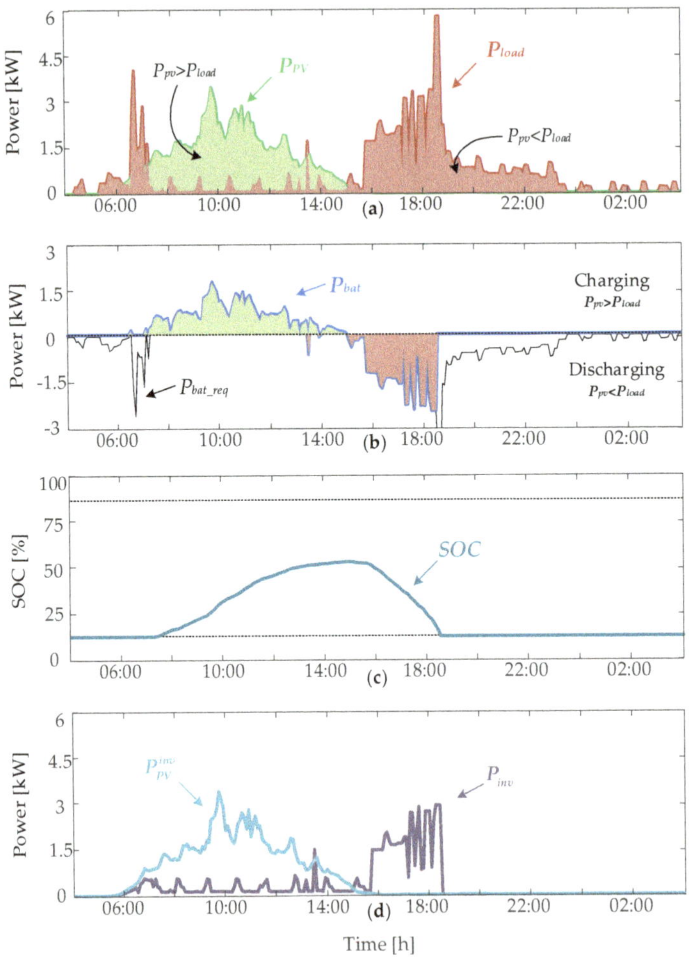

Figure 7. One-day power profile: (**a**) PV power P_{PV} and load demand P_{load}; (**b**) battery power P_{bat}, where positive power is absorbed power (charging periods) and negative power is delivered power (discharging periods); (**c**) battery SOC; and (**d**) input power of inverter of the DC-coupled configuration P_{inv} and PV inverter of the AC-coupled configuration P_{inv}^{PV}.

By examining the electrical loading of the components in the DC- and AC-coupled configuration, it can be concluded that the DC/AC interface electrical loading is depended on the input power profiles and cannot be universally connected to higher or lower reliability for a certain configuration.

The summary of the examined cases for the inverter in DC-coupled configuration and PV inverter in AC-coupled configuration is given in Table 1.

Table 1. Summary of The Inverter Loading.

	(PV) Inverter Loading	
	Battery Charging	**Battery Discharging**
DC coupled configuration	P_{PV}	$P_{PV} + P_{bat}$
AC coupled configuration	$P_{PV} + P_{bat}$	P_{PV}

3.2. Thermal Loading of Power Converters

Thermal loading of the power converters can be investigated by means of the thermal stresses of the power devices, i.e., Insulated Gate Bipolar Transistors (IGBT). The thermal stress of the power device can be translated from the electrical loading by the electro-thermal model. The thermal stress is represented by the variations in the junction temperature cycling T_j of the power device. Junction temperature is the stress factor which will cause the bond wire lift-off (representing the wear out failure of a power device). The detailed procedure for the electro-thermal modelling is presented in [18,19].

Thermal loading of the IGBT during one-day operation is investigated. Electrical loading profiles of the power converters in the DC-coupled configuration are shown in Figure 8. Furthermore, the associated junction temperature of the IGBTs (indicated in Figure 3) are also presented, where the variations in the junction temperature are correlated with the electrical loading profiles. The highest variations in the junction temperature occur in the power devices of the battery converter. This is aligned with the variations in the electrical lading of the battery converter.

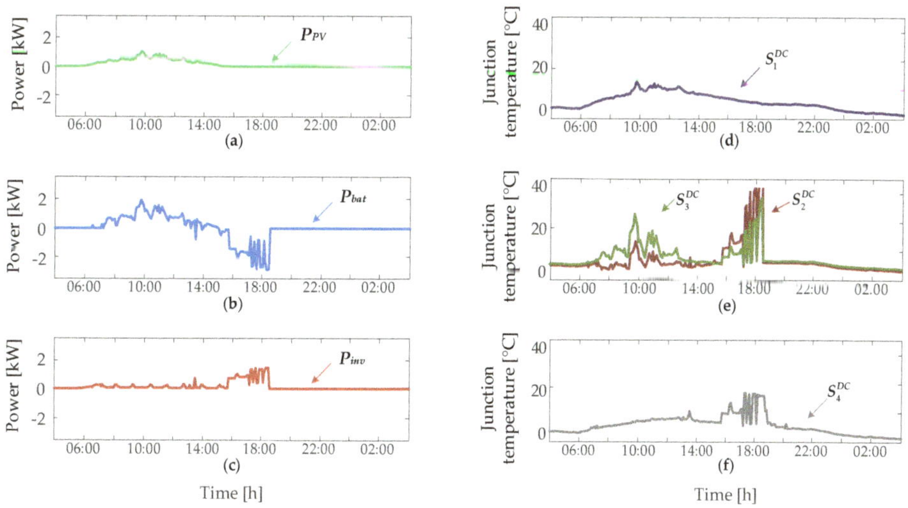

Figure 8. Electrical and thermal loading of the power electronic components in the DC-coupled configuration: (**a**) PV converter P_{PV}; (**b**) battery converter P_{bat}; (**c**) inverter P_{inv}; (**d**) junction temperature of S_1^{DC} in PV converter; (**e**) junction temperature of S_2^{DC} and S_3^{DC} in battery converter; and (**f**) junction temperature of S_4^{DC} in inverter.

Similarly, the electrical and the thermal loading profiles during one-day operation in the AC-coupled configuration are shown in Figure 9.

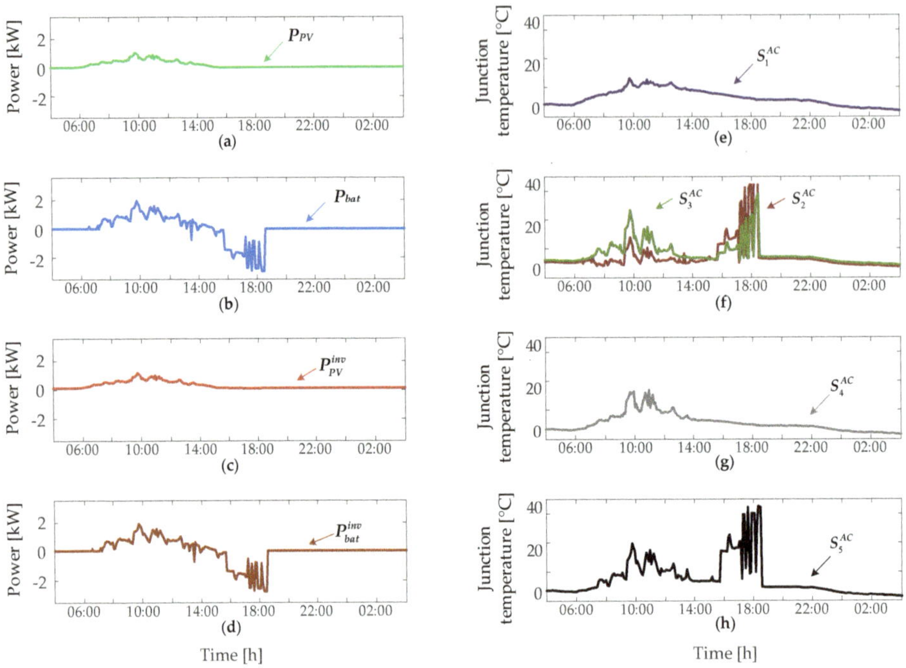

Figure 9. Electrical and thermal loading of the power electronic components in the AC-coupled configuration: (**a**) PV converter P_{PV}; (**b**) battery converter P_{bat}; (**c**) PV inverter P_{PV}^{inv}; (**d**) battery inverter P_{bat}^{inv}; (**e**) junction temperature of S_1^{AC} in PV converter; (**f**) junction temperature of S_2^{AC} and S_3^{AC} in battery converter; (**g**) junction temperature of S_4^{AC} in PV inverter; and (**h**) junction temperature of S_5^{AC} in battery inverter.

4. Reliability Evaluation

In this section, the reliability analysis is conducted by following the procedure presented in [17]. It involves the lifetime modeling of the system components and the (un)reliability analysis where the statistic of the failure rate is considered.

4.1. Lifetime Modelling

It is necessary to choose adequate lifetime model of the power device providing information on the number of cycles to failure for each set in the thermal stress matrix: the average junction temperature T_{jm}, the cycle amplitude ΔT_j and the cycle period t_{on}. An overview of the lifetime models predominately used for this purposes is provided in [18]. In this study, a lifetime model estimating the power cycling capability of IGBTs with junction temperature limitations of 150 °C is used [20]. The number of cycles to failure is then defined as shown in Equation (7). The input to the model is the matrix of n_i for T_{jm} and ΔT_j and t_{on} values. The rest of the model parameters is provided in Table 2. Then, the amount of device's consumed life is evaluated by means of Miner's rule shown in Equation (8).

$$N_f = K \cdot (\Delta T_j)^{\beta_1} \cdot e^{\frac{\beta_2}{T_{jm}+273}} \cdot (t_{on})^{\beta_3} \cdot (I)^{\beta_4} \cdot (V)^{\beta_5} \cdot (D)^{\beta_6} \quad (7)$$

$$LC = \sum_i \frac{n_i}{N_{fi}} \quad (8)$$

where N_f is the number of cycles to failure under the stress condition of the mean junction temperature T_{jm}, the cycle amplitude ΔT_j and the cycle period t_{on}. LC represents lifetime consumption (LC) which starts from the beginning-of-life. When the LC accumulates to unity, IGBT power device is considered to reach its end-of-life.

Table 2. Parameters of Insulated Gate Bipolar Transistors (IGBT) Lifetime Model.

Parameter	Value	Description
I	10 A	Current per wire bond
V	6 V/100	Blocking voltage
D	300 µm	Diameter of bonding wire
K	2.03×10^{14}	Technology factor
β_1	−4.416	Contribution of Coffin-Manson law
β_2	1285	Contribution of Arrhenius law
β_3	−0.463	Influence of transient thermal response on the chip
β_4	−0.716	Contribution of accelerated wire bonds failure close to end of life
β_5	−0.761	Accounted correlation between blocking voltage and chip thickness
β_6	−0.5	Considered impact of wire diameter on bond interface and thermal expansion

4.2. Unreliablity Function

From lifetime evaluation, a B15 lifetime is obtained based on the statistic from the lifetime model. It represents the amount of time during which 15% of the population (of the same IGBT type) will fail. In general, lifetime data of the wear out failure is following Weibull distribution [19], which probability density function is expressed as:

$$f(x) = \frac{\beta}{\eta^{\beta}} x^{\beta-1} \exp\left[-\left(\frac{x}{\eta}\right)^{\beta}\right] \tag{9}$$

where x represents the operation time, and η and β are the scale and the shape parameter respectively. η parameter represents time instance at which 63.2% of the population will fail. β parameter represents the failure mode, where the same failure mode will have similar β values. From the probability density function, it is, furthermore, possible to obtain cumulative density function by integration over operation time:

$$F(x) = \int_0^x f(x)dx = 1 - \exp\left[-\left(\frac{x}{\eta}\right)^{\beta}\right] \tag{10}$$

This cumulative density function is often referred as an unreliability function and it is representing the proportion of the accumulated failure population over time. In order to obtain the Weibull cumulative density function, it is necessary to determine η and β parameter values. For this analysis, the value of β parameter is taken from [21] and accounts for the failure mode related to the IGBT devices. η parameter can be calculated by using Equation (10) and previously obtained B15 lifetime information. Once η and β parameters are determined, it is possible to obtain the unreliability curve of power device under study.

The unreliability curve of each device gives information about the component-level reliability. To investigate system-level reliability, it is necessary to construct the Reliability Block Diagram (RBD) of the overall system. In general, the RBD is a representation of the reliability interaction between the components in the system. Parallel paths in the diagram are redundant i.e., the system fails once when all the parallel paths in the RBD fail. Contrary, if series paths are considered, the system fails when one of the components in a series path fails. With respect to that, in order to move from the component-level reliability (i.e., reliability of a single IGBT) to converter-level reliability (i.e., reliability of a single converter unit consisting of IGBTs), the RBD for each of the components in the system is created. For both configuration types, the series connection of all studied components is considered. The system unreliability is then calculated as:

$$F_{\text{system}}(x) = 1 - \prod_{i=1}^{n}(1 - F_i(x)) \tag{11}$$

where F_i is the unreliability function of the i-th component of the system and n is the number of components. The flow chart of the aforementioned process is shown in Figure 10.

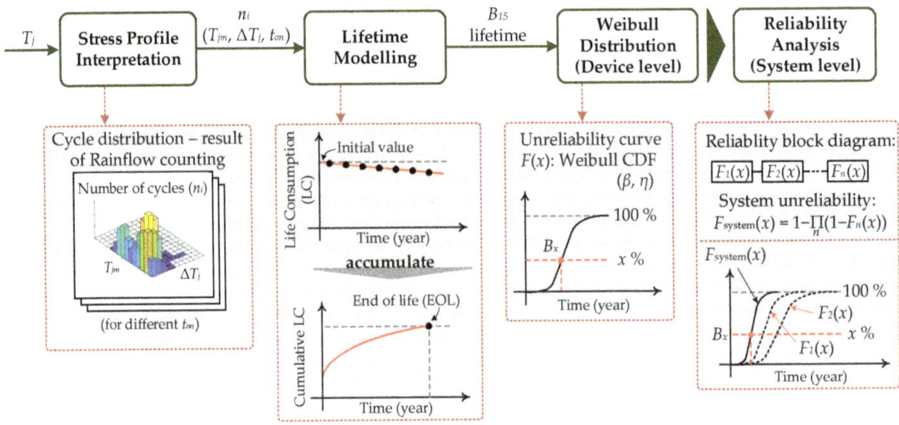

Figure 10. Flow diagram of the mission profile-based reliability evaluation of PV-battery system.

5. Case Study

A case study for the PV-battery system with the installation site in Germany is carried out in order to compare the reliability of the DC- and AC-coupled configurations. Firstly, a system description is provided, then a mission profile including real, on-site measurement data is presented. For the given one-year mission profile, reliability evaluation is performed by following the procedure presented in Section 4.

5.1. System Description

The system under study consist of a 6 kW PV system with a 3 kW/7.5 kWh battery system. The system sizes and relevant parameters are provided in Table 3 and correspond to the configurations outlined in Figures 3 and 4. As indicated, all the converter units are rated at 3 kW. A higher power rating is achieved with a modular structure, where e.g., 6 kW can be achieved by a parallel connection of two 3 kW units. The power devices in each of the converters are chosen to be the same, in order to ensure a fair comparison.

5.2. Mission Profile

A one-year mission profile from the installation site in Germany is chosen for the case study, as it is shown in Figure 11. It consists of a solar irradiance and ambient temperature profiles with a sampling rate of 5 min per sample [12]. The mission profile consists of the strong seasonal variations, e.g., mean solar irradiance during winter months is 34.98 W/m^2 which is 72% lower than the average value during summer months.The load profile is presenting a typical household load with the average of 4.8 kWh/year and without strong variations throughout the year [22].

Table 3. Parameters of the Single-Phase PV-Battery System.

Parameter	Value	
	DC Coupled Configuration	AC Coupled Configuration
PV array rated power (at STC)	6 kW	
Battery energy capacity	7.5 kWh	
PV converter rated power	6 kW (3 kW × 2 units)	
Battery converter rated power	3 kW	
Inverter rated power	6 kW (3 kW × 2 units)	-
PV inverter rated power	-	6 kW (3 kW × 2 units)
Battery inverter rated power	-	3 kW
DC-link capacitor	C_{dc} = 1100 μF	
LC-filter	L_{inv} = 4.8 mH, C_f = 4.3 μF	
Switching frequency	DC/DC Converters: f_{sw} = 20 kHz	
Switching frequency	DC/AC Converters: f_{sw} = 10 kHz	
DC-link voltage	V_{dc} = 450 V	
Grid nominal voltage (RMS)	V_g = 230 V	
Grid nominal frequency	$\omega_0 = 2\pi \times 50$ rad/s	

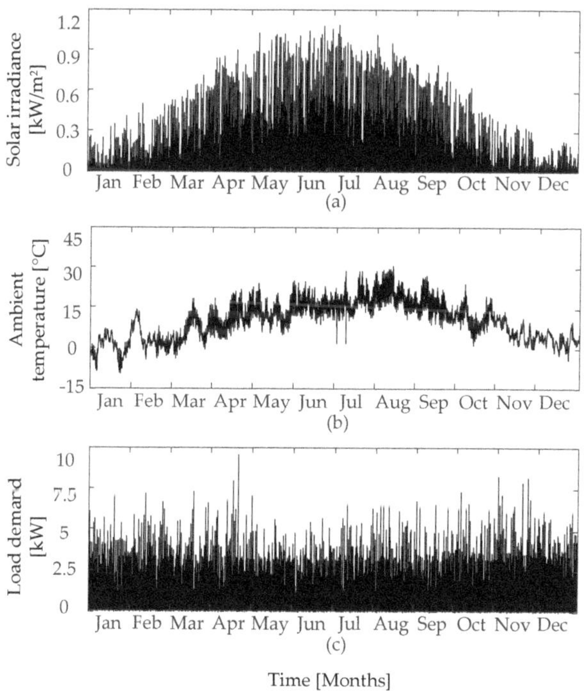

Figure 11. A one-year mission profile of the installation site in Germany (sampling rate of 5 min per sample): (**a**) solar irradiance; (**b**) ambient temperature; and (**c**) household load.

5.3. Results

Reliability assessment is carried out for the aforementioned mission profile and results are shown for both configurations. Firstly, the one-year stress profiles of the power devices are shown. Then, the unreliability curves of the power devices are given and finally, the reliability of the system and its components is investigated.

5.3.1. Reliability of the DC-Coupled Configuration

Stress profiles of the power devices are shown in Figure 12. By following the procedure presented in Figure 10, the stress profiles are decomposed to the cycles at different cycle depths and cycle averages, where it is possible to determine the range of T_{jm} and ΔT_j from the thermal stress profile.

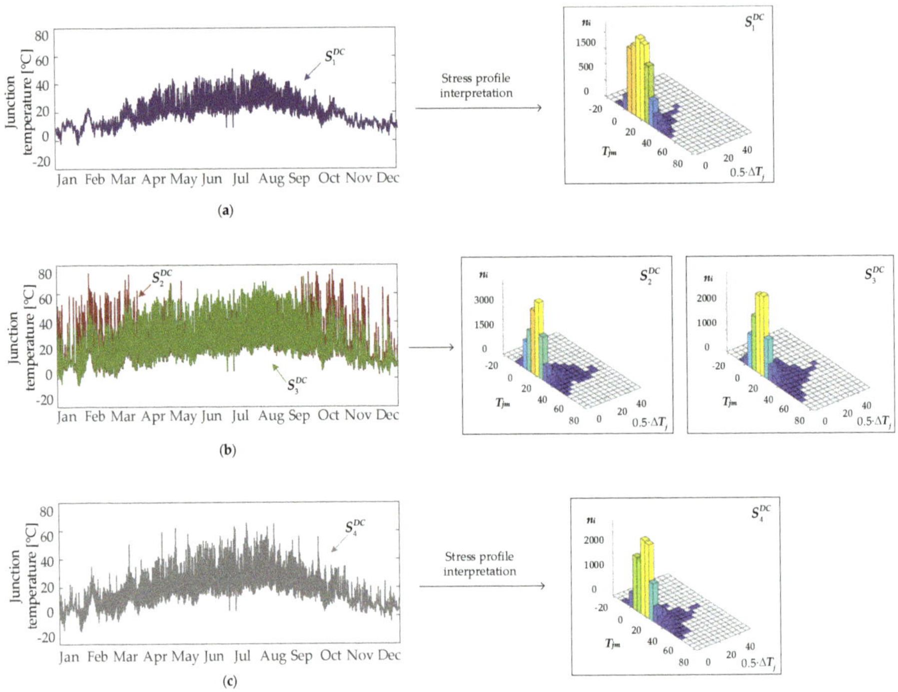

Figure 12. The resulting stress profiles and their cycling decomposition in the DC-coupled configuration for the input mission profile: (**a**) IGBT of PV converter S_1^{DC}; (**b**) IGBTs of battery converter S_2^{DC} and S_3^{DC} (**c**) IGBT of inverter S_4^{DC}.

The highest number of cycles for all IGBTs occurs for the mean junction temperature T_{jm} of 20 °C. However, the exact number of cycles differs for each IGBT based on the stress they are exposed to during the operation. By examining the stress profiles, it can be observed that the smallest cycle amplitude has IGBT S_1^{DC} of the PV converter. Additionally, this IGBT experiences the smallest number of cycles through out the operation. Hence, IGBT S_1^{DC} is the least stressed power device of all during one-year operation. This indicates that it will experience the highest lifetime. Stress of the battery converter is examined by means of IGBTs S_2^{DC} and S_3^{DC} junction temperature. In general, battery operation is more dynamic than one of the PV panels, meaning that electrical loading of the battery converter changes more often with battery charging and discharging periods. Hence, it is expected that the power devices of the battery converter will be more stressed than the rest of the devices of the system. Considering that, it is shown that IGBT S_2^{DC}, which is stressed by battery discharging, has the highest number of cycles out of all power devices. IGBT S_3^{DC} experiences less number of cycles, which is aligned with the fact that battery spent more time discharging during a year. Electrical loading of the inverter is based on power generated by PV and battery operation, as defined in Equation (2). Hence, S_4^{DC} stress profile is similar to the one of the S_1^{DC} of PV converter. However, the additional power provided by the battery operation is causing higher stress for the S_4^{DC}.

In order to examine the reliability of the system and its components, the RBD for the DC-coupled configuration is defined and shown in Figure 13. Corresponding unreliability functions are presented in Table 4 and the results are shown in Figure 14.

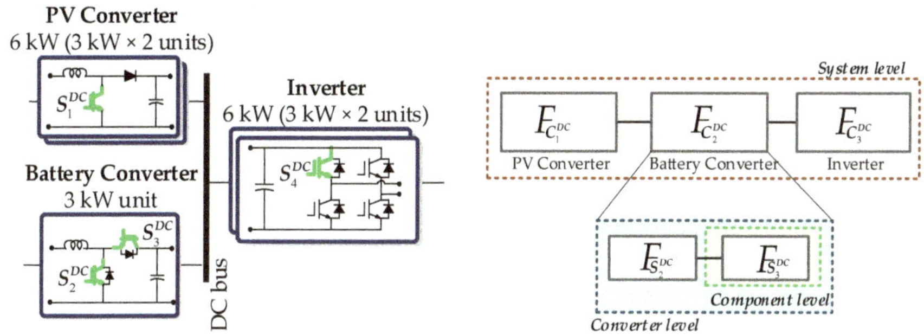

Figure 13. DC-coupled configuration: Power electronic interface and its reliability block diagram.

Table 4. DC-Coupled Configuration: Unreliability Functions of the System Components.

Component	Devices	Unreliability Function
C_1: PV converter	$2 \times S_1^{DC}$	$F_{C_1^{DC}} = 1 - (1 - F_{S_1^{DC}})^2$
C_2: Battery converter	S_2^{DC}, S_3^{DC}	$F_{C_2^{DC}} = 1 - (1 - F_{S_2^{DC}})(1 - F_{S_3^{DC}})$
C_3: Inverter	$2 \times 4 \times S_4^{DC}$	$F_{C_3^{DC}} = 1 - (1 - F_{S_4^{DC}})^8$
DC configuration	C_1, C_2, C_3	$F_{DC} = 1 - (1 - F_{C_1^{DC}})(1 - F_{C_2^{DC}})(1 - F_{C_3^{DC}})$

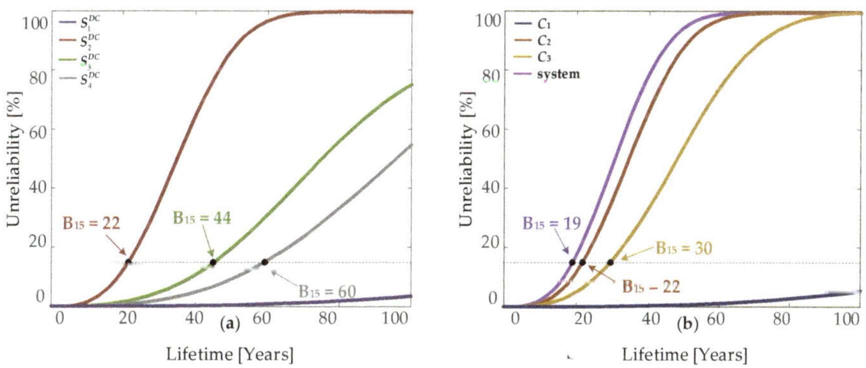

Figure 14. Unreliability curves of the DC-coupled configuration at the: (**a**) component-level; (**b**) converter-level.

On the component-level, the lowest B15 lifetime has IGBT S_2^{DC}. Contrary, the highest B15 lifetime has IGBT S_1^{DC} and it accounts for more than 100 years. The obtained reliability results on the component-level correspond to previously analyzed stress profiles of the IGBTs. On the converter-level, battery converter experiences lowest operational time with B15 lifetime of 21 years. Hence, it is the most reliability-critical component of the system. Overall system connected in DC-coupled configuration has B15 lifetime of 19 years.

5.3.2. Reliability of the AC-Coupled Configuration

For AC-coupled configuration, the stress profiles of the power devices are shown in Figure 15, together with the associated cycle distribution.

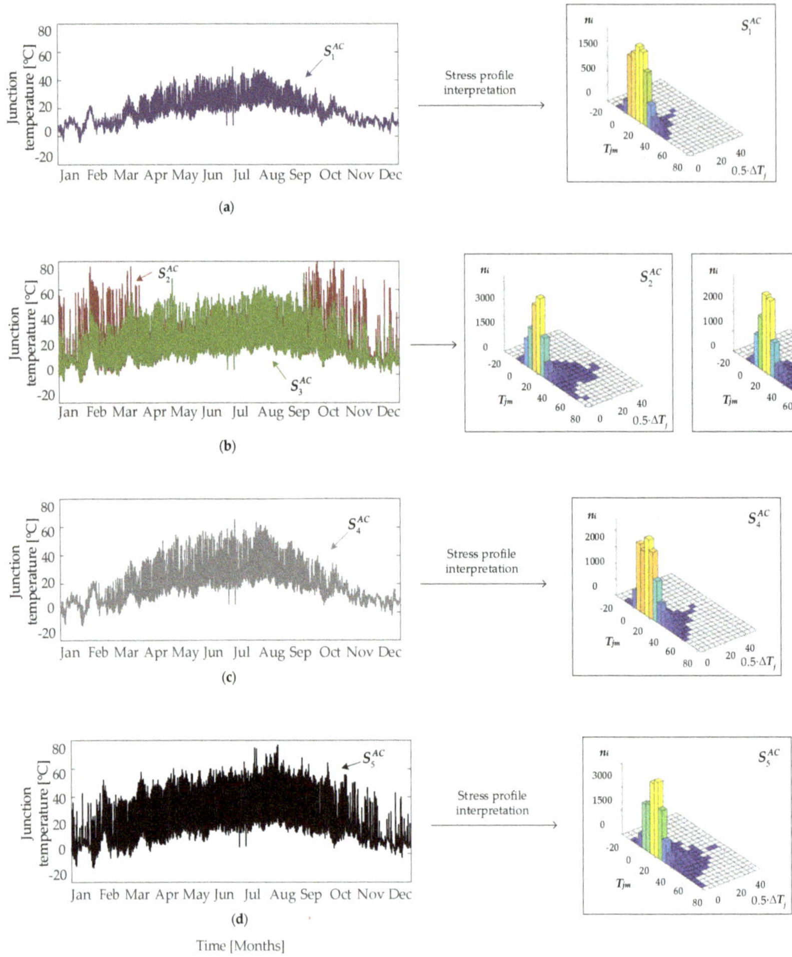

Figure 15. The resulting stress profiles and their cycling decomposition in the AC-coupled configuration for the input mission profile: (**a**) IGBT of PV converter S_1^{AC}; (**b**) IGBTs of battery converterS_2^{AC} and S_3^{AC}; (**c**) IGBT of pv inverter S_4^{AC}; (**d**) IGBT of inverter S_5^{AC}.

Similarly as for the DC-coupled configuration, the highest number of cycles is present for mean junction temperature of 20 °C for all IGBTs. The lowest cycle amplitude has the IGBT S_1^{AC} which is experiencing the lowest stress during operation. The highest number of cycles have IGBTs S_4^{AC} and S_4^{AC} of battery converter and battery inverter respectively. This is mainly due to the battery dynamics. However, in the case of AC-coupled configuration and its point of connection for the battery unit, two power electronics units are affected (compared to a single unit in DC coupled configuration). This will, consequently, influence the reliability of the system which is assessed by the means of the RBDs shown in Figure 16 and associated unreliability functions given in Table 5.

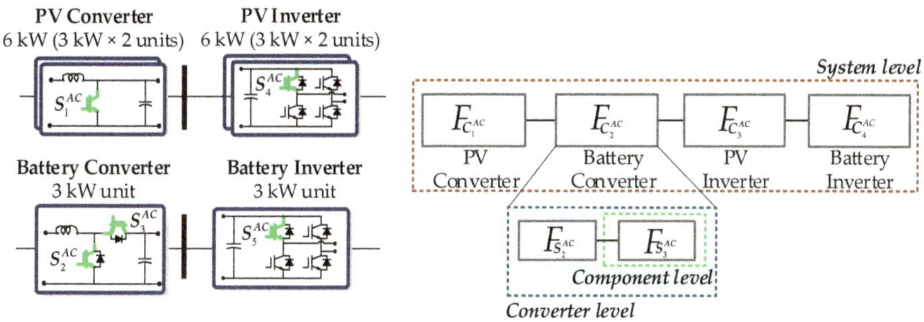

Figure 16. AC-coupled configuration: Power electronic interface and its reliability block diagram.

Table 5. AC-Coupled Configuration: Unreliability Functions of the System Components.

Component	Devices	Unreliability Function
C_1: PV converter	$2 \times S_1^{AC}$	$F_{C_1^{AC}} = 1 - (1 - F_{S_1^{AC}})^2$
C_2: Battery converter	S_2^{AC}, S_3^{AC}	$F_{C_2^{AC}} = 1 - (1 - F_{S_2^{AC}})(1 - F_{S_3^{AC}})$
C_3: PV Inverter	$2 \times 4 \times S_4^{AC}$	$F_{C_3^{AC}} = 1 - (1 - F_{S_4^{AC}})^8$
C_4: Battery Inverter	$4 \times S_5^{AC}$	$F_{C_4^{AC}} = 1 - (1 - F_{S_5^{AC}})^4$
AC configuration	C_1, C_2, C_3, C_4	$F_{AC} = 1 - (1 - F_{C_1^{AC}})(1 - F_{C_2^{AC}})(1 - F_{C_3^{AC}})(1 - F_{C_4^{AC}})$

The component-level unreliability curves are shown in Figure 17a. The highest unreliability has the IGBT S_5^{AC} of the battery inverter and its B15 lifetime accounts for 20 years. Similarly, B15 lifetime of the IGBT of the battery converter S_2^{AC} is 23 years. On the converter-level, the most reliability-critical unit is battery inverter. It means that this unit will develop highest rate of failure over time among all the units of the system. The overall B15 lifetime of the system connected in the AC-coupled configuration is 12 years.

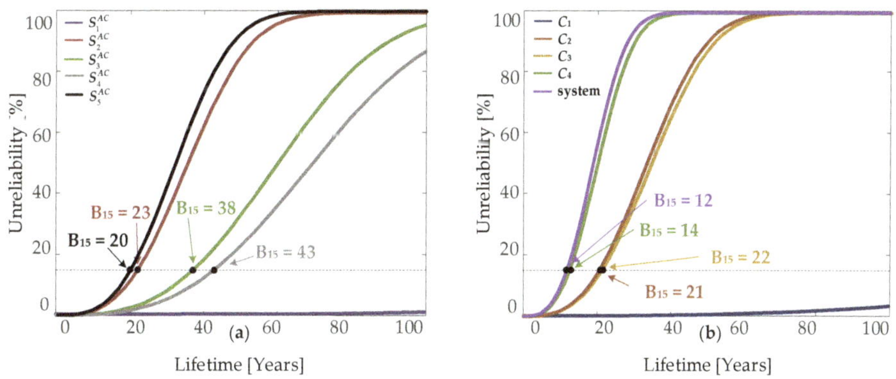

Figure 17. Unreliability curves of the AC-coupled configuration at the: (**a**) component-level; (**b**) converter-level.

5.3.3. Discussion

From the carried analysis, it is concluded that the PV-battery system connected in the DC-coupled configuration has a longer lifetime and a higher reliability. The summary of the B15 lifetime for the components of the both configurations is provided in Table 6. B15 lifetime of the DC-coupled configuration accounts for 19 years, which is 7 years longer than the one of the same system connected in the AC-coupled configuration.

Table 6. Summary of the Reliability Study Results—B15 Lifetime for the Components of the DC- and AC-Coupled Configurations.

	DC-Coupled Configuration	AC-Coupled Configuration
PV converter	>100	>100
Battery converter	22	21
(PV) inverter	30	22
Battery inverter	-	14
System	19	12

The main reasoning can be found in the fact that AC-coupled configuration has an additional unit—battery inverter. The excess PV power in AC-coupled configuration is transferred at the AC side, imposing thermal stress to the two units in the system. The same amount of power in DC-coupled configuration is transferred to the battery unit on the DC side and it is imposing thermal stress to a single unit. Furthermore, the reliability of the PV inverter in the AC-coupled configuration is not improved, when compared to the one in the DC-coupled configuration.

The benefit of the battery integration to the PV system in each of the configurations can be assessed by means of the self-consumption ratio (SCR). In general, SCR represents the ratio of the self-consumed electricity as a percentage of the total electricity generated by the PV. For the PV-battery system connected in the DC- and AC-coupled configuration, SCR accounts for 68.38% and 70.15% respectively. The minor difference in SCR indicate that both configuration types have a similar rate of the self-consumed power. It can be concluded that the addition of battery unit had the same benefit. Hence, for the presented case study, DC-coupled configuration is a more favorable option.

6. Conclusions

In this paper, a procedure for the reliability assessment of the PV-battery system is presented. Two configuration types are evaluated - DC- and AC-coupled configurations. This is done in order to assess how different point of connection of the battery unit is affecting the electrical loading of the power electronic units and, consequently, their reliability. Hence, by means of the presented reliability evaluation, a comparison on the component-, converter- and system-level reliability of the two configurations is studied. By performing the reliability analysis information on the most reliability-critical components among power electronic units of the each configuration is obtained. A mission profile-based reliability evaluation for a case study with the real measurement data from the installation site in Germany is conducted. The results have shown that the AC-coupled configuration has lower reliability, where the reliability-critical component of the system is battery inverter.

Author Contributions: Conceptualization, M.S. and A.S.; Formal analysis, M.S.; Investigation, M.S.; Methodology, M.S. and A.S.; Software, M.S.; Supervision, A.S. and F.B.; Validation, A.S.; Visualization, M.S.; Writing—original draft, M.S.; Writing—review & editing, A.S. and F.B.

Funding: This research was funded by the APETT grant number from the Innovation Foundation.

Conflicts of Interest: The authors declare no conflict of interest.

References

1. REN21. *Renewables 2018: Global Status Report (GRS)*; REN21: Paris, France, 2019.
2. Deloitte. *Supercharged: Challenges and Opportunities in Global Battery Storage Markets*; Deloitte: New York, NY, USA, 2018.
3. GTM. *DC-Coupled Solar-Plus-Storage Systems Are Gaining Ground*; Greentech Media: Boston, MA, USA, 2018.

4. Mashal, J.; Sloane, I. A Battery for Hire: AC vs. DC Coupling for Solar + Energy Storage Projects. 2018. Available online: https://blog.fluenceenergy.com/energy-storage-ac-dc-coupled-solar (accessed on 15 July 2019)
5. National Renewable Energy Laboratory (NREL). *U.S. Utility-Scale PhotovoltaicsPlus-Energy Storage System Costs Benchmark*; Technical Report NREL/TP-6A20-71714; National Renewable Energy Laboratory: Golden, CO, USA, 2018.
6. Faia, R.; Faria, P.; Vale, Z.; Spinola, J. Demand Response Optimization Using Particle Swarm Algorithm Considering Optimum Battery Energy Storage Schedule in a Residential House. *Energies* **2019**, *12*, 1645. [CrossRef]
7. Kosmadakis, I.E.; Elmasides, C.; Eleftheriou, D.; Tsagarakis, K.P. A Techno-Economic Analysis of a PV-Battery System in Greece. *Energies* **2019**, *12*, 1357. [CrossRef]
8. Meneghel, D.; da Costa Bortoni, E.; Karimi, A. Boosting DC/AC Ratio of PV Plant for BESS Integration on DC side. In Proceedings of the 2018 IEEE Conference on Technologies for Sustainability (SusTech), Long Beach, CA, USA, 11–13 November 2018; pp. 1–4.
9. Barchi, G.; Pierro, M.; Moser, D. Predictive Energy Control Strategy for Peak Switch and Shifting Using BESS and PV Generation Applied to the Retail Sector. *Electronics* **2019**, *8*, 526. [CrossRef]
10. Hofer, J.; Svetozarevic, B.; Schlueter, A. Hybrid AC/DC building microgrid for solar PV and battery storage integration. In Proceedings of the 2017 IEEE Second International Conference on DC Microgrids (ICDCM), Nuremburg, Germany, 27–29 June 2017; pp. 188–191.
11. National Renewable Energy Laboratory (NREL). *Installed Cost Benchmarks and Deployment Barriers for Residential Solar Photovoltaics with Energy Storage*; Technical Report NREL/TP-7A40-67474; National Renewable Energy Laboratory: Golden, CO, USA, 2016.
12. Sangwongwanich, A.; Angenendt, G.; Zurmuhlen, S.; Yang, Y.; Sera, D.; Sauer, D.U.; Blaabjerg, F. Enhancing PV inverter reliability with battery system control strategy. *CPSS Trans. Power Electron. Appl.* **2018**, *3*, 93–101. [CrossRef]
13. US Energy Information Administration. *European Residential Electricity Prices Increasing Faster than Prices in United States*; US Energy Information Administration: Washington, DC, USA, 2018.
14. National Renewable Energy Laboratory (NREL). *U.S. Solar Photovoltaic System Cost Benchmark*; Technical Report; National Renewable Energy Laboratory: Golden, CO, USA, 2016.
15. SMA Solar Technology AG. *Planning Guidelines—The System Solution for More Independence*; SMA Solar Technology AG: Niestetal, Germany, 2013.
16. Schneider-Electric. Rooftop for Self-Consumption with Storage. Available online: https://solar.schneider-electric.com/solution/residential-self-consumption/(accessed on 15 July 2019)
17. Yang, Y.; Sangwongwanich, A.; Blaabjerg, F. Design for reliability of power electronics for grid-connected photovoltaic systems. *CPSS Trans. Power Electron. Appl.* **2016**, *1*, 92–103. [CrossRef]
18. Reigosa, P.D.; Wang, H.; Yang, Y.; Blaabjerg, F. Prediction of Bond Wire Fatigue of IGBTs in a PV Inverter Under a Long-Term Operation. *IEEE Trans. Power Electron.* **2016**, *31*, 7171–7182.
19. Sangwongwanich, A.; Yang, Y.; Sera, D.; Blaabjerg, F.; Zhou, D. On the Impacts of PV Array Sizing on the Inverter Reliability and Lifetime. *IEEE Trans. Ind. Appl.* **2018**, *54*, 3656–3667. [CrossRef]
20. Bayerer, R.; Herrmann, T.; Licht, T.; Lutz, J.; Feller, M. Model for Power Cycling lifetime of IGBT Modules—Various Factors Influencing Lifetime. In Proceedings of the 5th International Conference on Integrated Power Electronics Systems, Nuremberg, Germany, 11–13 March 2008; pp. 1–6.
21. Sandia National Laboratories. *PV System Component Fault and Failure Compilation and Analysis*; Technical Report; Sandia National Laboratories: Albuquerque, NM, USA, 2018.
22. Bost, M.; Hirschl, B.; Aretz, A. *Effekte von Eigenverbrauch und Netzparität bei der Photovoltaik*; IOW GmbH: Berlin, Germany, 2011.

electronics

MDPI

Article

Validation of Forward Voltage Method to Estimate Cracks of the Solder Joints in High Power LED

Federica Pinti [1,*], Alberto Belli [1], Lorenzo Palma [1], Massimo Gattari [2] and Paola Pierleoni [1]

[1] Department of Information Engineering (DII), Università Politecnica delle Marche, 60131 Ancona, Italy; a.belli@univpm.it (A.B.); l.palma@univpm.it (L.P.); p.pierleoni@univpm.it (P.P.)

[2] iGuzzini Illuminazione S.p.A, Via Mariano Guzzini 37, 62019 Recanati, Italy; massimo.gattari@iguzzini.com

* Correspondence: f.pinti@pm.univpm.it; Tel.: +39-07-1220-4847

Received: 29 April 2020; Accepted: 27 May 2020; Published: 1 June 2020

check for updates

Abstract: The Light Emitting Diode (LED) has many advantages compared to traditional lamps, such as a long lifetime, color rendering and energy saving. It requires good thermal management, since as the temperature increases, the lifetime decreases. Furthermore, the presence of cracks in the Solder Joint of an LED (SJL) compromises the correct dispersion of heat and causes the joint fatigue. This can lead to a decrease in the lifetime of the assembled LED. In this study, we validated that an SJL can be considered faulty if the Forward Voltage (Vf) acquired before and after thermal cycles increases by more than 2%. The voltage measurement method was validated by comparing the results with the techniques commonly used to evaluate the defects of a solder joint as the X-ray analysis and the metallographic section. The failure analysis results present the probability of failure and the lifetime of the SJL achieved by analyzing the data using the Norris–Landberg Model. The lifetime calculated over 1800 SJLs considered in the validation process is greater than 20 years for 95.9% of the tested LEDs.

Keywords: LED; thermal cycling test; accelerated test; solder joint; cracks

1. Introduction

In the last few decades, the Light Emitting Diode (LED) has become an important competitor for traditional light sources such as incandescent or fluorescent lamps. LED lighting is currently the best light source with respect to its impact on the environment and less consumption of energy with respect to traditional light sources [1,2]. LED technology leads to innovative solutions that traditional lamps can not achieve. It allows the ability to create complex lighting systems for various scenarios, such as smart city [3,4] or smart lighting in the cultural heritage field [5]. Long-term analysis of lighting performance can be conducted by studying lighting parameters, such as luminous flux and the correlated color temperature, which can be controlled and monitored in real time [6]. Despite the investment costs, LEDs offer a longer life and the luminous flux is more constant over time compared to other technologies. In fact, LEDs represent the best solution in interior lighting applications such as homes and offices, and in the automotive sector for high brightness signals such as lights and traffic lights.

In general, different approaches can be used to predict the lifetime of electronic components [7,8]. In lighting, thermal management is the fundamental aspect to guarantee constant yields over time and a high performance of the LED. When temperature increases, several aspects are affected: the light color emitted changes, intensity decreases, and lifetime reduces [9]. Durability and reliability tests, such as thermal and power tests, are generally very expensive, therefore accelerated tests are widely used [10]. Accelerated tests reduce test and response times by accelerating the identification of design and fatigue defects. Among standard tests proposed to characterize the solder joint fatigue, the Thermal Cycling Test (TCT) is the most widely used by the scientific community [11].

One of the most relevant problems of LED behavior in lighting systems is the solder joint fatigue between Printed Circuit Board (PCB) and the LED [12]. In literature, many works deal with the Solder Joint of LED (SJL) reliability. The main methods of analyzing the solder joint lifetime are based on strain density [13], on junction temperature measurement [14], or on the effect of voids [15].

One of the main challenges in studying the solder joints fatigue is to detect cracks and monitor their propagation [16]. The SJL strongly depends on the different components of the device such as the LED type, the soldering paste, the substrate, and the finish. In particular, cracks develop due to the large variation of the Coefficient of Thermal Expansion (CTE) between LED and PCB. The presence of cracks is quite complicated to characterize as they are difficult to identify through a non-destructive method [17,18]. Scanning Electron Microscopy (SEM), cross scanning, and dye-and-pry cannot be used for in situ monitoring of the cracks because they all are destructive methods [17].

In the study of Zhao et al. [19] a model enables the analysis of the thermal–mechanical performance of solder interconnects. The experiment was conducted under cyclic temperature, taking into account that time and temperature depend on the creep behavior of the interconnected material. They assessed that the fatigue resistance is sensitive for changes in the thickness and meniscus of solder interconnects.

Elger et al. use the transient thermal analysis for the detection of cracks. They developed a method based on thermal analysis. The increase of the thermal resistance between the initial signal and the signal after cycles is correlated with the presence of cracks in the solder joint by cross sections [20]. Other researchers have so far relied on the detection of resistance variation to estimate the presence of cracks. In literature there are different criteria to define the resistance variation, for example the resistance change of 5 ohm [21], an increase of the resistance of 10 ohm [22], and a resistance threshold of 450 ohm [23]. The resistance variation defined in the JESD22-B111 standard is a 1000 ohm resistance threshold lasting 1 µs, or a 100 ohm resistance threshold, if the initial resistance value is less than 85. The solder joint failure described in the IPC/JEDEC-9702 standard is given by a 20% increase in resistance. In the study of Henshall et al. [24] three different electrical fault criteria are compared: 500 ohm threshold, 20% increase in resistance, and infinite resistance. In this study, we conclude that the use of the IPC-9701A standard failure criterion for 20% increase in resistance provides the most sensitive failure measure. Other studies assert that solder joint cracking can be defined by evaluating the increase in the differential voltage [10,25]. Osram Spa defines the 2% increase in the Forward Voltage (Vf), the failure criterion, which corresponds to the 20% increase in resistance [26].

In this paper, we aimed to validate the failure criterion of a 2% increase in the forward voltage as a method to detect cracks in the SJL. It is easy to implement, repeatable, non-destructive, and requires common measuring instruments present in many laboratories. The study presents the experimental protocol, the accelerated test carried out, the failure analysis, and the obtained SJL lifetime.

2. Materials and Methods

The experimental protocol section consists on the description of the circuits, which are subjected to the TCT and the voltage measurements method. TCT and the oscilloscope test, which is carried out to validate the voltage measurement method, are described. Finally, we present the traditional methods used to evaluate solder defects, the X-ray analysis for voids and cracks, and the metallographic section for a complete analysis of the physical structure of the components.

2.1. Experimental Protocol

Following the IPC-9701A standard and Osram's directives, an experimental protocol to detect cracks in SJL subjected to the thermal cycling test is proposed. A total of 42 circuits, of which 40 are subjected to thermal test, and 2 are used as a reference, were developed. Each circuit consisted of

45 LEDs arranged in 3 rows by 15 columns as can be seen in Figure 1. In total, 1800 LEDs were used in this study.

The composition of each circuit is defined below:

- LED Samsung LH181B.
- Senju solder paste.
- Substrate in thermal CEM3.
- Passivated copper finish.

The instrumentation used consisted of a Source Meter Unit (SMU), a thermometer, and a PC. A software capable of sending commands to the instrument to generate a current pulse and save the voltage measurement was implemented. The Keithley 2400 SMU [27,28] is the unit that sends the current impulse and measures the voltage across the Device Under Test (DUT). The advantage of this instrument is that it allows a simple and compact configuration compared to the use of a separate Digital Multimeter (DMM) and source, and allows for the 4-terminal measurement configuration. The fundamental principle of the 4-wire connection is to supply power to the device and perform the measurement of the Vf through two separate connection lines. Using this configuration, it is possible to connect the measurement cables directly to the ends of the device to measure voltage, thus avoiding the influence of resistive drops on the power supply wires. The connection between the SMU and the DUT consisted of a bed of nails. This is a traditional electronic test fixture made up of 4 spring contacts (spring-loaded pins), which create the electrical connection between the measuring instrumentation and the unit. Figure 1 shows the experimental setup used in this study.

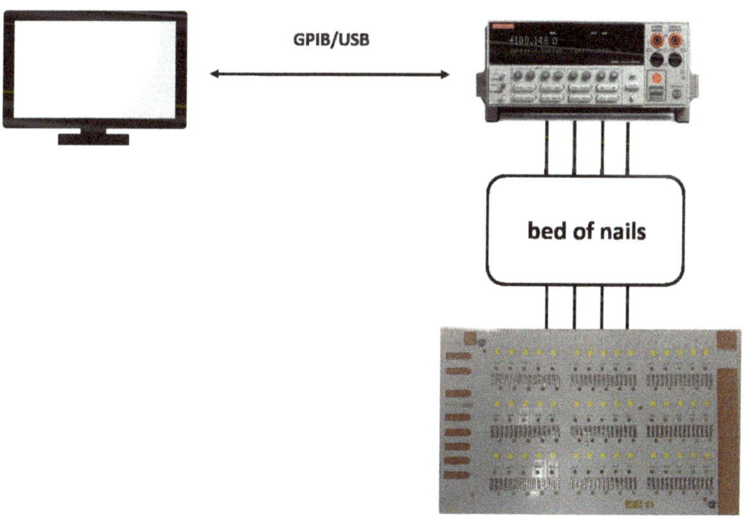

Figure 1. Instrumentation used in the experimental protocol. The PC was connected to the Source Meter Unit (SMU) with General Purpose Interface Bus—Universal Serial Bus (GPIB-USB), the SMU and bed of nails were linked with a 4-wire connection, and the bed of nails created the electrical connection with the DUT via 4 spring contacts.

Before subjecting the circuits to TCT, the voltage of each LED was measured at a controlled temperature (25 °C), inside a fixed temperature chamber. In order to monitor the temperature, a Fluke 51 thermometer with a type k thermocouple was used. This first measure was fundamental in the failure analysis because it represents the reference value.

The Vf was measured by sending a 20 ms current pulse at 10 mA to each LED. The instrument allowed us to set the integration time of the A/D converter that represents the period of time in which the input signal was measured. The integration time affects the usable digits, the amount of reading noise, and the ultimate reading rate of the instrument. The integration time was divided into three different speeds: the fastest integration time (fast), which results in increased reading noise and fewer usable digits, the slowest integration time (high accuracy), which provides the best noise immunity, and the default setting (normal), which is a compromise between speed and noise. In order to have both good speed and low noise, normal was used, which resulted in a 20 ms pulse. The current value was fixed at 10 mA in order to avoid heating the LED, and the consequently variation of Vf.

In total, 1900 thermal cycles were performed and the test, at every 100 cycles, was suspended to carry out the measurement of Vf [20]. Before measuring, we waited until the circuit temperature reached 25 °C inside the fixed temperature chamber.

The failure criterion was an increase of 2% in the forward voltage. For each 100 cycles, the forward voltage measured were compared with the reference value measured before the start of the TCT. A failure was recorded once the following condition was met:

$$Vf(x) - Vf(0) > 0.05 \tag{1}$$

where Vf (x) is the forward voltage measured at each cycle, and the Vf (0) is the reference value. The threshold value of 0.05 V is determined by calculating the 2% of the reference value, considering LEDs have an average Vf of 2.57 V.

The MATLAB programming environment was used to control the instrument, and in data analyzing.

2.2. Thermal Cycling Test

TCT is important because thermo-mechanical stress is one of the main reason for structural failures, causing the cracking of solder joints between the package and board. LED modules were exposed to temperature tests between −40 to 120 °C [29]. Due to differences in thermo-mechanical stress under hot and cold conditions, important information can be obtained. TCT is the process of testing the SJL by subjecting it to temperature conditions in excess of its normal service parameters to uncover faults in a short amount of time. The TCT was performed in a climatic chamber CST 27/2T of Angelantoni test technologies s.r.l. visible in Figure 2. It consisted of two chambers and a transfer carriage, which permits movement of circuits from one chamber to another. Proprietary software allowed us to start the test, set the specifications, and save the acquisition. Once the software started, the oven required 10 min to reach the set temperature. After that the test started. As shown in Figure 3, DUT were kept for 15 min in a cold chamber and 15 min in a hot chamber, the movement occurred through the transfer carriage with a 5 s transfer time. The test continued by alternating between the hot and cold chambers for up to 100 cycles.

In each chamber, an offset of 5 °C (recommended value of the oven manufacturer) was set, which guarantees keeping the oven temperature at −40 and 120 °C, respectively. This was an important setting because the movement from one chamber to another involves the variation of the temperature. It was possible to view the acquisition of the data through the proprietary software during the test.

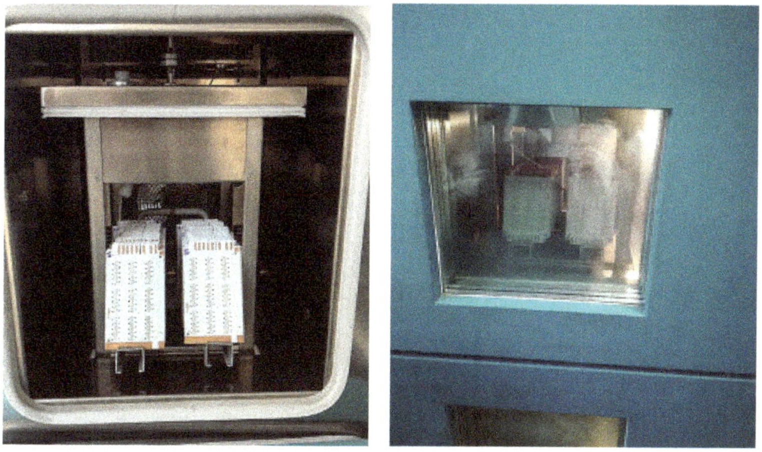

(**a**) (**b**)

Figure 2. (**a**) Circuits positioned in the climatic chamber CST 27/2T of Angelantoni test technologies. (**b**) A mechanical transfer carriage moves circuits from one chamber to another.

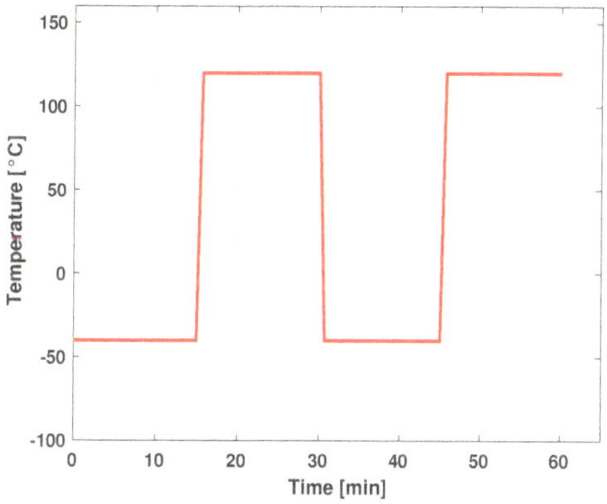

Figure 3. Thermal profile of the Thermal Cycling Test (TCT), which the circuits were subjected to.

2.3. Oscilloscope Test

In order to validate the experimental protocol, current and voltage were measured at the end of each acquisition to check that they were steady. The instrument used for the measurements is the Oscilloscope LeCroy waverunner LT322 Series (LeCroy, New York, NY, USA). The test was carried out by powering each LED at 10 mA in direct current. Typically, the oscilloscope has 2 type of probes: the voltage probes and the current probes. We connected the voltage probes to the LED and the current probes to the power cable. The oscilloscope displays electrical signal trends over the time. In order to obtain a stable signal, the function trigger set to single modality was used, in order to view only one scan of the signal at a time. The display of the oscilloscope shows in red the trend of the current and in blue the trend of the voltage. When the LED was powered, both signals increased: the current rises from 0 mA to 10 mA and the voltage rises from 0 V to 2.57 V. The transient time was 0.08 ms (80 µs). The delay of SMU between sending the impulse and the voltage measurement response was 1 ms. This value was bigger than the transitional time, therefore, the signals of current and voltage were in a steady state when the measurement was acquired.

2.4. X-ray Analysis

X-ray analysis is widely used in the electronic field to evaluate solder defects. SJL inspection allows for detecting structural defects of the electronic boards such as missing solder, the presence of voids, and short circuits. In this study, X-ray analysis was performed to obtain a complete analysis of the solder joint and to evaluate the presence of cracks. In order to validate the analysis method, the X-rays of LEDs that have not undergone the TCT were compared with the X-rays of SJL that reached failure.

2.5. Metallographic Section

The metallography consists in the study of the physical structures of crystals and metal alloys using a microscope. In this way, it is possible to determine the dimensions and the shape of crystals, the distribution of the phases, the direction of the slipping lines, the level of purity, and the presence of contaminations. The metallographic section was divided into six phases:

1. Sample sectioning: the sample was sectioned in order to allow the resin to be incorporated. Given the LED fragility, this operation was made manually.
2. Sample incorporation: the sample was incorporated into an acrylic resin to allow the smoothing and grinding phases and to simplify the microscopic observation.
3. Smoothing: this phase consisted of smoothing the sample using diamond grit abrasive paper.
4. Grinding: this phase was performed with diamond abrasive papers.
5. Polish: the samples ground were immersed into a corrosive chemical solution, which highlights the structures.
6. Microscopic observation: the observation of the sample was carried out with Axioplan Zeiss microscope.

Metallographic section was used to verify the presence of cracks in the SJL, and to evaluate length and thickness of the cracks. In order to validate the 2% increase of Vf as a method of estimating cracks in the SJL, the metallographic section of the unbroken SJL was compared with the metallographic section of failed SJL.

3. Physics Failure Analysis

3.1. Weibull Analysis

The Weibull distribution is a continuous probability distribution, described by two parameters: the scale parameter η and the shape parameter β. Weibull analysis is a methodology used for performing lifetime data analysis. Weibull distribution defines a failure curve, which is normally used in the industrial environment for all electronic applications.

The Weibull distribution is determined by the following equation:

$$F(t) = 1 - e^{[-(t/\eta)]^{\beta}} \tag{2}$$

where $F(t)$ is the probability of failure at t number of cycles or hours, β is the shape parameter or Weibull slope, and η the scale parameter or characteristic Weibull lifetime.

3.2. Norris–Landberg Model

The solder joint fatigue is a low-cycle failure and it originates from the Coffin–Manson model [30]. In fact, almost all lifetime prediction models are derived from Coffin–Manson. The Coffin–Manson model was used to evaluate the growth of cracks in the solder joint, the mechanical failures, the fatigue of the material, and the deformation of the material.

The model is described by the following equation:

$$N(\triangle\epsilon_p)^n = C \tag{3}$$

where N is the number of cycles to failure, $\triangle\epsilon_p$ is the plastic strain range per cycle, n is an empirical material constant, and C is a proportionality factor. The number of cycles obtained with the Coffin–Manson model depends on the plastic deformation, which increases as the cycles increase. The measurement of this deformation is often difficult to obtain, therefore, other models that consider temperature variation as a fundamental parameter were developed. The Norris–Landberg model assumes that the plastic deformation range is proportional to the temperature excursion range, and introduces two other factors taking into account the cycling frequency and the maximum temperature of the solder paste. In this way, the Acceleration Factor (AF) can be defined as:

$$AF = \frac{N_{field}}{N_{test}} = \left(\frac{f_{field}}{f_{test}}\right)^{-m}\left(\frac{\triangle T_{field}}{\triangle T_{test}}\right)^{-n}\left(e^{\frac{Ea}{K}\left(\frac{1}{T_{max,field}}-\frac{1}{T_{max,test}}\right)}\right) \tag{4}$$

where *field* and *test* indicate the real condition and the test condition, respectively. f is the cycling frequency expressed by the cycles per day at the *field* and *test* condition, m is a cycling frequency exponent and a typical value is about 0.33. $\triangle T$ is the temperature range during cycles at *field* and *test*, T_{max} is the maximum temperature of the joint in *field* and *test* condition expressed in Kelvin, n is the exponent of the temperature range, a typical value is about 1.9. Ea is the activation energy and K is the Boltzmann constant, $\frac{Ea}{K}$ is 1414 [30]. SJL lifetime represents the period of time during which the joint guarantees the correct operation. The lifetime can be expressed in different measurement units depending on the system application. In lighting applications, it is common to express the lifetime in hours [31]. To simplify the understanding of the data it is advisable to express lifetime in years, especially in the industrial and commercial environment. The equation below computes the lifetime starting from N_{field}:

$$Lifetime_{years} = \frac{N_{field}}{f_{field}} \times \frac{1}{365} = \frac{AF \times N_{test}}{f_{field}} \times \frac{1}{365} \tag{5}$$

4. Results

The presence of cracks and the failure of SJL were evaluated considering the 2% increase in Vf with respect to the reference value. In total, at the end of the 1900 cycles, there are 74 faulty SJLs.

Figure 4 shows the number of SJLs that failed as the thermal cycles increase. The first failure was at 1400 cycles with one broken SJL. At 1500 cycles one more SJL failed, and at 1600 cycles, another 10 SJLs failed. At 1900 cycles, the amount of broken SJLs increased to 37. In the last cycle, the number of failure assumed an exponential trend.

Figure 4 was obtained by fitting the data of failed SJLs. The analytical formula, which described the curve, is presented below:

$$N_{failed}(x) = 0.00015 \times e^{0.0065x} \tag{6}$$

Coefficients are calculated with 95% confidence bound and the R^2 value obtained is 0.94.

In order to evaluate the SJL performance it is necessary to determine the probability of reliability and the probability of failure. The reliability R(x) indicates the probability that solder joint has a correct functioning during TCT. The probability of failure determines the probability that solder joint presents cracks. The reliability and probability of failure are described by Equations (7) and (8):

$$R(x) = \frac{N_{total} - N_{failed}(x)}{N_{total}} \tag{7}$$

$$F(x) = 1 - R(x) \tag{8}$$

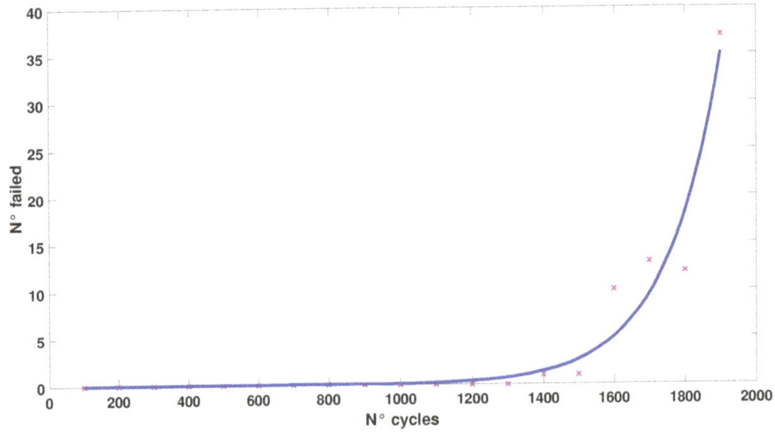

Figure 4. Trend of the failures at each cycle. Markers represent the number of failed determined at each cycle.

As thermal cycles increase, the percentage of reliability decreases, from 1% to 0.96%. Instead, the probability of failure $F(x)$ increases when thermal cycles increase. At the end of thermal cycles, the unreliability increases up to 0.035% . Figure 5 shows the unreliability assessed through the SJLs that reach the failure during test. This curve is obtained fitting the data of unreliability and the analytical formula of the curve is represented by:

$$F(x) = 0.00013 \times e^{3.5x} \tag{9}$$

Coefficients are calculated with 95% confidence bound and the R^2 value obtained is 0.99.

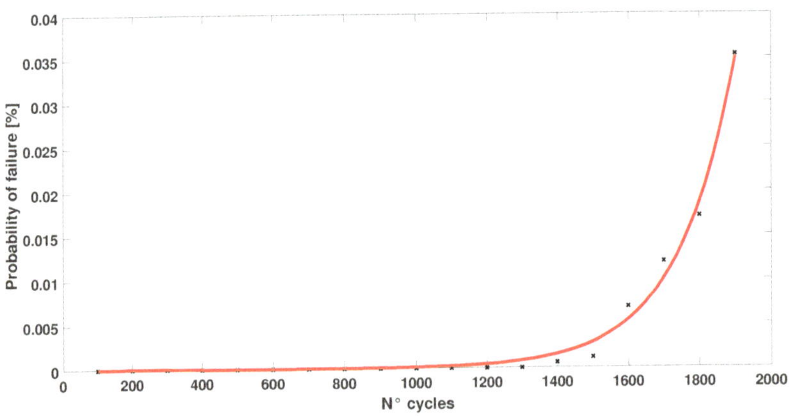

Figure 5. Variation of the probability of failure related to the thermal cycles variation.

As we said in the previous chapter, the Weibull distribution is the most used function in the electronic environment to represent faults analysis. Figure 6 represents the acquired data using the Weibull distribution, including Weibull parameters.

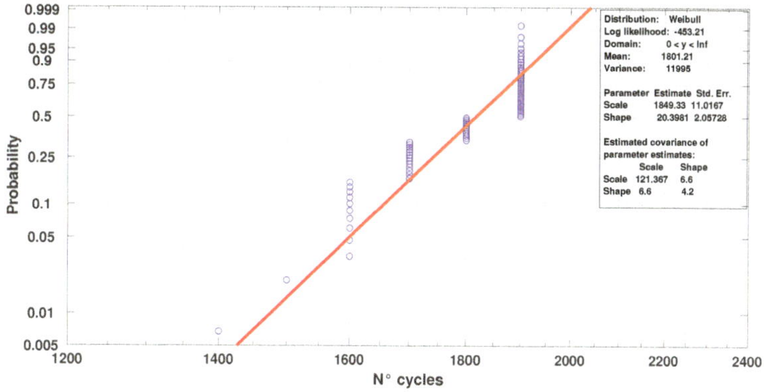

Figure 6. Graphical representation of data using the Weibull distribution. Purple dots represent the Solder Joints of an LED (SJLs) that reached the failure at every thermal cycle; the red line describes the distribution curve of the faults.

Results refer to the relationship between real condition and test condition. We have chosen a frequency cycling of 6 on–off per day on the basis of the data provided by the company according to the type of LED and the final application of the product. TCT frequency cycling depends on the type of the accelerated test used. In this case, the frequency cycling is of 48 on–off per day. TCT requires that devices remain 15 min inside hot chamber and 15 min inside cold chamber. Each cycle ended in approximately 30 min, therefore, 48 cycles were performed in 24 h corresponding to the number of daily on–offs. Table 1 shows the results obtained from the TCT.

Table 1. The first column presents the thermal cycles, the second the number of failed SJLs at each cycle, and the third the total number of failed at each cycle. The fourth column presents the average increase in Forward Voltage (Vf) and the related standard deviation. Last column reports the SJL lifetime.

N° Cycles	SJLs Failed (Samples)	Total SJLs Failed (Samples)	Vf Increase (Mean ± std) (%)	SJL Lifetime (*AF* = 28.8) (Years)
100	0	0	0	N/A
200	0	0	0	N/A
300	0	0	0	N/A
400	0	0	0	N/A
500	0	0	0	N/A
600	0	0	0	N/A
700	0	0	0	N/A
800	0	0	0	N/A
900	0	0	0	N/A
1000	0	0	0	N/A
1100	0	0	0	N/A
1200	0	0	0	N/A
1300	0	0	0	N/A
1400	1	1	3.6 ± 0.0	17.1
1500	1	2	3.3 ± 0.6	18.4
1600	10	12	3.6 ± 1.4	19.7
1700	13	25	4.1 ± 1.8	21.1
1800	12	37	4.3 ± 2.1	22.4
1900	37	74	4.5 ± 2.8	23.7

SJL lifetime represents the estimated number of years in a real installation during which the joint guarantees the correct operation.

At the ends of TCT, the 4.1% of the SJLs have reached failure still providing a SJL lifetime between 17 and 24 years, the remaining 95.9% of the SJLs present a joint lifetime longer than 24 years.

In order to validate the methodology used to identify cracks, X-ray analysis and metallographic section were performed. The X-ray analysis permits us to evaluate the presence and the size of voids inside the solder joint. This type of analysis allows us to visualize the air trapped in the soldering paste. Voids are observable as they have a different color than the surrounding environment, as we can see in Figure 7a. This type of visualization also allows us to distinguish the cracks formed inside the joint [32]. The crack represents a fracture inside the solder paste, which is equivalent to an air passage, and it is easily distinguishable by X-rays. Cracks cause a malfunction of the SJL as electrical continuity is compromised. X-ray analysis is a non-destructive method that allows us to detect cracks which compromise the correct dispersion of heat and cause an increase of Vf. Figure 7a shows the X-ray analysis of a LED not used in TCT. In this figure, it is possible to notice the presence of voids in the SJL. It does not lead to malfunction because the percentage of voids is lower than the percentage defined in the IPC-A-610D standard. Figure 7b, presents the X-ray analysis of a SJL submitted to the TCT, and that reached the failure. Through analysis of the image, it is possible to identify cracks in the solder joint.

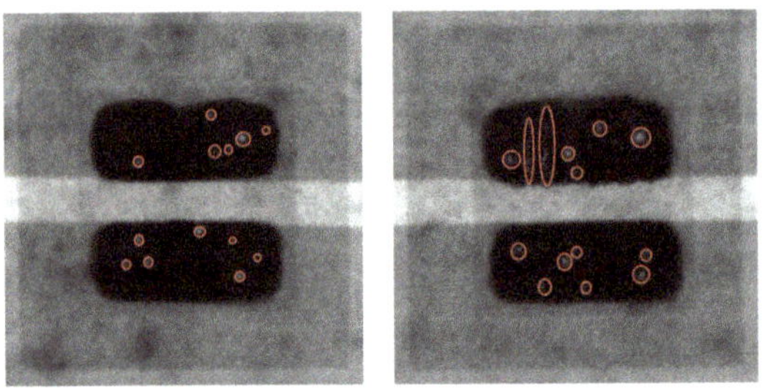

(a) (b)

Figure 7. (**a**) X-ray analysis of an LED that has not undergone TCT, red circles underline the voids present in the joint. (**b**) X-ray analysis of a failed SJL with voids and cracks pointed out by red circles.

The metallographic section is a complete analysis of the physical structure of all components in a circuit. This technique is widely used to evaluate the reliability of the solder joint but has the disadvantage of being a destructive method. In our workn this analysis was used to validate the results obtained by the voltage measurement method. The analysis was made and analyzed using the Axioplan Zeiss microscope. The metallographic section shows the solder joint between LED and PCB. In Figure 8a, a metallographic section of a non-failed SJL is presented. The SJL has no cracks. It guarantees the correct electrical operation without any increase in forward voltage.

Conversely, in failed SJL, the metallographic section shows cracks formed during TCT [17]. Figure 8b presents an example of a broken SJL. Cracks are along the entire joint.

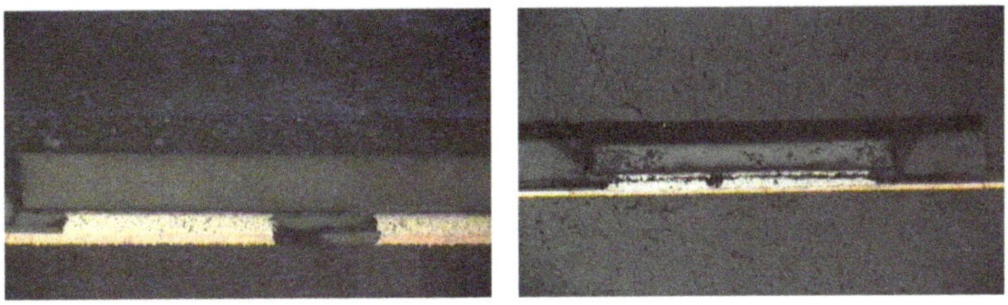

(a) (b)

Figure 8. (**a**) A metallographic section of an unbroken SJL with no cracks. (**b**) A metallographic section of a broken SJL with evident transverse cracks.

5. Discussion

The aim of this work was not to characterize the lifetime of the LED but to provide a testing method usable to furnish significant evaluation in lighting applications. In this paper, a validation of 2% failure criterion proposed by Osram was carried out. Results of our paper check the validity of this criterion to detect the presence of cracks in the SJL. At 1900 cycles 4.1% of the considered SJL have an increase of 2% of Vf.

In the analysis of failure, the reliability and the probability of failure were calculated. The analysis demonstrates that the probability of failure grows exponentially related to the thermal cycles variation. As can be seen from Table 1, the Vf increase grows at each cycle with the number of thermal cycles. Results show the correlation between the solder joint fatigue and the number of thermal cycles. The AF was determined using the Norris–Landberg model by which the SJL lifetime is calculated.

The lifetime variation according to the number of cycles and the number of failed SJLs was analyzed. The SJL lifetime increases linearly with the number of thermal cycles, as can be deduced from Equation (5). On the contrary, the trend of the lifetime related to the number of failed SJLs was approximated with the normal distribution.

6. Conclusions

This paper proposes to validate the criterion of 2% increase in Vf to detect cracks in the SJL. In order to reduce test and response time, an accelerated test was carried out. TCT was chosen because it is the most used test in the scientific community. In total, 1900 thermal cycles were performed and the comparison between Vf and the reference value was calculated. A difference greater than 2% represents the failure criterion for SJL.

In the experimental protocol, 40 circuits with 1800 LEDs were subjected to the TCT and 4.1% of the SJLs failed. The probability of failure obtained has an exponential trend. The SJL lifetime was determined using the Norris–Landberg Model. From obtained results, 95.9% of the considered SJL had a lifetime greater than 20 years. This method was validated by comparing the results of Vf increase with the X-ray analysis and the metallographic section. The comparison allows us to conclude that all the SJLs that showed an increase in Vf actually present cracks.

Obviously, this is not the only parameter to consider in calculating the lifetime of a product, where the overall duration is given by the component with the shortest lifetime; however, the proposed parameter represents a validated method that allows companies to estimate the SJL duration in a short time and to calibrate the guarantees also based on these results.

Author Contributions: Formal analysis, F.P.; funding acquisition, M.G. and P.P.; investigation, A.B. and L.P.; methodology, F.P. and A.B.; project administration, M.G. and P.P.; resources, F.P. and L.P.; supervision, P.P.; writing—original draft, F.P.; writing—review and editing, L.P. All authors have read and agreed to the published version of the manuscript.

Funding: This research received no external funding.

Acknowledgments: The authors are grateful to the iGuzzini Illuminazione S.p.A. (www.iguzzini.com), Silga S.p.a. (www.silga.com), Gitronica S.p.a. (www.gitronica.com) for the opportunity to carry out this research work, they also acknowledge for the technical support of the materials used for experiments.

Conflicts of Interest: The authors declare no conflicts of interest.

References

1. Corral-García, J.; Lemus-Prieto, F.; González-Sánchez, J.L.; Pérez-Toledano, M.Á. Analysis of Energy Consumption and Optimization Techniques for Writing Energy-Efficient Code. *Electronics* **2019**, *8*, 1192. [CrossRef]
2. Mataloto, B.; Ferreira, J.C.; Cruz, N. LoBEMS—IoT for Building and Energy Management Systems. *Electronics* **2019**, *8*, 763. [CrossRef]
3. Leccese, F.; Cagnetti, M.; Trinca, D. A smart city application: A fully controlled street lighting isle based on Raspberry-Pi card, a ZigBee sensor network and WiMAX. *Sensors* **2014**, *14*, 24408–24424. [CrossRef] [PubMed]
4. Costa, D.G.; Duran-Faundez, C. Open-source electronics platforms as enabling technologies for smart cities: Recent developments and perspectives. *Electronics* **2018**, *7*, 404. [CrossRef]
5. Pierleoni, P.; Belli, A.; Palma, L.; Valenti, S.; Raggiunto, S.; Incipini, L.; Ceregioli, P. The Scrovegni Chapel Moves Into the Future: An Innovative Internet of Things Solution Brings New Light to Giotto's Masterpiece. *IEEE Sens. J.* **2018**, *18*, 7681–7696. [CrossRef]
6. Raggiunto, S.; Belli, A.; Palma, L.; Ceregioli, P.; Gattari, M.; Pierleoni, P. An Efficient Method for LED Light Sources Characterization. *Electronics* **2019**, *8*, 1089. [CrossRef]
7. Chiodo, E.; Mazzanti, G. A new reliability model for power system components characterized by dynamic stress and strength. In Proceedings of the 2006 IEEE International Symposium on Power Electronics, Electrical Drives, Automation and Motion (SPEEDAM 2006), Taormina, Italy, 23–26 May 2006; pp. 840–845.
8. Chiodo, E.; Di Noia, L.; Mottola, F. Electrical insulation components reliability assessment and practical Bayesian estimation under a Log-Logistic model. *Int. J. Eng Technol.* **2018**, *7*, 1072–1082. [CrossRef]
9. Abbing, F.D.R.; Pertijs, M.A. Light-emitting diode junction-temperature sensing using differential voltage/current measurements. In Proceedings of the 10th IEEE SENSORS Conference, Limerick, Ireland, 28–31 October 2011; pp. 861–864.
10. Chang, M.H.; Das, D.; Varde, P.; Pecht, M. Light emitting diodes reliability review. *Microelectron. Reliab.* **2012**, *52*, 762–782. [CrossRef]
11. Magnien, J.; Rose, J.; Pfeiler-Deutschmann, M.; Hammer, R.; Mitterhuber, L.; Defregger, S.; Schrank, F.; Kraker, E. Accelerated thermo-mechanical test method for LED modules. In Proceedings of the 2016 17th IEEE International Conference on Thermal, Mechanical and Multi-Physics Simulation and Experiments in Microelectronics and Microsystems (EuroSimE), Montpellier, France, 18–20 April 2016; pp. 1–6.
12. Engelmaier, W. Fatigue life of leadless chip carrier solder joints during power cycling. *IEEE Trans. Compon. Hybrids Manuf. Technol.* **1983**, *6*, 232–237. [CrossRef]
13. Fulmek, P.; Langer, G.; Wenzl, F.P.; Nemitz, W.; Schweitzer, S.; Hoschopf, H.; Nicolics, J. Direct junction temperature measurement in high-power leds. In Proceedings of the 2014 37th IEEE International Spring Seminar on Electronics Technology, Dresden, Germany, 7–11 May 2014; pp. 58–63.
14. Ekpu, M.; Bhatti, R.; Okereke, M.I.; Mallik, S.; Otiaba, K. Fatigue life of lead-free solder thermal interface materials at varying bond line thickness in microelectronics. *Microelectron. Reliab* **2014**, *54*, 239–244. [CrossRef]
15. Yunus, M.; Srihari, K.; Pitarresi, J.M.; Primavera, A. Effect of voids on the reliability of BGA/CSP solder joints. *Microelectron. Reliab.* **2003**, *43*, 2077–2086. [CrossRef]
16. Zhao, X.; Caers, J.; Noijen, S.; Zhong, Y.; De Jong, M.; Gijsbers, H.; Elger, G.; Willwohl, H. Potential interconnect technologies for high power LEDs assemblies. In Proceedings of the 2012 4th IEEE Electronic System-Integration Technology Conference, Amsterdam, The Netherlands, 17–20 September 2012; pp. 1–8.
17. Pan, J.; Silk, J. A study of solder joint failure criteria. In Proceedings of the International Symposium on Microelectronics. International Microelectronics Assembly and Packaging Society, Long Beach, CA, USA, 9–13 October 2011; pp. 000694–000702.

18. Lee, W.; Nguyen, L.; Selvaduray, G.S. Solder joint fatigue models: review and applicability to chip scale packages. *Microelectron. Reliab.* **2000**, *40*, 231–244. [CrossRef]
19. Zhao, X.; Watte, P.; de Vries, H.; van Hees, G. Challenges in Predicting the Solder Interconnect Lifetime of High Power Electronics. In Proceedings of the 2018 19th IEEE International Conference on Electronic Packaging Technology (ICEPT), Shanghai, China, 8–11 August 2018; pp. 1511–1517.
20. Elger, G.; Kandaswamy, S.V.; Derix, R.; Conti, F. Detection of solder joint cracking of high power leds on AI-IMS during temperature shock test by transient thermal analysis. In Proceedings of the 20th IEEE International Workshop on Thermal Investigations of ICs and Systems, London, UK, 24–26 September 2014; pp. 1–6.
21. Suhling, J.C.; Gale, H.; Wayne Johnson, R.; Nokibul Islam, M.; Shete, T.; Lall, P.; Bozack, M.J.; Evans, J.L.; Seto, P.; Gupta, T.; et al. Thermal cycling reliability of lead-free chip resistor solder joints. *Solder. Surf. Mount Technol.* **2004**, *16*, 77–87. [CrossRef]
22. Farooq, M.; Goldmann, L.; Martin, G.; Goldsmith, C.; Bergeron, C. Thermo-mechanical fatigue reliability of Pb-free ceramic ball grid arrays: Experimental data and lifetime prediction modeling. In Proceedings of the 53rd IEEE Electronic Components and Technology Conference, New Orleans, LA, USA, 27–30 May 2003; pp. 827–833.
23. Lau, J.; Hoo, N.; Horsley, R.; Smetana, J.; Shangguan, D.; Dauksher, W.; Love, D.; Menis, I.; Sullivan, B. Reliability testing and data analysis of lead-free solder joints for high-density packages. *Solder. Surf. Mount Technol.* **2004**, *16*, 46–68. [CrossRef]
24. Henshall, G.; Bath, J.; Sethuraman, S.; Geiger, D.; Syed, A.; Lee, M.; Newman, K.; Hu, L.; Kim, D.; Xie, W.; et al. Comparison of thermal fatigue performance of SAC105 (Sn-1.0 Ag-0.5 Cu), Sn-3.5 Ag, and SAC305 (Sn-3.0 Ag-0.5 Cu) BGA components with SAC305 solder paste. In Proceedings of the IPC APEX EXPO, Bannockburn, IL, USA, 31 March–2 April 2009.
25. Soltani, M.; Freyburger, M.; Kulkarni, R.; Mohr, R.; Groezinger, T.; Zimmermann, A. Reliability study and thermal performance of LEDs on molded interconnect devices (MID) and PCB. *IEEE Access* **2018**, *6*, 51669–51679. [CrossRef]
26. GmbH, O. New Package Technology for SSL High Power. Available online: https://fhi.nl/app/uploads/sites/32/2017/09/RUTRONIK-BELGIE.pdf (accessed on 13 May 2020).
27. Tek. Keithley 2400 SourceMeter User's Manual. Available online: https://www.tek.com/datasheet/series-2400-sourcemeter-instruments (accessed on 13 May 2020).
28. Tek. Keithley 2400 Specifications. Available online: https://download.tek.com/manual/2400S-900-01_K-Sep2011_User.pdf (accessed on 13 May 2020).
29. Chen, Q.; Chen, Q.; Luo, X. Fast estimation of LED's accelerated lifetime by online test method. In Proceedings of the 2014 64th Electronic Components and Technology Conference (ECTC), Orlando, FL, USA, 27–30 May 2014; pp. 1992–1995.
30. Vasudevan, V.; Fan, X. An acceleration model for lead-free (SAC) solder joint reliability under thermal cycling. In Proceedings of the 2008 58th IEEE Electronic Components and Technology Conference, Lake Buena Vista, FL, USA, 27–30 May 2008; pp. 139–145.
31. Alfarog, A.O.; Qu, X.; Wang, H.; Blaabjerg, F.; Li, Z. Lifetime prediction of led lighting systems considering thermal coupling between led sources and drivers. In Proceedings of the 2017 43rd IEEE Annual Conference of the Industrial Electronics Society, Beijing, China, 29 October–1 November 2017; pp. 1273–1278.
32. Jiang, C.; Fan, J.; Qian, C.; Zhang, H.; Fan, X.; Guo, W.; Zhang, G. Effects of voids on mechanical and thermal properties of the die attach solder layer used in high-power LED chip-scale packages. *IEEE Trans. Compon. Packag. Manuf. Technol.* **2018**, *8*, 1254–1262. [CrossRef]

electronics

Review

Two Decades of Condition Monitoring Methods for Power Devices

Giovanni Susinni [1], Santi Agatino Rizzo [1,*] and Francesco Iannuzzo [2]

[1] Department of Electrical Electronic and Computer Engineering (DIEEI), University of Catania, Viale Andrea Doria, 6, 95125 Catania, Italy; giovanni.susinni@unict.it
[2] Department of Energy Technology, Aalborg University, 9220 Aalborg, Denmark; fia@et.aau.dk
* Correspondence: santi.rizzo@unict.it; Tel.: +39-095-738-2308

Abstract: Condition monitoring (CM) of power semiconductor devices enhances converter reliability and customer service. Many studies have investigated the semiconductor devices failure modes, the sensor technologies, and the signal processing techniques to optimize the CM. Furthermore, the improvement of power devices' CM thanks to the use of the Internet of Things and artificial intelligence technologies is rising in smart grids, transportation electrification, and so on. These technologies will be widespread in the future, where more and more smart techniques and smart sensors will enable a better estimation of the state of the health (SOH) of the devices. Considering the increasing use of power converters, CM is essential as the analysis of the data obtained from multiple sensors enables the prediction of the SOH, which, in turn, enables to properly schedule the maintenance, i.e., accounting for the trade-off between the maintenance cost and the cost and issues due to the device failure. From this perspective, this review paper summarizes past developments and recent advances of the various methods with the aim of describing the current state-of-the-art in CM research.

Keywords: condition monitoring; junction temperature; power device; reliability; power electronics

check for updates

Citation: Susinni, G.; Rizzo, S.A.; Iannuzzo, F. Two Decades of Condition Monitoring Methods for Power Devices. *Electronics* **2021**, *10*, 683. https://doi.org/10.3390/electronics10060683

Academic Editor: Ahmed Abu-Siada

Received: 1 January 2021
Accepted: 24 February 2021
Published: 15 March 2021

1. Introduction

Nowadays, power electronics is widespread in a huge number of daily applications that improve services for the collective [1]. Furthermore, power electronics has a key role in renewable energy systems [2,3], lighting [4,5], electric mobility [6,7], and other systems that enable sustainable development [8].

A crucial aspect is the reliability and lifetime prediction of the whole power conversion system. The warrant of the highest robustness level while minimizing the product and maintenance cost is extremely mandatory. For example, the devices used in avionic and automotive applications must have a fault rate close to zero that imposes stringent requirements during the system design. In the same way, wind farms must guarantee normal operations without interruption, but this is extremely difficult because of the expensive access to farms for easier maintenance. In this context, many approaches to forecasting the lifetime of power electronic systems and the single power device have been intensively studied.

To guarantee a high level of reliability, it is important to comply with several standards [9–11], and different strategies are usually performed such as the use of the fault-tolerant topologies with redundant components [12,13], and the advanced reliable design of power electronic devices using innovative materials [14,15].

Unfortunately, the enhanced system's robustness does not prevent failure and, indeed, it is never completely foreseeable. Therefore, a maintenance operation before a failure is necessary. Considering the costs of maintenance operations, lifetime prediction combined with condition monitoring approaches [16,17] are very useful tools to choose when a maintenance operation has to be carried out.

Some studies have shown that capacitors are fragile with a failure of 30%, whereas the failure of the PCB and connectors is around 36% [18–20]. The remaining part is related to the semiconductor and soldering failures in device modules that consist of the most important area of concern for converter system failures. In this context, countless condition monitoring (CM) methods for the evaluation of the semiconductor state of health have been widely explored in the literature.

The target of this work is to provide an overview of various CM methods that have been used to evaluate the state of the health of power devices. More specifically, the first part of the proposed review was focused on the impact of the Internet of Things (IoT) and artificial intelligence (AI) technologies used for CM of power devices, with a chronological overview of the main CM methods over the last two decades. Then, it a first CM method based on acoustic emission was presented and used to detect any physical damage in a power module packaging. It is worth underlining that it enables one to estimate the state of aging of a power module.

Then, CM methods based on the optical properties of the semiconductor power devices were presented, including temperature estimation. These methods are usually based on an optical beam that is reflected or scattered back from the semiconductor lattice. There is an inherent dependence between the temperature and the energy related to the photoemission. More specifically, such energy is a function of the junction temperature (JT); hence, in turn, the energy variation can be used to estimate the temperature of the chip. After that, a depth-analysis of the several approaches to extract the junction temperature in the semiconductor devices based on the physical or electrical properties was performed. The early works using a physical CM method for the JT measurement were done by directly contacting the chip surface with a thermo-sensitive material such as a point contact system (such as thermocouples and liquid crystal). On the other hand, electrical methods for the junction temperature measurement are often the preferred choice for CM of power devices because the temperature estimation can be carried out through the measurement of electrical quantities. Among various electrical CM methods, thermal test chips (TTCs) are directly fabricated on the die surface of the device, and the voltage drop can be used to estimate the temperature variations. On the other hand, the temperature sensitive electrical parameters (TSEPs) are based on the measurement of the voltage drop during the converter operations. Generally, the measurement of the voltage drop can be carried out using some voltage probes connected to the device terminals.

Although the measurement of the power devices' junction temperature was widely treated in many ways, it is still an active relevant topic owing to the current trade-off between the advantages and limitations of the methods proposed so far. Therefore, all these methods have been compared in terms of their main aspects such as sensitivity, linearity, cost, and online monitoring operations.

2. Conditioning Monitoring Methods and Their Future Application

In the literature, the research topic based on CM methods has been gaining interest as the various maintenance strategies allow to increase the lifetime of the overall power conversion system. In many applications, it is becoming crucial to monitor the state of the health (SOH) of power devices to prevent a failure, that is, the possibility for the operators to obtain a lifetime estimation, thus properly scheduling any maintenance operations.

Figure 1 depicts a timeline of the various CM methods for power devices that have been widely studied in the literature and used by the industry in the last two decades. The first use of CM methods for power devices dates back to the 2000s, where the measure of on-state voltage of the devices was used as a well-known parameter to monitor the device condition. Later, the measure of the threshold voltage or the gate turn-off voltage of a device was used to estimate the temperature dependence of the power devices by measuring a low voltage; then, the acquisition system was developed to be more simple and less bulky. In the last 5 years, the estimation of the SOH has been carried out using contactless approach such as the acoustic method or the photodiode approach, where the acoustic emission or the light emission, respectively, of a particular device have been taken into account.

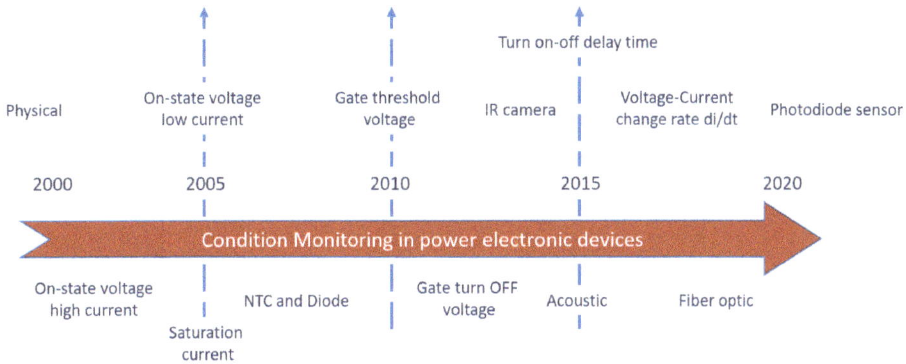

Figure 1. Timeline of the main condition monitoring (CM) methods in the last two decades. IR, infrared; NTC, negative temperature coefficient.

The thermal stresses such as the increasing of the mean temperature and abrupt temperature fluctuations are the main failure mechanisms. Consequentially, the temperature is an index of the power device SOH. Other CM methods focus on other quantities. Table 1 summarizes the physical or electrical quantities measured for each CM method and, in the following subsections, a brief overview is presented. It is worth noting that, among the various CM methods in the literature, the acoustic one is used to detect the state of the aging of the power device without any estimation of the working temperature, while all the other strategies are focused on the estimation of the junction temperature (JT). More specifically, some CM methods perform such estimation by directly measuring the temperature, such as the infrared (IR) camera, a negative temperature coefficient (NTC) resistor, fiber optic cable, and photodiodes sensors. Other methods provide an indirect estimation by mapping electrical quantities in a temperature value, such as the thermo-sensitive electrical parameters (TSEPs).

Table 1. Measured quantities for the different condition monitoring (CM) methods. IR, infrared; NTC, negative temperature coefficient; TTC, thermal test chip; TSEP, thermo-sensitive electrical parameter.

Method	Measured Quantity
Acoustic	Acoustic waves
OPTICAL—Fiber optic	Light wavelength
OPTICAL—Photodiode sensor—internal	Light intensity
OPTICAL—Photodiode sensor—external	Light intensity
OPTICAL—IR camera print	Light wavelength
OPTICAL—IR camera	Light wavelength
Physical	Resistance
TTCs—NTC	Resistance
TTCs—Diode	Voltage
TSEP—On-state voltage, low current	V_{ce}, V_{ds}
TSEP—On-state voltage, high current	Device current, V_{ce}, V_{ds}
TSEP—Saturation current	V_{ce}–V_{ds}
TSEP—Gate threshold voltage	V_{ge}—V_{gs}
TSEP—Gate turn OFF voltage	V_{ge}–V_{ce}, V_{gs}–V_{ds}, Gate resistance
TSEP—Turn on-off delay time	Device current, V_{ce}–V_{ds}
TSEP—Voltage–current change rate di/dt	Device current, V_{ce}–V_{ds}
TSEP—Peak gate current	V_{ge}–V_{gs}

According to recent research, the emerging trend of IoT and AI technologies are gaining more and more interest and they are expanding rapidly in the field of CM methods [21].

AI aims to assist electronic systems with intelligence that is capable of human-like learning and reasoning. This technology possesses countless advantages and has been widely applied in numerous industrial and research areas such as maximum power point tracking (MPPT) control for Photovoltaic (PV) plants, anomaly operation detection for inverter, and prediction of the SOH of a power converter.

The use of enabling AI technologies allows the power converter systems to be embedded with capabilities of self-awareness and self-adaptability, thus system autonomy can be enhanced. Similarly, the development of data science, including sensor technology, IoT, and big data analytics, provides a wide variety of data for power electronic systems throughout different stages of its life-cycle. Furthermore, AI technology can exploit data to estimate the system health status with high sensitivity in condition monitoring for aging detection of power devices. Only in a few works [22–25] has the condition monitoring and fault detection in power electronics AI-based fault detection been presented.

Figure 2 shows a proof-of-concept of the integration between IA and IoT technologies for CM of power devices. More specifically, Figure 2a depicts an example of a user (red box), which consists of a specific power converter and its power devices that have to be monitored, and an example of a provider of CM services (PS) (yellow block). More specifically, the devices of a single-user power application are connected with several sensors that enable the monitoring of the SOH of each device (such as the measurement of the TJ or the state of aging of a device). Then, the controller system interface (CSI) manages the sensors and collects all the measured data. The CSI block plays an important role for the CM of the power system because it is able to provide control signals and it may exchange data with different users in the IoT framework (see Figure 2b).

The PS (sometimes also the CSI) uses AI technologies that act as "intelligent agents", exploiting sensors able at perceiving the environment of the power converter. Indeed, during some power converter operations, the JT can be dangerously increased over a threshold maximum value, hence the AI is able to suddenly shut down the power system. Moreover, the AI-integrated system may enable a power derating in order to avoid any failure of the converter and, hence, the power devices. On the other hand, AI should perform more complex tasks, such as detecting any change in the device behavior that could lead to future malfunctioning; interacting with the CSI "to propose" solutions, i.e., fault preventing strategies, to current anomalies or potential future ones; and so on. Regardless of whether AI is located in the PS, in the CSI, or both, a local backup control system is necessary. The local controller should be simple (without complex hardware and control strategies) and highly reliable. The local controller should have less stringent limits than the AI, thus the local controller intervenes only when the system is close to a dangerous situation. For example, the local converter shuts down the converter when the JT passes over a preset limit and this situation occurs owing to the lack of a smarter command (load reduction, switching frequency variation, and so on) from the AI of the PS because of different causes (e.g., temporary lack of connection, unexpected AI actions or reasoning).

It is worth noting that the PSs are able to contribute to the decision making in both offline and real-time analysis using data acquired in the past and saved into a dedicated database (green silos). From this perspective, all the collected data can be used to enhance the predictive model of the SOH of the power devices.

Moreover, the PS may interact with enterprises, authorities, public companies, and so on. For example, real-time use of CM occurs when a PS notices a critical functioning of the converters of a large PV system. The PS informs the PV owner as well as interacts with the PV management system to enable them to promptly and properly operate. On the other hand, the PS also informs the electric network operator (ENO) of the smart grid where the PV is installed about a potential disconnection of the PV system. This information enables the ENO to adopt countermeasures that mitigate the impact of such a disconnection. In turn, the ENO informs sensitive customers, e.g., hospitals, of a potential lack of service or asks them to turn on a backup system in order to reduce the absorbed power.

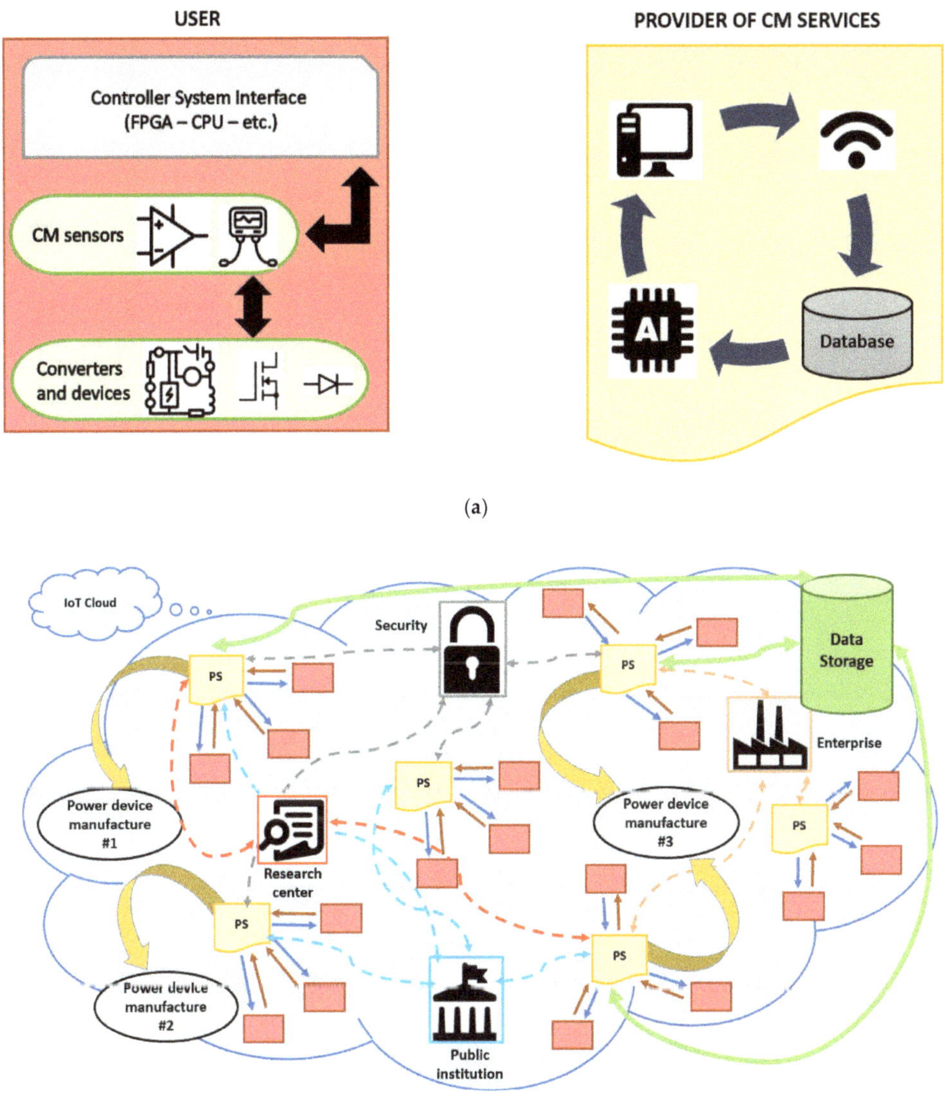

(a)

(b)

Figure 2. (**a**) User and provider of CM services (PS). (**b**) Proof of concept of Internet of Things (IoT) and artificial intelligence (AI) technologies for CM of power devices.

Furthermore, the measurements of the JT and other parameters of the power devices can be used by some public institutions such as universities or research centers (blue pale and red dashed lines) for an off-line analysis of the data to carry out some models of the SOH of the power devices. Moreover, the PS may be connected to a security organization (grey dashed lines) that may collect data and it should be able to interrupt an electrical service in the case of failure.

It is worth underlining that the PSs could provide data to the power device manufacturers with the aim to share information related to the state of aging of the power devices.

The manufacturers can use the data to improve their power devices as well as to obtain more accurate power device models. Indeed, manufacturers can act as PSs.

3. Acoustic Methods

Acoustic emission has been widely investigated in the literature as a CM quantity useful in different application fields such as pumps, industrial electrical machines, and so on. Moreover, in the field of power electronics, acoustic monitoring has been extensively used to detect any defects or damage in transformers and capacitors [25–27]. Only in the last decade, a few works have been focused on the acoustic phenomenon such as a measurement method for monitoring the SOH of the power semiconductor devices [28–33].

Acoustic emission has been used to detect any physical damage in a power module packaging using an acoustic microscope. Furthermore, from the experimental evidence, it has been proven that acoustic emissions are related to the switching operations of power devices. In the case of fast switching operation (tens of nanoseconds), a certain amount of current is switched, which causes a large di/dt, which involves magnetic interaction within the module packaging. This means that the magnetic force could be the source of the acoustic emission, such as the mechanical breaking of the structure inside the component package. However, the physical phenomena causing the acoustic emission are not definitively understood.

Thereby, the device under test (DUT) is monitored contactless with an acoustic sensor that is usually placed in the proximity of the package. It intrinsically eliminates the issues related to contact directly with the voltage probes.

A correlation between the SOH of the power module and the analysis of its acoustic emission during the switching process has been analyzed [28–30]. It has been demonstrated that the acoustic peak in an aged device is smaller in comparison with a new one. However, in these works, only the acoustic emission of an Insulated Gate Bipolar Transistor (IGBT) connected into short-circuit has been investigated. Meanwhile, the authors of [32] present an early experimental setup used to prove that acoustic emission is related to the switching of power semiconductor components. Furthermore, the authors have proposed an analysis based on propagation delays to assess the source of the acoustic emission. The authors in [33] have investigated the acoustic emission as a CM method to measure the fatigue mechanisms in the power module. More specifically, they have investigated the physical degradation, observing the aging process of the whole power module by measuring the frequency spectrum of acoustic emission. The authors in [34] have measured the acoustic emission during converter operations to estimate the aging of a power semiconductor module due to power cycling. However, a spectrum analysis has been conducted to process the acquired data. The experimental results have shown a correlation between acoustic emission and the drain-source voltage, which is a common indicator of degradation of the bond wires of the power module.

As for disadvantages, the acoustic method needs an expensive and complex sensing circuit to correctly decode the acoustic emission. Furthermore, the system has to be shielded against Electromagnetic interference (EMI) and the superposition of noise contributions.

4. Optical Methods

Temperature variation, especially the sudden increase of the JT, plays a significant role in terms of power device reliability [34,35]. CM methods performing on-line JT monitoring raise great interest in terms of planning maintenance operations because the working conditions of a power converter are extremely unpredictable. From this perspective, the CM methods based on the optical properties of the semiconductor power devices were studied in depth because they are useful for temperature estimation. These methods are usually based on an optical beam that is reflected or scattered back from the semiconductor lattice. There is an inherent dependence between temperature and the energy related to the photoemission. More specifically, such energy is a function of the JT, hence, in turn, the energy variation can be used to estimate the temperature of the chip. It is worth remembering that these solutions based on optical quantities have some other drawbacks, such as the high cost and the impracticality in high-voltage converters.

There are various techniques for thermal mapping based on the use of an IR sensor [36–40], IR microscope [41], 2D radiometry [41,42], and the laser deflection technique [43–45], while fiber optic [46–48] and the IR camera [49–54] can both be used to obtain a thermal mapping or the JT value. In the following, the aforementioned optical techniques are briefly discussed.

4.1. Infra-Red/Visible Emission

An IR sensor is able to detect changes in the amount of infrared radiation of an object, which may vary depending on the temperature and surface characteristics of the objects in front of the sensor. The use of the IR sensors as a CM method for the measurement of the JT in a power device [36–39] is almost inexpensive as well as not very intrusive. However, these sensors have a low response time and, furthermore, the IR sensors average out the junction temperature value of the power device, and hence the accuracy is very low.

On the other hand, the emerging trend of the Wide-bandgap (WBG) power devices, such as the SiC power MOSFETs, are more and more diffusing devices as they concurrently enable high switching frequency, high voltage, and high-temperature operations. Therefore, the study of the electroluminescence proprieties of the SiC material for on-line CM has started to attract wide interest. For different reasons, the electroluminescence proprieties have already been studied back in 1907 [55], while, in the last three years, the inadvertent light emission phenomenon in the intrinsic body diode has drawn attention. While the body diode is in forward conduction mode, the chip glows a visible blue light [56]. The light brightness of the SiC body diode strongly depends on both the injected current magnitude and the JT. Hence, the measurement of light brightness can be used as a novel CM method for temperature detection, where only a few works have already focused on this topic [56–58]. The first proof of concept of the SiC light emission in a commercial power module has already proven the potentiality of this CM method [57]. An inexpensive passive sensing circuit, such as a silicon photodiode and a resistor, was adopted and the photodiode output voltage was correlated to the light emission intensity as a function of the temperature. It is worth noting that the system is small enough such that it can be easily embedded in the package. Another approach considers a light circuit sensing using two commercial photodiodes with an active signal conditioning circuit [58]. This approach has been adopted for JT estimation in a real application such as a pulse-width modulation (PWM) driven converter. The temperature-dependent changes in the spectrum of the light emission from the body diode of a SiC module have also been investigated [59]. The method has been proven through static characterization and dynamic double pulse measurement using two silicon photomultipliers, which can detect the peak intensity and, consequentially, the temperature dependence. Different from the previous CM methods, the last one, based on the light intensity of the SiC body diode, enables high-voltage operations and, even more importantly, the JT can be estimated during on-line operations.

4.2. Optical Fibers

The use of optical fiber as a CM method for the power modules has been discussed in very few works [46–48]. It is worth underlining that this CM method can be used without removing the dielectric gel on the power module surface, and the JT can be measured by placing the fiber optic cable in direct contact with the power chip. On the other hand, the measure is able to give only a local temperature; further, for almost all fiber optic methods, the measurement response time is generally high. Furthermore, it requires an external conditioning circuit unit that may be bulky for a specific application.

As an example, in [49], an optical fiber sensor has been used to measure the die temperature of an IGBT power module to estimate the thermal impedance (see Figure 3). The module top lid has been removed because the optical system has to be placed close to the die. Printing the die and bond-wires to increase the emissivity of the chip is usually preferable, but causes a cost increment and severely limits the on-line use of the method.

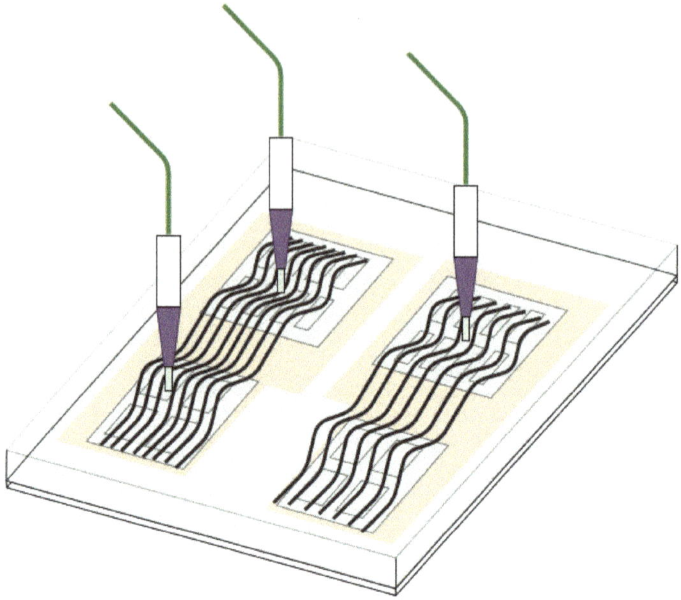

Figure 3. Example of a power module with optical fiber thermal sensors (Based on [49]).

4.3. IR-Detection Apparatuses

IR-detection apparatuses are not really used as a CM method for power devices, but are extremely useful in laboratory testing (e.g., under power cycling) [49–53]. More specifically, the use of an IR camera allows to display the thermal map of the whole surface of the power module under test, as shown in Figure 4. As expected, the module temperature is not uniformly distributed; the temperature gradient between the center and the edge of the module can be greater than 40 °C. Usually, an IR camera is used to carry out a spatial thermal mapping on the device surface, but it is not able to provide an accurate measure of the device JT. It is worth remembering that some temperature measurement errors can be done using an IR camera because of the surface degradation of materials and the intrinsic low emissivity of aluminum. Even in this case, the IR temperature measurements are usually conducted by varnishing the surface of the DUT with a particular solution that increases the thermal emissivity on the surface.

4.4. Other Techniques

The IR microscope [40], 2D radiometry [41,42], and the laser deflection technique [43–45] have also been used as CM methods for power devices. All the aforementioned methods are able to provide a very precise JT estimation of a semiconductor device, but, on the other hand, the devices under test are to be driven with a specific testing sequence, not matching with the real operation in a power converter. They are also very expensive solutions and are not easily embedded in a real application.

Among the various sensors that use the 2D radiometry and laser deflection technique, it is worth remembering the InSb photovoltaic detector, which is a high-speed, low-noise infrared detector that delivers high sensitivity, and with an optical microsensor whose operating principle is based on detecting the absorption, deflection, and phase shift of an optical beam.

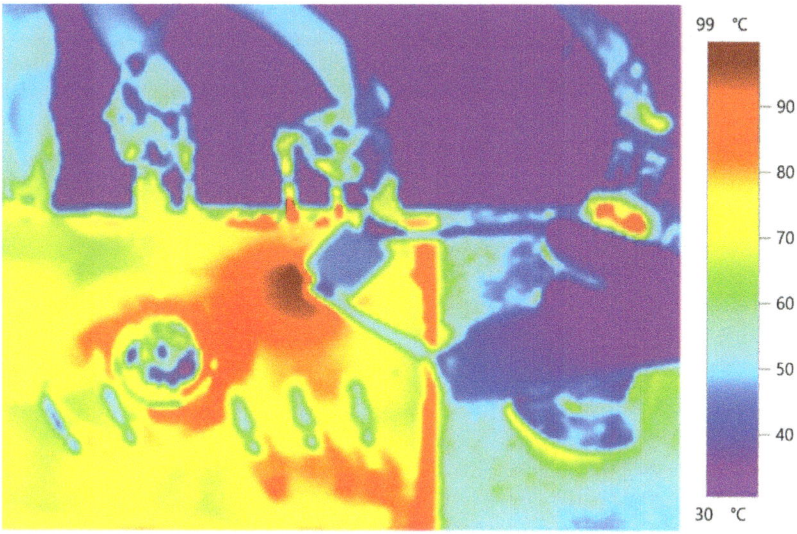

Figure 4. Thermal map of a SiC power module in the case of a current injection in the body diode.

5. Physical Methods

The early works treating CM methods for the JT measurement have been done by directly contacting the chip surface with a thermo-sensitive material such as a point contact system. In this case, direct access to the semiconductor chip is necessary and, consequentially, the package must be removed.

Various equipment has been used for the physical contact measurement, including thermocouples, thermistors, scanning thermal probes, and multiple contact or blanket coatings such as liquid crystals and thermographic phosphors [59–65]. The aforementioned equipment relies on the transfer of thermal energy from the DUT to the thermal sensors. In this case, the spatial resolution related to the contact measurements strictly depends on the size and the thermal capacitance of thermo-sensible materials. The ability to provide a temperature map utilizing a matrix of sensors and a wide spatial resolution (can reach less than 100 nm) are the main advantages. In the following, the aforementioned physical techniques have been briefly discussed.

5.1. Thermocouples

The physical contact methods that rely on the use of thermocouples are not widespread in practical applications as the chip of the power module must be accessible to the thermal probe and, from this perspective, the on-line measurements and high voltage operations are strongly limited. Furthermore, the measurement of the thermal variation of the power module strictly depends on the time response of the probe, which may be considerably slower (few seconds) than the variation of the module JT.

Nowadays, only in a few cases [60,61], the JT of an IGBT module has been experimentally measured during on-line converter operation. More specifically, the temperature has been determined using several thermocouples physically connected to the chip (see Figure 5). On the other hand, several works use the measurement of some thermocouples as target values to prove the accuracy and effectiveness of new on-line junction temperature estimation models [62–65]. For example, the effectiveness of a model carried out for a three-phase power module IGBT by considering the transient thermal impedance has been proven using several thermocouples [63]. An experimental setup and an on-line control system that includes a microcontroller and a matrix of K-type thermocouples have been built up to verify a numerical thermal model for IGBT devices [64]. In [65], an electrical-thermal model has been carried out in terms of both the transient and steady-state responses. To validate the model, an array of thermocouple has been installed on the chip surface. A thermal model based on the Fourier series solution of heat conduction equations has also been validated using several thermocouples placed on the surface of the silicon die, on the base plate, and on the heat sink, in order to characterize the transient electrothermal behavior of an IGBT module [66].

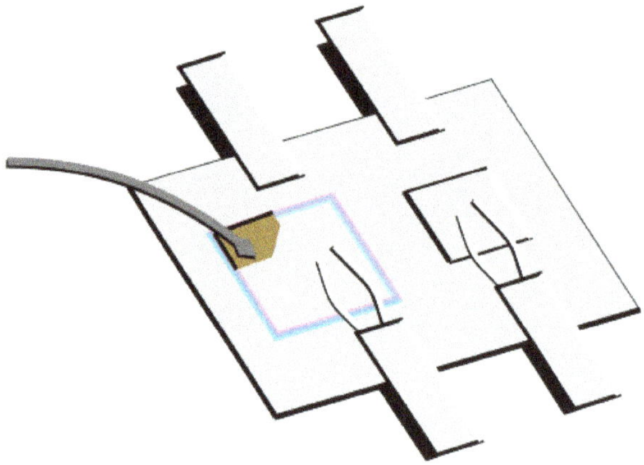

Figure 5. Measurement of the junction temperature (JT) of a device under test (DUT) with a thermocouple probe on the die surface (Based on [62]).

5.2. Liquid Crystals

The earliest physical CM methods for the measurement of the JT in a power device have been obtained by using the scanning thermal probes, as well as multiple contact or blanket coatings, such as liquid crystals and thermographic phosphors [60].

More specifically, the thermochromic liquid crystals consist of a thermal imaging tool for mapping surface and spatial temperature distributions. It is worth remembering that the molecular structure and optical properties of the liquid crystals vary with the temperature. Hence, the JT measurement of a power device can be done by measuring the wavelength of the reflected light. These CM methods have a very good spatial resolution, but on the other hand, they are extremely highly invasive and cannot be used in a real power converter application owing to the bulky sensing circuits.

6. Electrical Methods

Electrical methods for JT measurement are often the preferred choice for CM of power devices because the temperature estimation can be carried out through the measurement of electrical quantities. More specifically, it is worth remembering that the proprieties of the semiconductor materials are temperature dependent and, hence, the measurement of the voltage drop or the current that flows into the device can be used as a valid temperature estimator. Among the various electrical CM methods, thermal test chips (TTCs) are directly fabricated on the die surface of the device, and the voltage drop can be used to estimate the temperature variations. On the other hand, the TSEPs are based on the measurement of the voltage drop (or current) during the converter operations. Generally, the measurement of the voltage drop can be carried out using some voltage probes that are connected to the device terminals. The TSEPs are usually the preferred choice for CM because of their user-friendliness, fast response time to the temperature transients, and good accuracy.

6.1. Thermal Test Chips

TTCs were originally developed for the thermal characterization of device packages [67], and since then have also been used in IGBT power modules. TTCs act as thermal sensors to monitor the JT and they are fabricated on the proximity of the silicon chip.

TTCs can be suitable for on-line temperature measurements. Various types of TTCs have been realized, such as integrated diodes and resistance temperature detectors (RTDs) [67–75]. As the forward voltage of the diodes strongly depends on the temperature variation, the measure of the voltage drop can be used for temperature estimation. It is worth remembering that the temperature presents an exponential dependence on the forward voltage. Likewise, RTDs are also used as the temperature-sensitive parameter because the voltage drop is related to the resistance variations. The variable resistance, R_t, can be expressed as follows:

$$R_t = R_0(1 + \alpha_0 \Delta T) \tag{1}$$

where R_0 is the value of the resistance at 0 °C, α_0 is the resistance temperature coefficient that strictly depends on the material, and ΔT is the temperature variation.

To use TTCs, a modified IGBT power module layout with an accessible on-chip temperature terminal has been proposed in [68]. A string of diodes on the top of the chip has been fabricated and the measurement of the JT has been performed by measuring the forward voltage drop. Instead, in [69] a thin-film RTD placed on the top of the IGBT chip has been realized to measure the average temperature of the die. A similar solution where an NTC thermistor has been embedded in the IGBT power module has been also investigated [70]. Innovative use of a kelvin-emitter resistor, placed directly on the IGBT die surface, as a junction temperature sensor has been also adopted [71]. It provides only a local temperature measurement. Meanwhile, in [72], a chain of integrated diodes has been fabricated on the die surface to investigate the JT variations during a power cycling test.

The widespread nature of SiC power modules in different power electronics applications has also driven forward the research of innovative control techniques that require real-time monitoring or estimation of the module's JT. From this perspective, several works [72–75] have been focused on the development of electrical models of the devices in which several NTC thermistors have been integrated on the die surface. The measurement of the temperature variation enables the estimation of aging of a device and, consequentially, the device model can be continuously updated.

The main drawbacks of the TTCs are the production cost and manufacturing complexity of the embedded sensors. Indeed, such layout modification complexity of the power module packaging can considerably increase and also requires additional terminals for the temperature measurements. Furthermore, it is worth remembering that the diodes and RTDs can be affected by degradations along the lifetime of the device that may affect the accuracy of the measurement. These issues have limited the spread of TTCs in commercial power devices.

6.2. Methods Using the Thermo-Sensitive Electrical Parameters (TSEPs)

The CM methods outlined so far require visual or physical access to the chip. To overcome this limitation, the temperature measurement by thermo-sensitive electrical parameters (TSEPs) has been used as a valid alternative for the estimation of the JT of a power device. The key point consists of correlating the temperature of the semiconductor material with the electrical quantities during the normal operation of the converter. More specifically, the semiconductor devices have an intrinsic dependence on the temperature related to different parameters, such as the mobility of the carriers $\mu(T)$, intrinsic concentration $n_i(T)$, and the bandgap energy $E_g(T)$. It is worth remembering that the $E_g(T)$ and $n_i(T)$ increase at higher temperatures, while $\mu(T)$ has a complex dependence with the temperature that is related to the doping concentration and traps in the gate oxide and silicon interface.

Therefore, the temperature dependence on the aforementioned parameters may be written as follows [76–78]:

$$E_g(T) = E_g(T_0) - \alpha_1 \frac{T^2}{T + \beta_1} \tag{2}$$

$$n_i(T) = N^{\alpha_2} e^{\frac{-\gamma}{T}} \tag{3}$$

$$\mu(T) = \mu_0 \frac{\beta_2 \left(\frac{T}{T_0}\right)^{\alpha_3}}{1 + \beta_2 \left(\frac{T}{T_0}\right)^{\alpha_4}} \tag{4}$$

where α_1, α_2, α_3, β_1, β_2, and γ are empirical coefficients; N is the number per unit volume of effectively available levels states; and T_0 is the room temperature.

Consequentially, the measurement of the electrical quantities measured at the device terminal can be used as a temperature estimator.

Therefore, TSEPs methods use passive voltage or current probes that measure the electrical quantities at the device electrodes, without direct access to the chip device, then the JT is estimated from these measurements. Furthermore, the TSEPs are the preferred approaches to easily obtain JT measurements on packaged devices with a fast time response (less than 100 microseconds). On the other hand, the TSEPs methods do not provide a thermal map of the DUT and, hence, the JT peak is often hard to evaluate [79]. Such an issue is more severe in multichip devices where the voltage or current measurements only provide a rough temperature of the whole device, without the possibility to know the effective temperature distribution among several paralleled chips [80]. In the following subsection, the main TSEPs methods are briefly discussed.

6.2.1. On-State Voltage Measurement

Among the different TSEPs methods, on-state voltage measurement under low current injection has been the most used in many industrial and academic applications. In this case, the TSEP is the voltage drop across the device. The advantage of using this CM method lies in the easy calibration procedure and the negligible self-heating of the DUT.

This CM method is widely employed when the devices have a PN junction in their structure. More specifically, bearing in mind a vertical diffusion MOSFET power device, the temperature variation can be evaluated as the on-resistance $R_{ds,on}$ fluctuations during the converter operations. For the sake of simplicity, the $R_{ds,on}$ can be approximated as follows (see Figure 6):

$$R_{ds,on} \approx R_{ch} + R_d + R_{sub} + R_{cs} + R_{cd} + R_s + R_a + R_{jfet} \tag{5}$$

where R_{ch} is the channel resistance, R_d is the drift region resistance, R_{sub} is the substrate resistance, R_{cs} and R_{ds} are the source and drain contact resistance, R_s is the source resistance, R_{jfet} is the JFET resistance, and Ra is the accumulation resistance. Furthermore, the R_{ch} and R_d can be evaluated as follows [78]:

$$R_{ch} = \frac{L_{ch}}{W_{ch}\mu_{ch}C_{ox}(V_{gs} - V_{th})} \tag{6}$$

$$R_d = \frac{L_d}{q\mu_d N_d A_d} \tag{7}$$

where L_{ch} and W_{ch} are the channel length and width, respectively; C_{ox} is the gate capacitance; L_d and A_d are the drift region length and area, respectively; N_d is the doping concentration of the drift region; and μ_{ch} and μ_d are the channel and drift region mobility, respectively. It is worth noting that R_{ch} decreases at higher temperatures because both μ_{ch} and V_{th} decrease at higher temperatures. On the other hand, R_d acts as a positive temperature coefficient thermistor owing to the temperature dependence of μ_d, which decreases at higher temperatures.

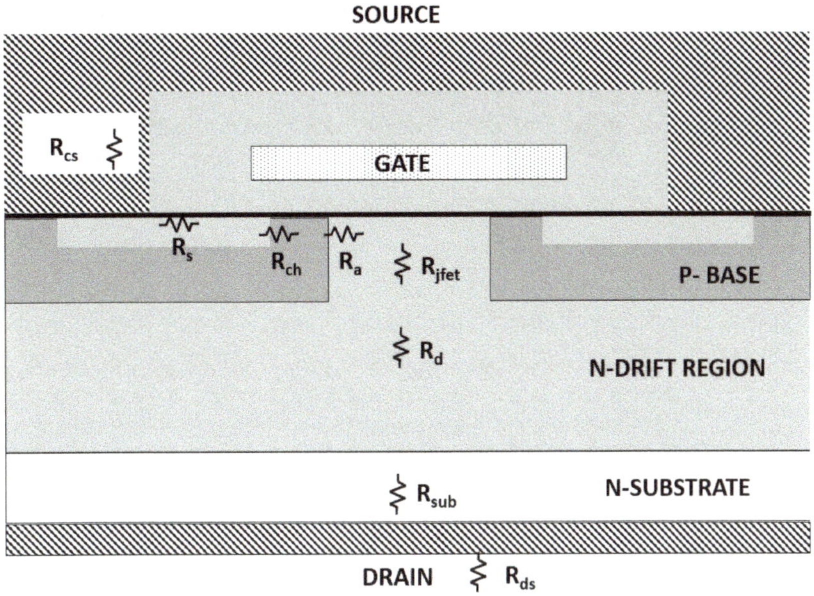

Figure 6. Power vertical diffused MOSFET structure with its internal resistances.

It is worth underlining that the temperature coefficient of $R_{ds,on}$ may differ for the power devices from different vendors, which is mainly caused by the different design of the device. Therefore, notwithstanding an easier calibration procedure, the CM method must be calibrated when a different device is adopted.

Firstly, the calibration procedure is mandatory, which is used to find the relationship between the JT and the TSEP. Typically, the calibration step consists of the use of a current source I_{Cal}, in a range from 1 mA to tens of A. It is worth noting that, during the calibration procedure, the device temperature can usually be fixed by a temperature-controlled heat sink. Then, the temperature measurement can be carried out during the dissipation stage, where the TSEP is measured in a typical converter application. In this case, a current source, *Id*, feeds the DUT to increase its temperature by means of power dissipations. Therefore, the voltage drop across the device, under known electrical conditions, is measured as a function of the temperature.

A simplified schematic of the circuits for the measurement of the voltage under low current is depicted in Figure 7 for an IGBT (Figure 7a) and a MOSFET (Figure 7b). The measurement can be carried out for both the on-state and off-state voltage. A voltmeter is usually connected in parallel to the DUT for the measurement of the voltage drop. It is worth noting that the current *Ical* must be at least hundreds of mA to guarantee a linear relationship between the voltage drop and the temperature [80,81]. In the literature, many works [80–96] have focused on voltage measurement under low current injection in power diodes during forward polarization [84–87], in IGBT power modules [90–95], as well as in power BJTs. Some works [94,95] have focused on the JT estimation in an IGBT power module whose on-state voltage (i.e., collector-emitter voltage, V_{CEon}) has a negative temperature coefficient. The main drawback of this method is the high dependence on the collector current during the measurement of $V_{CE,on}$. Hence, the load current should be diverted during the measurement and this momentary interruption limits the use of this method in real-time applications.

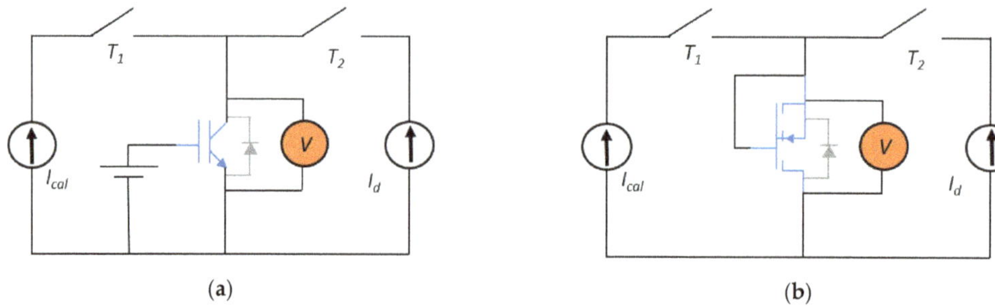

Figure 7. Electrical circuits for the static measurement of the voltage under a low current: (**a**) IGBT and (**b**) MOSFET.

The principle of operation of the CM methods based on the on-state voltage measurement at high current injection is almost similar to that of the low current injection methods. The measurement of the $V_{CE,on}$ (or $V_{ds,on}$) voltage drop across the device is used as a TSEP, as described for the low current injection mode. The main difference with respect to the previous CM method lies in the calibration procedure. More specifically, a higher current is used for the calibration procedure and it produces a non-negligible self-heating. From this perspective, the relation between the voltage drop on the DUT and the temperature also depends on the value of the injected current.

The experimental setup for the temperature measurement is depicted in Figure 8. A high current generator feeds the DUT with a pulsed current, I_H, and a voltmeter is connected in parallel to the DUT. It is important to point out that the measurement of the JT can be obtained during the heating process.

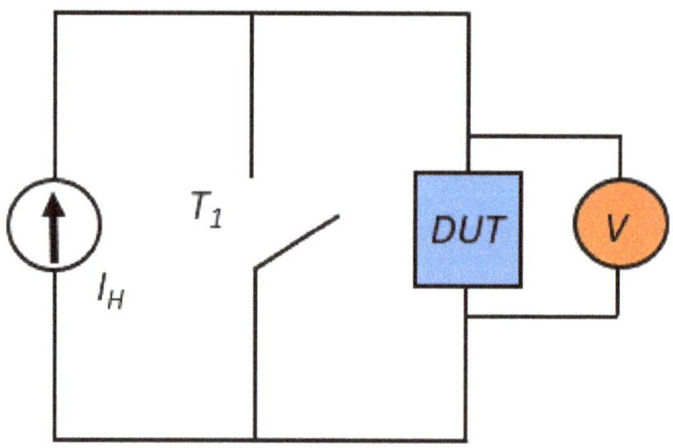

Figure 8. Electrical circuits of a DUT for the static measurement of the voltage under high current.

The TSEPs are usually the MOSFET on-state drain-source voltage [97], the power diodes forward voltage [91], and the IGBT on-state emitter-collector voltage [91,97]. The sensitivity of the aforementioned TSEP is strictly related to the on-state current value, regardless of the specific device. The JT estimation is only practicable for current values greater than tens of Ampere [98]. Hence, this method appears to be very useful, especially for on-line JT measurement during the normal converter operation. Several circuit solutions to measure the V_{CEon} of the power device have been devised [81,90,96–99].

This approach also presents some limitations owing to the voltage swing between the on-state and off-state of the device. This implies the use of advanced electronic sensing circuits, thus increasing the complexity of the system. Innovative and compact sensing circuits to face these issues have been proposed [97,98].

Another issue is the contact resistances of the voltage probes, which cause an undesired voltage drop that may produce an overestimation of the JT measurement [99]. This issue has been partially mitigated with a correction factor based on the layout of the power module [98,99]. Unfortunately, the introduced correction factor has to be calibrated as the device aging progresses.

6.2.2. Saturation Current

The measurement of the saturation current, I_{sat}, has also been used as a TSEP [88,93,100,101] in power modules with IGBTs or MOSFETs. This current can be measured using a current probe or a voltage probe (by adding a shunt resistor). The electrical quantities measured provide a JT estimation due to the dependence on the chip temperature of the channel electron mobility, μ_{ch}; of the threshold, V_{th}; and of the PNP transistor current gain β for the IGBT [101]. It is worth remembering that the current I_{sat} shows a complex temperature dependence, but under the assumption that all the devices are at the same temperature and by neglecting the self-heating, the current I_{sat} in a device can be simply approximated as follows:

$$I_{sat} = \frac{1}{2}\frac{\mu_{ch}(T)W_{ch}C_{ox}}{L_{ch}}(V_{GS} - V_{th}(T))^2 \quad (8)$$

The measurement setup consists of a voltage source, V_{GT}, connected between the gate-emitter (or gate-source) terminals of the DUT and a DC source voltage, V_D, connected between the drain-source or collector-emitter terminals of the DUT. Figure 9 shows the setup of an IGBT device. The voltage value of V_{GT} is usually higher than the threshold voltage V_{th} of the device and a pulsed current is injected into the DUT by controlling the switch T_1. The saturation current can be measured through the voltage drop on the R_{shunt}. The setup demonstrates that the thermal characterization of the device cannot be performed during the on-line converter operation.

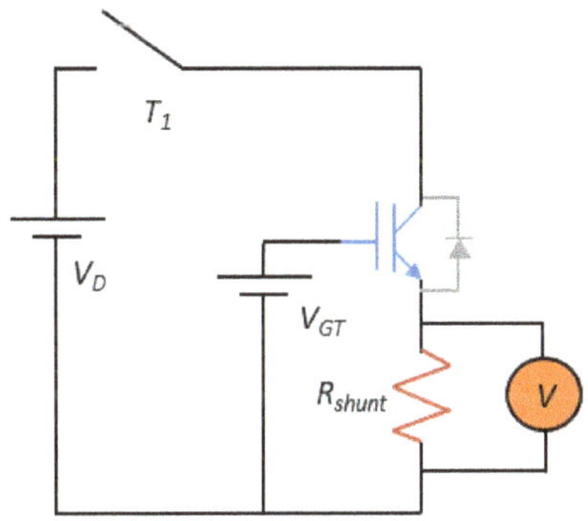

Figure 9. Electrical circuits for the measurement of the saturation current in an IGBT device.

The first procedure is the calibration step, where the DUT is usually placed in a controlled hot plate that overheats the device and, hence, the I_{sat} is measured at varying plate temperatures. Then, the measurement procedure (see Figure 9) consists of performing a non-destructive short-circuit to produce a significant channel temperature variation over a short period of time. From this perspective, the measurement variation of the current I_{sat} can be associated with a specific temperature value. Moreover, the temperature calibration may not be performed without power losses that influence the device self-heating [100–102]. Furthermore, it has been demonstrated that JT measurement is more accurate only for high temperatures.

6.2.3. Gate Threshold Voltage

The threshold voltage V_{th} is defined as the voltage to be applied to the gate-source terminals to have a given current, which is the minimum current that must flow into the device channel to assume the device is turned on. Instead, from the standpoint of power electronic devices, V_{th} is defined as the level of gate bias needed to observe a transition from weak inversion to strong inversion. For a MOS transistor structure, the V_{th} can be approximated as follows [103]:

$$V_{th} \approx 2\varphi_F(T) - \frac{Q_{SS}}{C_O} + \varphi_{ms}(T) + \sqrt{\frac{2\varepsilon q N_A}{C_O}}\sqrt{2\varphi_F(T)} \tag{9}$$

where φ_F is the Fermi potential, Q_{SS} is the extrinsic change due to surface states, C_O is the gate oxide capacitance, φ_{ms} is the metal-semiconductor work function difference, ε is the oxide dielectric constant, q is the elementary charge unit, and N_A is the body doping.

By referring to (9), it can be demonstrated that the voltage V_{TH} decreases with the increasing temperature [103], and it is a TSEP useful for temperature monitoring of MOSFETs [90,103] and IGBTs [93,94,104,105]. A potential measurement setup for the calibration procedure and the measurement of the V_{th} as TSEP in the case of an IGBT device is depicted in Figure 10.

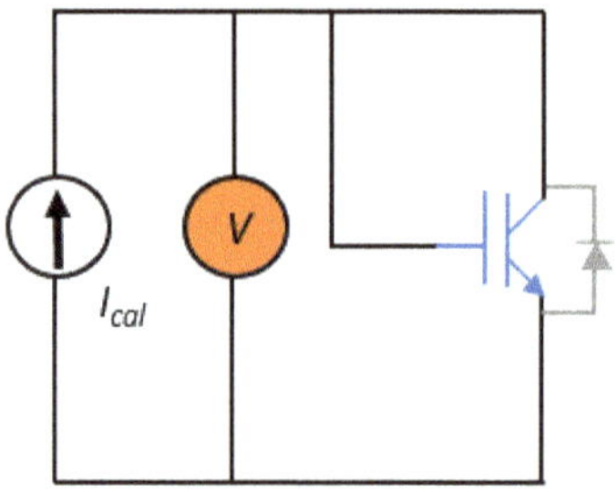

Figure 10. Circuit for the calibration step of the threshold voltage method.

The gate and drain (collector) terminals are short-circuited and a current source, I_{cal}, feeds the DUT, while a voltmeter measures the V_{th}. It is worth noting that the calibration step is based on the low current injection method and, thereby, the self-heating is negligible.

Some works [93,94] have focused on the temperature dependence of the V_{th} measured by varying the collector-emitter voltage and the current collector value for an IGBT device. The I_{cal} value has to be higher than 5 mA to have a correct calibration step for high temperatures and high sensitivity [93,94]. Other works [104,105] have focused on the temperature measurements after the power dissipation of the device. More specifically, a current source with two different current levels, one for dissipation (high current injection) and the other for JT measurement (low current injection), has been proposed. This CM method is not suitable for on-line condition monitoring [106,107].

6.2.4. Gate-Source or Gate-Emitter Voltage Turn ON-OFF

The gate-emitter (source) voltage, V_{ge} (or V_{gs}), is used as a TSEP during the turn-on and turn-off of the switch [108,109]. The high sensitivity and the linear dependence of V_{ge} (or V_{gs}) with the temperature are the strengths of this method. Similarly to the threshold voltage method, the V_{ge} (or V_{gs}) TSEP method cannot be used for on-line JT estimation in a power converter application, because the gate and collector (drain) terminal has to be shorted. The experimental setup of the gate-source or gate-emitter voltage as the CM method is very similar to that of the threshold voltage (see Figure 10). In this case, the current injected into the DUT is higher than the current used in the threshold voltage method and, consequentially, the self-heating is not negligible.

Figure 11 depicts the simulation of the gate-emitter voltage V_{ge} of an IGBT during the turn-off while varying the device temperature. It is worth noting that the following analysis can be done by considering the turn-on of a device. The Miller plateau becomes wider as the temperature increases. In other terms, the time shift Δt in the figure is strictly related to the temperature of the chip and can be detected using a time counter that triggers from the first falling edge to the second one after the Miller plateau.

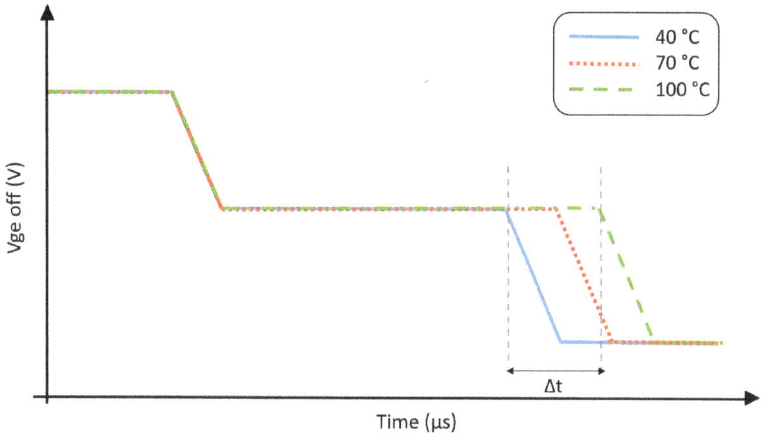

Figure 11. Simulated V_{ge} waveform during the turn off of an IGBT by varying the temperature working operations (figure based on [110]).

The Miller plateau width t_d can be approximated as follows [109]:

$$t_d = \frac{R_{Gint}(T) \cdot C_{rss}(T) \cdot (V_{DD} - V_{ON})}{\left(\frac{I_{load}(T)}{g_m(T)} + V_{th}(T)\right)} \tag{10}$$

where R_{Gint} is the internal gate resistance, C_{rss} is the Miller capacitance, V_{DD} is the DC-link voltage, V_{ON} is the on-state voltage, I_{load} is the load current, and g_m is the transconductance. Equation (10) shows that t_d is directly proportional to C_{rss} and R_{Gint}. It is worth noting that the impact of temperature variation on V_{ON} and V_{DD} is negligible, while the temperature variations of the terms (I_{load}/g_m) and V_{TH} partly neutralize each other. The internal gate resistance depends on the temperature as the electron mobility decreases at higher temperatures. Therefore, t_d increases at higher temperatures owing to the temperature dependence of C_{rss} and R_{Gint}. Therefore, the time interval t_d of the Miller plateau in the V_{ge} (or V_{gs}) voltage can be used as a TSEP to estimate the JT of IGBTs (or MOSFETs).

The authors in [109] have proved the temperature independence of the collector-emitter voltage. Instead, the calibration step measurement has been improved in [53], where an auxiliary sensing circuit has been added to the gate driver to reduce undesirable oscillations during the turn off of the device. Meanwhile, in [110], the linear dependence of t_d with respect to the temperature of the chip has been demonstrated, and a parametric analysis by varying the JT, I_{load}, and DC-link voltage has been performed.

6.2.5. Turn On-Off Delay Time

The switching behavior of the power devices has been also adopted as a CM method [110–116]. In this case, the TSEPs are the voltage and current waveforms during the turn-on and turn-off of the DUT. This method is quite similar to the V_{ge} (or V_{gs}) TSEP method, but the JT monitoring can be performed on-line during the converter operations. More specifically, the delay, ΔD, at turn-on, between the collector current i_c and the gate-emitter voltage V_{ge} for an IGBT device (see Figure 11), is used as a TSEP [110–112], as well as the delay between the drain current i_d and the gate-source voltage V_{gs} for a MOSFET device. Bearing in mind the IGBT devices, the turn-on delay is of great interest because ΔD increases linearly with the temperature [113], it only depends on the dc-link voltage, and it is not influenced by the value of i_c. More specifically, during the switching on time interval

t_{on}, the gate current charges the gate-emitter capacitance C_{GE} that is connected in series with the gate resistance R_{Gint}.

Therefore, the zero state waveform of the $v_{ge}(t)$ can be written as follows [109]:

$$v_{ge}(t) = V_G \cdot \left(1 - e^{\frac{t}{\tau}}\right) \; \tau \approx R_{Gint}\,(T) \cdot C_{GE}(T) \;\; t_{ON} \approx \tau(T) \cdot \ln\left(1 - \frac{V_{th}(T)}{V_G}\right) \quad (11)$$

where V_G is the driver gate-emitter voltage.

The dependence on the temperature of the turn-on delay ΔD can be analyzed by combining both (9) and (11). V_{th} decreases as the temperature increases and the value of the time constant τ depends on the temperature variations too. It is worth underlining that the gate charge (the intrinsic gate capacitances) has a weak dependence on temperature, while the internal gate-resistance R_{Gint} has a stronger dependence on temperature owing to the channel mobility μ, which decreases at higher temperatures.

Figure 12 depicts the simulation of an ideal IGBT device during the turn-on at varying working operation temperatures (40 °C, 70 °C, and 100 °C). The shift on the right of the waveforms is strictly related to the aforementioned temperature dependence. The previous method is also valid for MOSFETs.

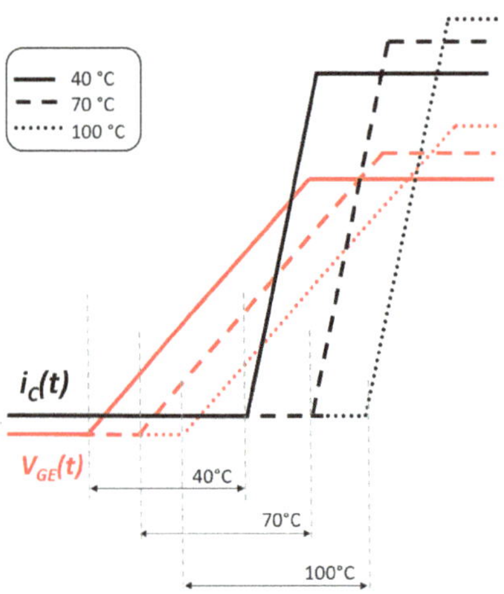

Figure 12. Simplified i_c–V_{ge} turn ON waveforms of an IGBT at different temperature working operations.

An advanced sensing circuit (voltage probes, Field Programmable Gate Array (FPGA), Analog to Digital Converter (ADC)) that records the transient evolutions of both V_{ge} and i_c waveforms has been proposed in [112]. The delay is calculated as the time interval between the time instant the rising edge of the V_{ge} is detected, and the rising edge of the current i_c (Figure 12). This method allows a sensitivity close to 2 ns/°C. Moreover, because a gate resistor with a large resistance improves the accuracy of the temperature measurements during the switching behavior of the converter, but worsens the efficiency, a variable gate resistor has been proposed to set a higher value exclusively when the JT is measured [113].

Similarly, the turn-off delay can also be used as a TSEP, reaching a sensitivity level close to the one obtained with the turn-on delay method [114]. Other works have proposed an alternative sensing circuit for the JT estimation during the turn OFF [115]. However, the turn-off delay method does not attract interest because it is not linear at high-temperature operations, and the time delay depends greatly on both the i_c current and the DC link voltage [115,116].

In general, the turn-on and turn-off TSEPs methods require high bandwidth sensors and an advanced sampling circuit for temperature measurement, which considerably increase the cost of the overall system. Furthermore, these methods usually require an external circuit to trigger a counter for the estimation of the turn-on and turn-off delay time.

6.2.6. Current and Voltage Change Rate

In the last decade, the research has moved from the study of the electrical quantities (such as the voltage and current waveforms) to their derivative functions, which are observed during the device commutation, called dynamic thermo-sensitive electrical parameter (DTSEP) methods. More specifically, the collector-emitter voltage change rate (dv_{ce}/dt) and the collector current change rate (di_c/dt) have been used as temperature estimators [117–121]. The temperature dependence of both dv_{ce}/dt and di_c/dt has been explored theoretically as well as confirmed experimentally [119–121].

As an example, the dV_{ce}/dt in an IGBT device can be approximated as follows [121]:

$$\frac{dv_{ce}}{dt} \approx \frac{1}{\tau_{gc}(T)}\left(\frac{V_{GE,ON} - V_{GE,OFF}}{1 + \left(\frac{C_O}{g_m(T)\tau_{gc}(T)}\right)}\right) \tau_{gc} \approx R_{Gint}(T) \cdot C_{GC}(T) \quad (12)$$

where C_O is the charge extraction capacitance and $V_{GE,ON}$ and $V_{GE,OFF}$ are the on-off gate driver voltages, respectively.

It is worth noting that the term dV_{ce}/dt depends on the physical parameters of the IGBT device and the temperature dependence is not easy to obtain. More specifically, the JT affects the dV_{ce}/dt through the MOS channel parameters such as the L_{ch}, W_{ch}, emitter recombination parameter, channel mobility, and so on. A detailed discussion of all the temperature parameter dependencies is given in [121]. The dependence of many parameters influencing the derivative quantities on the temperature strongly limits the use of this CM method for on-line JT measurement in practical power converter applications. A wide investigation of the IGBT maximum dv_{ce}/dt for the JT estimation has revealed the severe limits owing to the influence of the control method, the DC link voltage, and the load current [121]. Likewise, the maximum di_c/dt during turn-off as a TSEP has been also investigated in [122]. Even in this case, the measurement of the current change rate has been performed using an additional circuit able to capture the current and voltage transient dynamics, which require both high bandwidth sensors and the use of voltage probes and Rogowski coil probes. Furthermore, this sensing circuit should be designed to avoid any disturbance, and it has to be insensitive to the temperature variation of the system.

In the recent generations of IGBT and SiC high power modules, the Kelvin emitter pin has been introduced. Such an additional pin involves in the package an integrated inherent parasitic inductance L_{eE} between the Kelvin pin and power emitters pin [122–124], as shown in Figure 13. The transient collector current characteristic during the turn OFF process has been introduced as a potential DTSEP, called the maximum collector current falling rate $-dI_C/dt_{max}$ [125]: the collector current I_C flows in the inductance L_{eE} and the resulting voltage drop enables an easier investigation of the JT measurement.

Moreover, in [122], both the static and dynamic behaviors of the stored carriers in the IGBT collector current during the falling rate have been analyzed. Furthermore, the influences of the physical parameters of the device on the temperature sensitivity of $-dI_C/dt_{max}$ have been fully investigated. However, several drawbacks of these methods are related to the strong dependence of the applied voltage and the gate resistance, and the thermal characterization can only be done off-line.

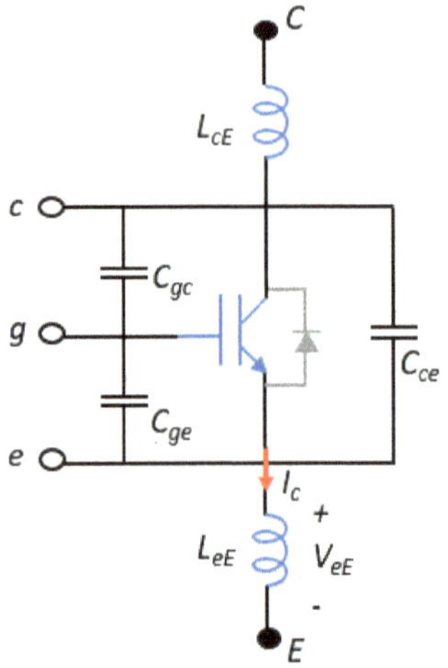

Figure 13. IGBT module equivalent circuit.

6.2.7. Peak Gate Current

An innovative method for JT measurement in IGBTs and MOSFETs, based on the temperature dependence of the internal gate resistance, has been studied in the last years. Firstly, the measure of R_{Gint} in a power module has already been investigated using a standard RLC meter [126], where a common approach is to consider the equivalent series resistance (ESR) of both the gate-emitter and gate-collector capacitance (see Figure 14a). Another method to estimate the R_{Gint} variation has been related to the measurement of the gate charge during the turn-on of the DUT [127]. Therefore, the peak gate current during the turn-on switching behavior has been assumed as a valid TSEP.

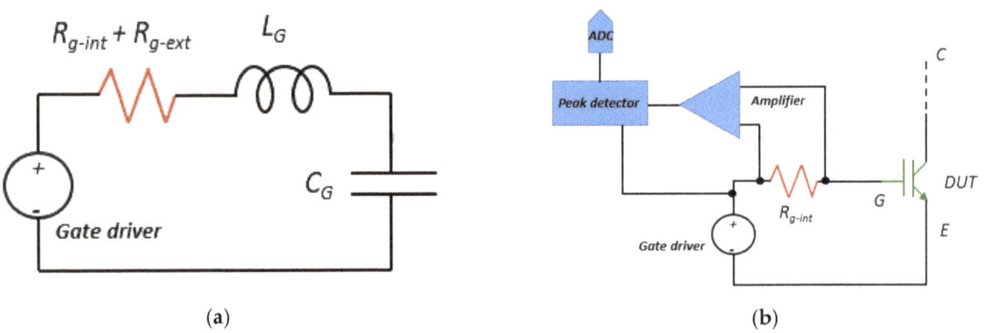

Figure 14. Gate driver RLC network. (**a**) Peak detector schematic to detect peak voltage over the external gate resistor (**b**).

JT measurement via the peak gate current can be studied during the standard charging cycles of the gate terminal. Considering an IGBT device, the turn-on process starts when the gate driver output voltage changes from a negative value to a positive one. Therefore, the gate current can be computed as the step response of a second-order RLC circuit [128] (see Figure 14a). The parasitic inductance L_G can be neglected and the peak current can be estimated by simply using the Ohm's law, provided that the RLC circuit is overdamped. It is worth noting that the external gate resistance R_{Gext} does not have a significant temperature dependence. Therefore, the temperature variation of R_{Gint} can be carried out by the measurement of the peak current variation. In other words, the maximum value of the gate current provides a suitable strategy for the measurement of the chip JT. The measurement circuit is shown in Figure 14b. The peak voltage on the external gate resistor during turn-on is measured with a peak detector circuit (a differential amplifier and a peak detector). Then, the acquired data are processed by an analog to digital converter to the microcontroller. This measurement circuit can be integrated into the gate driver, and the JT monitoring can be operated during the on-line operation of the converter. This method does not require calibration steps and, more importantly, the voltage peak has a linear relationship with the temperature.

Only a few works have focused on the peak gate current as a TSEPs method. More specifically, the sensing circuit depicted in Figure 14b has been proposed in [129–131], where the JT has been esteemed in an IGBT power module. The authors have asserted that the proposed method has better accuracy for JT measurement compared with other TSEP methods in the literature. However, this method requires additional complex trigger circuits for the measurement of JT, which may introduce additional disturbance into the system. It is worth remembering that the aging of the power module may affect the internal gate resistances. Hence, a correction factor should be introduced for calibration with the aging of the device.

7. Comparison of the CM Methods

Table 2 summarizes all the CM methods discussed previously. A comparison between the advantages and disadvantages of each approach is outlined.

Among the aforementioned optical CM methods, the use of fiber optic shows the highest accuracy and sensitivity. On the other hand, the device package has to be removed to carry out the temperature measurement. The optical methods based on the photodiode sensors and the use of the IR camera are able to operate contactless, without the lift-off of the device package. It is worth underlying that all the optical methods can be used during the on-line converter operations.

The acoustic method has been studied as a CM method for power devices in recent years. The strength of the proposed solution is owing to the ability to estimate the state of aging a power module and prevent any mechanics fatigue. It is worth remembering that it may be used during the on-line converter operations. In the literature, only a few papers have been focused on this CM method. Thus, the technology is not yet well mature to be widespread in commercial solutions.

It is worth highlighting that the physical and TTCs–NTC methods can be adopted for on-line JT measurements in a real power converter application. Furthermore, both CM methods show a strong linear dependence with the voltage and the temperature. As a drawback, method is almost obsolete and requires a device package modification. Meanwhile, the use of method needs a layout modification and it is strongly aging sensitive, hence the measurement setup has to be frequently calibrated. Moreover, the method based on the TTCs diode requires device layout modifications and shows poor linearity owing to a nonlinear dependence between the voltage drop of the diode and the temperature.

Finally, TSEPs methods have been widely used as CM methods for the estimation of the JT of the power devices, where the key point consists of correlating the temperature of the semiconductor material with the electrical quantities during the switching operation of the power device. More specifically, the TSEPs CM methods such as the on-state voltage under high current injection, the gate turn–off voltage, the turn on-off delay time, and the peak gate current enable the estimation of the JT during the on-line converter operations. Furthermore, the aforementioned CM methods exhibit high linearity between the voltage measurement and the temperature of the device.

On the other hand, the on-state voltage under high-level current injection, the gate threshold voltage, the saturation current, the gate turn-off voltage, and the voltage-current change rate require to switch off the power converter for the JT estimation. On the other hand, the aforementioned CM methods highlight the highest accuracy among the various TSEPs in the literature.

Table 2. Summary of different CM methods.

Method	Advantages	Disadvantages
OPTICAL—Fiber optic [46–48]	On-line measurements High sensitivity and accuracy	Package modification High cost
OPTICAL—Photodiode sensor—internal	On-line measurements Contactless	Technology not mature Package modification
OPTICAL—Photodiode sensor—external [56–58]	On-line measurements Contactless	Package modification
OPTICAL—IR camera [48–53]	Spatial resolution Contactless	Package modification Poor response time Poor accuracy
OPTICAL—IR camera print [49]	Spatial resolution Contactless	Poor response time Poor accuracy
Acoustic [28–33]	On-line measurements Contactless	Technology not mature Noise sensitive
Physical [59–65]	On-line measurements High linearity and sensitivity	Package modification Poor response time
TTCs—NTC [67,69,70,72,75]	On-line measurements High linearity	Layout modification Aging sensitive
TTCs—Diode [68,71]	On-line measurements High sensitivity	Layout modification Poor linearity
TSEP—On-state voltage, low current [80–95]	High sensitivity, linearity Easy calibration	High-cost sensing Off-line measurements
TSEP—Gate threshold voltage [89,93,94,103–105]	High sensitivity and linearity	Off-line measurements Unplug DUT
TSEP—On-state voltage, high current [96–99]	On-line measurements High linearity	High-cost sensing Aging sensitive
TSEP—Saturation current [87,93,100,101]	High sensitivity	Off-line measurements Poor linearity
TSEP—Gate turn OFF voltage [108,109]	High linearity	Off-line measurements High-cost sensing
TSEP—Turn on-off delay time [110–116]	On-line measurements High linearity	Aging sensitive High-cost sensing
TSEP—Peak gate current [129,130]	On-line measurements High linearity	Aging sensitive High-cost sensing
TSEP—Voltage-current change rate [117–121]	High sensitivity and linearity	Off-line measurements Gate resistance dependence

8. Conclusions

In this work, the main CM methods used to estimate the SOH of the semiconductor power devices were discussed and compared. The analysis has highlighted that the method based on the TSEP on-state voltage, measured under low currents injection, is the best one both for silicon and WBG power devices. Indeed, this CM method can involve a significant reduction of the experimental time duration of the calibration steps in comparison with other solutions. Furthermore, the experimental setup does not impact the device under test, i.e., the measurement does not degrade the electrical connections, the metallization, and the wire bonding. On the other hand, this method is not able to measure the junction temperature during the on-line converter operations. From this perspective, the TSEP methods based on the threshold voltage can be used during the on-line converter operations with comparable sensitivity and accuracy of the junction temperature estimation.

Finally, from the analysis of the literature arose the lack of studies of CM intrusiveness. Many CM methods have been presented so far, but only in a few cases do they discuss the intrusiveness of the proposed method, and very rarely do these works compare the intrusiveness of the proposed CM method with others. CM methods requiring the removal of the device package for temperature monitoring are intrusive for the device and this intrusiveness could make these CM methods impracticable in dusty or moist environments or in applications where atmospheric agents could damage the device. CM methods that need to shut down the converter are very intrusive for the converter operations and cannot be used in any application where the converter cannot shut down. Finally, a CM method adopting tools for the measurement, conditioning, elaboration, and so on is more of an encumbrance on the conversion system. This intrusiveness impedes their use in applications requiring high power density or, more in general, where the weight and encumbrance of the conversion system must be minimized. Therefore, accurate studies focusing on the CM intrusiveness, which also provide some figure of merits based on the previous aspects as well as the specific application, are strongly recommended.

Funding: This research received no external funding.

Conflicts of Interest: The authors declare no conflict of interest.

References

1. Blaabjerg, F.; Dragicevic, T.; Davari, P. Applications of Power Electronics. *Electronics* **2019**, *8*, 465. [CrossRef]
2. Abadi, I.; Imron, C.; Bachrowi, M.M.; Fitriyanah, D.N. Design and implementation of battery charging system on solar tracker based standalone PV using fuzzy modified particle swarm optimization. *AIMS Energy* **2020**, *8*, 142–155. [CrossRef]
3. García Salvador, R.M.; Novas, N.; Alcayde, A.; El Khaled, D.; Montoya, F.G. Electronics and Its Worldwide Research. *Electronics* **2020**, *9*, 977. [CrossRef]
4. Gritti, G.; Adragna, C. Analysis, design and performance evaluation of an LED driver with unity power factor and constant-current primary sensing regulation. *AIMS Energy* **2019**, *7*, 579–599. [CrossRef]
5. Musumeci, S.; Bojoi, R.; Armando, E.; Borlo, S.; Mandrile, F. Three-Legs Interleaved Boost Power Factor Corrector for High-Power LED Lighting Application. *Energies* **2020**, *13*, 1728. [CrossRef]
6. Yilmaz, M.; Krein, P.T. Review of the Impact of Vehicle-to-Grid Technologies on Distribution Systems and Utility Interfaces. *IEEE Trans. Power Electron.* **2013**, *28*, 5673–5689. [CrossRef]
7. Konara, K.M.S.Y.; Kolhe, M.L.; Sharma, A. Power dispatching techniques as a finite state machine for a standalone photovoltaic system with a hybrid energy storage. *AIMS Energy* **2020**, *8*, 214–230. [CrossRef]
8. Faraci, G.; Raciti, A.; Rizzo, S.A.; Schembra, G. Green Wireless Power Transfer System for a Drone Fleet Managed by Reinforcement Learning in Smart Industry. *Appl. Energy* **2020**, *259*, 114204. [CrossRef]
9. Ding, Y.; Loh, P.C.; Tan, K.K.; Wang, P.; Gao, F. Reliability evaluation of three-level inverters. In Proceedings of the 25th Annual IEEE Applied Power Electronics Conference and Exposition, Palm Springs, CA, USA, 21–25 February 2010; pp. 1555–1560.
10. De Leon-Aldaco, S.E.; Calleja, H.; Chan, F.; Jimenez-Grajales, H.R. Effect of the mission profile on the reliability of a power converter aimed at photovoltaic applications-a case study. *IEEE Trans. Power Electron.* **2013**, *28*, 2998–3007. [CrossRef]
11. Alam, M.K.; Khan, F.H. Reliability analysis and performance degradation of a boost converter. In Proceedings of the IEEE Energy Conversion Congress and Exposition, Denver, CO, USA, 15–19 September 2013; pp. 5592–5597.
12. Tajfar, A.; Mazumder, S.K. A fault-tolerant switching scheme for a photovoltaic high-frequency-link inverter. In Proceedings of the 27th Annual IEEE Applied Power Electronics Conference and Exposition, Orlando, FL, USA, 5–9 February 2012; pp. 2087–2094.
13. Madhukar Rao, A.; Umesh, B.S.; Sivakumar, K. A fault tolerant dual inverter configuration for islanded mode photovoltaic generation system. In Proceedings of the 1st International Future Energy Electronics Conference, Tainan, Taiwan, 3–6 November 2013; pp. 816–821.
14. Haumann, S.; Becker, M.; Rudzki, J.; Eisele, R.; Osterwald, F. Novel bonding and joining technology for power electronics—Enabler for improved lifetime, reliability, cost and power density. In Proceedings of the 28th Annual IEEE Applied Power Electronics Conference and Exposition, Long Beach, CA, USA, 17–21 March 2013; pp. 622–626.
15. Krebs, T.; Duch, S.; Schmitt, W.; Kotter, S.; Prenosil, P.; Thomas, S. A breakthrough in power electronics reliability–new die attach and wire bonding materials. In Proceedings of the IEEE 63rd Electronics Components and Technology Conference, La Vegas, NV, USA, 28–31 May 2013; pp. 1746–1752.
16. Mohagheghi, S.; Harley, R.G.; Habetler, T.G.; Divan, D. Condition monitoring of power electronic circuits using artificial neural networks. *IEEE Trans. Power Electron.* **2009**, *24*, 2363–2367. [CrossRef]
17. Xiang, D.; Ran, L.; Tavner, P.; Bryant, A.; Yang, S.; Mawby, P. Monitoring solder fatigue in a power module using case-above-ambient temperature rise. *IEEE Trans. Ind. Appl.* **2011**, *47*, 2578–2591. [CrossRef]

18. Li, D.; Li, X. Study of degradation in switching mode power supply based on the theory of PoF. In Proceedings of the International Conference Computing Science and Service Systems, Nanjing, China, 11–13 August 2012; pp. 1976–1980.
19. Bhargava, C.; Sharma, P.K.; Senthilkumar, M.; Padmanaban, S.; Ramachandaramurthy, V.K.; Leonowicz, Z.; Blaabjerg, F.; Mitolo, M. Review of Health Prognostics and Condition Monitoring of Electronic Components. *IEEE Access* **2020**, *8*, 75163–75183. [CrossRef]
20. Jiang, N.; Zhang, L.; Liu, Z.Q.; Sun, L.; Long, W.M.; He, P.; Xiong, M.Y.; Zhao, M. Reliability issues of lead-free solder joints in electronic devices. *Sci. Technol. Adv. Mater.* **2019**, *20*, 876–901. [CrossRef] [PubMed]
21. Takamiya, M.; Miyazaki, K.; Obara, H.; Sai, T.; Wada, K.; Sakurai, T. Power electronics 2.0: IoT-connected and AI-controlled power electronics operating optimally for each user. In Proceedings of the 2017 29th International Symposium on Power Semiconductor Devices and IC's (ISPSD), Sapporo, Japan, 28 May–1 June 2017; pp. 29–32.
22. Mellit, A.; Kalogirou, S.A. Artificial intelligence techniques for photovoltaic applications: A review. *Progr. Energy Combust. Sci.* **2008**, *34*, 574–632. [CrossRef]
23. Butler, S.W. Enabling a Powerful Decade of Changes [Flyback]. *IEEE Power Electron. Mag.* **2019**, *6*, 18–26. [CrossRef]
24. Balda, J.C.; Mantooth, A.; Blum, R.; Tenti, P. Cybersecurity and Power Electronics: Addressing the Security Vulnerabilities of the Internet of Things. *IEEE Power Electron. Mag.* **2017**, *4*, 37–43. [CrossRef]
25. Jasperneite, J.; Sauter, T.; Wollschlaeger, M. Why We Need Automation Models: Handling Complexity in Industry 4.0 and the Internet of Things. *IEEE Ind. Electron. Mag.* **2020**, *14*, 29–40. [CrossRef]
26. Levikari, S.; Kärkkäinen, T.J.; Andersson, C.; Tammminen, J.; Silventoinen, P. Acoustic Detection of Cracks and Delamination in Multilayer Ceramic Capacitors. *IEEE Trans. Ind. Appl.* **2019**, *55*, 1787–1794. [CrossRef]
27. Cheraghi, M.; Karimi, M.; Booin, M.B. An investigation on acoustic noise emitted by induction motors due to magnetic sources. In Proceedings of the 9th Annual Power Electronics, Drives Systems and Technologies Conference (PEDSTC), Tehran, Iran, 13–15 February 2018; pp. 104–109.
28. Smulko, J.; Józwiak, k.; Olesz, M.; Hasse, L. Acoustic emission for detecting deterioration of capacitors under aging. *Microelectron. Reliab.* **2011**, *51*, 624–627. [CrossRef]
29. Kärkkäinen, T.J.; Talvitie, J.P.; Kuisma, M.; Silventoinen, P.; Mengotti, E. Measurement challenges in acoustic emission research of semiconductors. In Proceedings of the 17th European Conference on Power Electronics and Applications (EPE'15 ECCE-Europe), Geneva, Switzerland, 8–10 September 2015; pp. 1–6.
30. Kärkkäinen, T.J.; Talvitie, J.P.; Kuisma, M.; Silventoinen, P.; Mengotti, E. Acoustic emission caused by the failure of a power transistor. In Proceedings of the IEEE Applied Power Electronics Conference and Exposition (APEC), Charlotte, NC, USA, 15–19 March 2015; pp. 2481–2484.
31. Kärkkäinen, T.J.; Talvitie, J.P.; Ikonen, O.; Kuisma, M.; Silventoinen, P.; Mengotti, E. Sounds from semiconductors—Acoustic emission experiment with a power module. In Proceedings of the 16th European Conference on Power Electronics and Applications, Lappeenranta, Finland, 26–28 August 2014; pp. 1–6.
32. Kärkkäinen, T.J.; Talvitie, J.P.; Kuisma, M.; Hannonen, J.; Ström, J.P.; Mengotti, E.; Silventoinen, P. Acoustic Emission in Power Semiconductor Modules—First Observations. *IEEE Trans. Power Electron.* **2014**, *29*, 6081–6086. [CrossRef]
33. Müller, S.; Drechsler, C.; Heinkel, U.; Herold, C. Acoustic emission for state-of-health determination in power modules. In Proceedings of the 13th International Multi-Conference on Systems, Signals & Devices (SSD), Leipzig, Germany, 21–24 March 2016; pp. 468–471.
34. Davari, P.; Kristensen, O.; Iannuzzo, F. Investigation of acoustic emission as a non-invasive method for detection of power semiconductor aging. *Microelectron. Reliab.* **2018**, *88–90*, 545–549. [CrossRef]
35. Oh, H.; Han, B.; McCluskey, P.; Han, C.; Youn, B.D. Physics-of failure, condition monitoring, and prognostics of insulated gate bipolar transistor modules: A review. *IEEE Trans. Power Electron.* **2015**, *30*, 2413–2426. [CrossRef]
36. Luo, H.; Wang, X.; Zhu, C.; Li, W.; He, X. Investigation and emulation of junction temperature for high-power IGBT modules considering grid codes. *IEEE J. Emerg. Sel. Top. Power Electron.* **2018**, *6*, 930–940. [CrossRef]
37. Scheuermann, U.; Schuler, S. Power cycling results for different control strategies. *Microelectron. Reliabil.* **2010**, *50*, 1203–1209. [CrossRef]
38. Zarebski, J.; Gorecki, K. The electro thermal large-signal model of power MOS transistors for SPICE. *IEEE Trans. Power Electron.* **2010**, *25*, 1265–1274. [CrossRef]
39. Breglio, G.; Irace, A.; Spirito, P.; Letor, R.; Russo, S. Fast transient infrared thermal analysis of smart Power MOSFETS in permanent short circuit operation. In Proceedings of the 18th International Symposium on Power Semiconductor Devices IC's, Naples, Italy, 4–8 June 2006; pp. 1–4.
40. Hunger, T.; Schilling, O. Numerical investigation on thermal crosstalk of silicon dies in high voltage IGBT modules. In Proceedings of the PCIM International Exhibition & Conference for Power Electronics, Intelligent Motion, Power Quality, Nuremberg, Germany, 27–29 May 2008.
41. Hillkirk, L.-M. Dynamic surface temperature measurements in SiC epitaxial power diodes performed under single-pulse self-heating conditions. *Solid State Electron.* **2004**, *48*, 2181–2189. [CrossRef]
42. Breglio, G.; Rinaldi, N.; Spirito, P. Thermal mapping and 3D numerical simulation of new cellular power MOS affected by electro-thermal instability. *Microelectron. J.* **2000**, *31*, 741–746. [CrossRef]

43. Spirito, P.; Breglio, G.; D'Alessandro, V.; Rinaldi, N. Thermal instabilities in high current power MOS devices: Experimental evidence electro-thermal simulations and analytical modeling. In Proceedings of the 23rd International Conference on Microelectronics, Nis, Yugoslavia, 12–15 May 2002; pp. 23–30.
44. Thalhammer, R.-K.; Wachutka, G. Physically rigorous modeling of internal laser-probing techniques for micro structured semiconductor devices. *IEEE Trans. Comput. Aided Des. Integr. Circuits Syst.* **2004**, *23*, 60–70. [CrossRef]
45. Furbock, C.; Thalhammer, R.; Litzenberger, M.; Seliger, N.; Pogany, D.; Gornik, E.; Wachutka, G. A differential backside laser probing technique for the investigation of the lateral temperature distribution in power devices. In Proceedings of the 11th International Symposium on Power Semiconductors Devices ICs, Toronto, ON, Canada, 26–28 May 1999; pp. 193–196.
46. Werber, D.; Wachutka, G. Interpretation of laser absorption measurements on 4 H-SiC bipolar diodes by numerical simulation. In Proceedings of the International Conference on Simulation of Semiconductor Processes and Devices, Yokohama, Japan, 9–11 September 2008; pp. 89–92.
47. Hamidi, A.; Coquery, G.; Lallemand, R.; Vales, P.; Dorkel, J.M. Temperature measurements and thermal modeling of high power IGBT multichip modules for reliability investigations in traction applications. *Microelectron. Reliabil.* **1998**, *38*, 1353–1359. [CrossRef]
48. Carubelli, S.; Khatir, Z. Experimental validation of a thermal modelling method dedicated to multichip power modules in operating conditions. *Microelectron. J.* **2003**, *34*, 1143–1151. [CrossRef]
49. Khatir, Z.; Carubelli, S.; Lecoq, F. Real-time computation of thermal constraints in multichip power electronic devices. *IEEE Trans. Compon. Packag. Technol.* **2004**, *27*, 337–344. [CrossRef]
50. Schmidt, R.; Scheuermann, U. Using the chip as a temperature sensor—The influence of steep lateral temperature gradients on the Vce(T)-measurement. In Proceedings of the 13th European Conference on Power Electronics and Applications, Barcelona, Spain, 8–10 September 2009; pp. 1–9.
51. Brckner, T.; Bernet, S. Estimation and measurement of junction temperatures in a three-level voltage source converter. *IEEE Trans. Power Electron.* **2007**, *22*, 3–12. [CrossRef]
52. Mermet-Guyennet, M.; Perpina, X.; Piton, M. Revisiting power cycling test for better life-time prediction in traction. *Microelectron. Reliabil.* **2007**, *47*, 1690–1695. [CrossRef]
53. Avenas, Y.; Dupont, L. Comparison of junction temperature evaluations in a power IGBTs module using an IR camera and three thermo-sensitive electrical parameters. In Proceedings of the Applied Power Electronics Conference and Exposition, Orlando, FL, USA, 5–9 February 2012.
54. Dupont, L.; Avenas, Y.; Jeannin, P.-O. Comparison of junction temperature evaluations in a power IGBT module using an IR camera and three thermosensitive electrical parameters. *IEEE Trans. Ind. Appl.* **2013**, *49*, 1599–1608. [CrossRef]
55. Round, H.J. A Note on Carborundum. In *Semiconductor Devices: Pioneering Papers*; World Scientific: Singapore, 1991.
56. Winkler, J.; Homoth, J.; Kallfass, I. Utilization of parasitic luminescence from power semiconductor devices for current sensing. In Proceedings of the PCIM Europe 2018, International Exhibition and Conference for Power Electronics, Intelligent Motion, Renewable Energy and Energy Management, Nuremberg, Germany, 5–7 June 2018; pp. 1–8.
57. Ceccarelli, L.; Luo, H.; Iannuzzo, F. Investigating SiC MOSFET body diode's light emission as temperature-sensitive electrical parameter. *Microelectron. Reliab.* **2018**, *88–90*, 627–630. [CrossRef]
58. Li, C.; Luo, H.; Li, C.; Li, W.; Yang, H.; He, X. Online Junction Temperature Extraction of SiC Power MOSFET with Temperature Sensitive Optic Parameter (TSOP) Approach. *IEEE Trans. Power Electron.* **2019**, *34*, 10143–10152. [CrossRef]
59. Winkler, J.; Homoth, J.; Kallfass, I. Electroluminescence-Based Junction Temperature Measurement Approach for SiC Power MOSFETs. *IEEE Trans. Power Electron.* **2020**, *35*, 2990–2998. [CrossRef]
60. Parsley, M. The use of thermochromics liquid crystals in research applications, thermal mapping and non-destructive testing. In Proceedings of the 7th IEEE Semiconductor Thermal Measurement and Management Symposium, Phoenix, AZ, USA, 12–14 February 1991; pp. 53–58.
61. Brekel, W.; Duetemeyer, T.; Puk, G.; Schilling, O. Time Resolved in situ Tvj Measurements of 6.5 kV IGBTs during Inverter Operation. In Proceedings of the PCIM Europe 2009: International Exhibition & Conference for Power Electronics Intelligent Motion Power Quality, Nurember, Germany, 12–14 May 2009.
62. Salem, T.E.; Ibitayo, D.; Geil, B.R. A Technique for Die Surface Temperature Measurement of High-Voltage Power Electronic Components using Coated Thermocouple Probes. In Proceedings of the IEEE Instrumentation and Measurement Technology Conference, Sorrento, Italy, 24–27 April 2006; pp. 651–654.
63. Sathik, M.H.M.; Prasanth, S.; Sasongko, F.; Padmanabhan, S.K.; Pou, J.; Simanjorang, R. Online junction temperature for off-the-shelf power converters. In Proceedings of the IEEE Applied Power Electronics Conference and Exposition (APEC), San Antonio, TX, USA, 4–8 March 2018; pp. 2769–2774.
64. Ikonen, M.; Häsä, H.; Rauma, K.; Silventoinen, P. A system for thermal model verification of a power switch. In Proceedings of the 37th IEEE Power Electronics Specialists Conference, Jeju, Korea, 18–22 June 2006; pp. 1–4.
65. Bonsbaine, A.; Trigkidis, G.; Benamrouche, N. An integrated electro-thermal model of IGBT devices (experimental validation). In Proceedings of the 44th International Universities Power Engineering Conference (UPEC), Glasgow, UK, 1–4 September 2009; pp. 1–5.
66. Du, B.; Hudgins, J.L.; Santi, E.; Bryant, A.T.; Palmer, P.R.; Mantooth, H.A. Transient Electrothermal Simulation of Power Semiconductor Devices. *IEEE Trans. Power Electron.* **2010**, *25*, 237–248.

67. Claassen, A.; Shaukatullah, H. Comparison of diodes and resistors for measuring chip temperature during thermal characterization of electronic packages using thermal test chips. In Proceedings of the13th Annual IEEE. Semiconductor Thermal Measurement and Management Symposium, Austin, TX, USA, 28–30 January 1997; pp. 198–209.
68. Motto, E.R.; Donlon, J.F. IGBT module with user accessible on-chip current and temperature sensors. In Proceedings of the Twenty-Seventh Annual IEEE Applied Power Electronics Conference and Exposition (APEC), Orlando, FL, USA, 5–9 February 2012; pp. 176–181.
69. Ka, I.; Avenas, Y.; Dupont, L.; Vafaei, R.; Thollin, B.; Crebier, J.C.; Petit, M. Instrumented chip dedicated to semiconductor temperature measurements in power electronic converters. In Proceedings of the IEEE Energy Conversion Congress and Exposition (ECCE), Milwaukee, WI, USA, 18–22 September 2016; pp. 1–8.
70. Zhou, Y.; Shi, W.; Tang, J.; Wang, X.; Li, W.; He, X.; Zhang, C.; Li, Z. Dynamic junction temperature estimation via built-in negative thermal coefficient (NTC) thermistor in high power IGBT modules. In Proceedings of the IEEE Applied Power Electronics Conference and Exposition (APEC), Tampa, FL, USA, 26–30 March 2017; pp. 772–775.
71. Baker, N.; Iannuzzo, F.; Beczkowski, S.; Kristensen, P.K. Proof-of-Concept for a Kelvin-Emitter On-Chip Temperature Sensor for Power Semiconductors. In Proceedings of the 21st European Conference on Power Electronics and Applications (EPE '19 ECCE Europe), Genova, Italy, 3–5 September 2019; pp. P.1–P.8.
72. Kempiak, C.; Lindemann, A.; Thal, E.; Idaka, S. Investigation of the usage of a chip integrated sensor to determine junction temperature during power cycling tests. In Proceedings of the CIPS 2018, 10th International Conference on Integrated Power Electronics Systems, Stuttgart, Germany, 20–22 March 2018; pp. 1–6.
73. Baker, N.; Iannuzzo, F.; Li, H. Impact of Kelvin-Source Resistors on Current Sharing and Failure Detection in Multichip Power Modules. In Proceedings of the 20th European Conference on Power Electronics and Applications (EPE'18 ECCE Europe), Riga, Latvia, 17–21 September 2018; pp. 1–7.
74. Kempiak, C.; Lindemann, A.; Idaka, S.; Thal, E. Investigation of an Integrated Sensor to Determine Junction Temperature of SiC MOSFETs during Power Cycling Tests. In Proceedings of the 10th International Conference on Power Electronics and ECCE Asia (ICPE 2019—ECCE Asia), Busan, Korea, 27–31 May 2019; pp. 3084–3089.
75. Liu, P.; Chen, C.; Zhang, X.; Huang, S. Online junction temperature estimation method for SiC modules with built-in NTC sensor. *CPSS Trans. Power Electron. Appl.* **2019**, *4*, 94–99. [CrossRef]
76. Liu, P.; Zhang, X.; Yin, S.; Tu, C.; Huang, S. Simplified Junction Temperature Estimation using Integrated NTC Sensor for SiC Modules. In Proceedings of the IEEE International Power Electronics and Application Conference and Exposition (PEAC), Shenzhen, China, 4–7 November 2018; pp. 1–4.
77. O' Donnell, K.P.; Chen, X. Temperature dependence of semiconductor band gaps. *Appl. Phys. Lett.* **1991**, *58*, 2924–2926. [CrossRef]
78. Baliga, B.J. *Fundamentals of Power Semiconductor Devices*; Springer International Publishing: Cham, Switzerland, 2013.
79. Blackburn, D.-L. An electrical technique for the measurement of the peak junction temperature of power transistors. In Proceedings of the 13th Annual Reliability Physics Symposium, Las Vegas, NV, USA, 1–3 April 1975; pp. 143–150.
80. Farjah, E.; Perret, R. Application and analysis of thermosensitive parameters in the case of hybrid power modules. In Proceedings of the IEEE IndustryApplcations Society Annual Meeting, Orlando, FL, USA, 8–12 October 1994; pp. 1284–1289.
81. Choi, U.; Blaabjerg, F.; Iannuzzo, F.; Jørgensen, S. Junction temperature estimation method for a 600V, 30A IGBT module during converter operation. *Microelectron. Reliab.* **2015**, *55*, 2022–2026. [CrossRef]
82. Oettinger, F.F.; Blackburn, D.L.; Rubin, S. Thermal characterization of power transistors. *IEEE Trans. Electron. Devices* **1976**, *23*, 831–838. [CrossRef]
83. Held, M.; Jacob, P.; Nicoletti, G.; Scacco, P.; Poech, M.-H. Fast power cycling test for insulated gate bipolar transistor modules in traction application. *Int. J. Electron.* **1999**, *86*, 1193–1204. [CrossRef]
84. Blackburn, D.L. A review of thermal characterization of power transistors. In Proceedings of the Fourth Annual IEEE Semiconductor Thermal and Temperature Measurement Symposium, San Diego, CA, USA, 10–12 February 1988; pp. 1–7.
85. Khatir, Z.; Dupont, L.; Ibrahim, A. Investigations on junction temperature estimation based on junction voltage measurements. *Microelectron. Reliabil.* **2010**, *50*, 1506–1510. [CrossRef]
86. Barnes, C.-M.; Tuma, P.-E. Practical considerations relating to immersion cooling of power electronics in traction systems. *IEEE Trans. Power Electron.* **2010**, *25*, 2478–2485. [CrossRef]
87. Sofia, J.-W. Electrical measurement using semiconductors. *Electron. Cool.* **1997**, *3*, 22–25.
88. Nowak, M.; Rabkowski, J.; Barlik, R. Measurement of temperature sensitive parameter characteristics of semiconductor silicon and silicon carbide power devices. In Proceedings of the 13th Power Electronics and Motion Control Conference, Poznan, Poland, 1–3 September 2008; pp. 84–87.
89. Blackburn, D.-L.; Berning, D.-W. Power MOSFET temperature measurements. In Proceedings of the Annual Power Electronics Specialists Conference, Cambridge, MA, USA, 14–17 June 1982; pp. 400–407.
90. Jakopovic, Z.; Bencic, Z.; Kolonic, F. Important properties of transient thermal impedance for MOS-gated power semiconductors. In Proceedings of the IEEE International Symposium on Industrial Electronics, Bled, Slovenia, 12–16 July 1999; pp. 574–578.
91. Perpiñà, X.; Serviere, J.-F.; Saiz, J.; Barlini, D.; Mermet-Guyennet, M.; Millán, J. Temperature measurement on series resistance and devices in power packs based on on-state voltage drop monitoring at high current. *Microelectron. Reliab.* **2006**, *46*, 1834–1839. [CrossRef]

92. Meysenc, L.; Saludjian, L.; Bricard, A.; Rael, S.; Schaeffer, C. A high heat flux IGBT micro exchanger setup. *IEEE Trans. Compon. Packag. Manuf. Technol. A* **1997**, *20*, 334–341. [CrossRef]
93. Cova, P.; Ciappa, M.; Franceschini, G.; Malberti, P.; Fantini, F. Thermal characterization of IGBT power modules. *Microelectron. Reliabil.* **1997**, *37*, 1731–1734. [CrossRef]
94. Ammous, A.; Allard, B.; Morel, H. Transient temperature measurements and modeling of IGBT's under short circuit. *IEEE Trans. Power Electron.* **1998**, *13*, 12–25. [CrossRef]
95. Duong, S.; Rael, S.; Schaeffer, C.; De Palma, J.F. Short circuit behaviour for PT and NPT IGBT devices—Protection against explosion of the case by fuse. In Proceedings of the European Conference on Power Electronics and Applications, Seville, Spain, 19–21 September 1995; pp. 249–254.
96. Forest, F.; Rashed, A.; Huselstein, J.-J.; Martiré, T.; Enrici, P. Fast power cycling protocols implemented in an automated test bench dedicated to IGBT module ageing. *Microelectron. Reliab.* **2015**, *55*, 81–92. [CrossRef]
97. Koenig, A.; Plum, T.; Fidler, P.; De Doncker, R.-W. On-line junction temperature measurement of CoolMOS devices. In Proceedings of the 7th International Conference on Power Electronics and Drive Systems, Bangkok, Thailand, 27–30 November 2007; pp. 90–95.
98. Kim, Y.-S.; Sul, S.-K. On-line estimation of IGBT junction temperature using on-state voltage drop. In Proceedings of the 1998 IEEE Industry Applications Conference, St Louis, MO, USA, 12–15 October 1998; pp. 853–859.
99. Dupont, L.; Avenas, Y. Evaluation of thermo-sensitive electrical parameters based on the forward voltage for on-line chip temperature measurements of IGBT devices. In Proceedings of the Energy Conversion Congress and Exposition (ECCE-2014), Pittsburgh, PA, USA, 14–18 September 2014; pp. 4028–4035.
100. Ghimire, P.; Pedersen, K.B.; Trintis, I.; Munk-Nielsen, S. Online chip temperature monitoring using U_{ce}-load current and IR thermography. In Proceedings of the Energy Conversion Congress and Exposition (ECCE-2015), Montreal, QC, Canada, 20–24 September 2015; pp. 6602–6609.
101. Castellazzi, A.; Wachutka, G. Low-voltage Power MOSFETs used as dissipative elements: Electrothermal analysis and characterization. In Proceedings of the 37th IEEE Power Electronics Specialists Conference, Jeju, Korea, 18–22 June 2006; pp. 1–7.
102. Ayadi, M.; Fakhfakh, M.; Moez, G.; Neji, R. Electro-Thermal Simulation of a Three Phase Inverter with Cooling System. *J. Model. Simul. Syst.* **2010**, *1*, 163–170.
103. Baliga, J. *Power Semiconductor Devices*; International Thomson Publishing: Boston, MA, USA, 1996.
104. Chen, H.; Pickert, V.; Atkinson, D.J.; Pritchard, L.S. On-line monitoring of the MOSFET device junction temperature by computation of the threshold voltage. In Proceedings of the 3rd IET International Conference on Power Electronics, Machines and Drives, Dublin, Ireland, 4–6 April 2006; pp. 440–444.
105. Cao, X.; Wang, T.; Lu, G.-Q.; Ngo, K.D.T. Characterization of lead-free solder and sintered nano-silver die-attach layers using thermal impedance. In Proceedings of the International Power Electronics Conference, Sapporo, Japan, 21–24 June 2010; pp. 546–552.
106. Huang, X.Y.; Lu, C.; Xie, X.; Fan, Y.; Zhang, J.; Meng, X. A study of test system for thermal resistance of IGBT. In Proceedings of the 2010 Asia Pacific Conference on Postgraduate Research in Microelectronics Electronics (PrimeAsia), Shanghai, China, 22–24 September 2010; pp. 312–315.
107. Strauss, B.; Lindemann, A. Indirect measurement of junction temperature for condition monitoring of power semiconductor devices during operation. In Proceedings of the PCIM Europe, International Exhibition and Conference for Power Electronics, Intelligent Motion, Renewable Energy and Energy Management, Nuremberg, Germany, 19–20 May 2015; pp. 1–6.
108. Butron Ccoa, J.A.; Strauss, B.; Mitic, G.; Lindemann, A. Investigation of Temperature Sensitive Electrical Parameters for Power Semiconductors (IGBT) in Real-Time Applications. In Proceedings of the PCIM Europe, International Exhibition and Conference for Power Electronics, Intelligent Motion, Renewable Energy and Energy Management, Nuremberg, Germany, 20–22 May 2014; pp. 1–9.
109. Berning, D.; Reichl, J.; Hefner, A.; Hernandez, M.; Ellenwood, C.; Lai, J.-S. High speed IGBT module transient thermal response measurements for model validation. In Proceedings of the 38th IAS Annual Meeting on Industry Applications Conference, Salt Lake City, UT, USA, 12–16 October 2003; pp. 1826–1832.
110. Sundaramoorthy, V.; Bianda, E.; Bloch, R.; Nistor, I.; Knapp, G.; Heinemann, A. Online estimation of IGBT junction temperature (Tj) using gate-emitter voltage (Vge) at turn-off. In Proceedings of the 15th European Conference on Power Electronics and Applications (EPE), Lille, France, 3–5 September 2013; pp. 1–10.
111. Kuhn, H.; Mertens, A. On-line junction temperature measurement of IGBTs based on temperature sensitive electrical parameters. In Proceedings of the 13th European Conference on Power Electronics and Applications, Barcelona, Spain, 8–10 September 2009; pp. 1–10.
112. Barlini, D.; Ciappa, M.; Castellazzi, A.; Mermet-Guyennet, M.; Fichtner, W. New technique for the measurement of the static and of the transient junction temperature in IGBT devices under operating conditions. *Microelectron. Reliabil.* **2006**, *46*, 1772–1777. [CrossRef]
113. Du, M.; Xin, J.; Wang, H.; Ouyang, Z.; Wei, K. Estimating Junction Temperature of SiC MOSFET Using Its Drain Current during Turn-On Transient. *IEEE Trans. Electron. Devices* **2020**, *67*, 1911–1918. [CrossRef]
114. Li, L.; Ning, P.; Wen, X.; Li, Y.; Ge, Q.; Zhang, D.; Tai, X. A turn-off delay time measurement and junction temperature estimation method for IGBT. In Proceedings of the Applied Power Electronics Conference and Exposition (APEC-2017), Tampa, FL, USA, 26–30 March 2017; pp. 2290–2296.

115. Zhang, Z.; Wang, F.; Costinett, D.J.; Tolbert, L.M.; Blalock, B.J.; Wu, X. Online junction temperature monitoring using turn-off delay time for silicon carbide power devices. In Proceedings of the IEEE Energy Conversion Congress and Exposition (ECCE), Milwaukee, WI, USA, 18–22 September 2016; pp. 1–7.
116. Luo, H.; Chen, Y.; Sun, P.; Li, W.; He, X. Junction Temperature Extraction Approach with Turn-Off Delay Time for High-Voltage High-Power IGBT Modules. *IEEE Trans. Power Electron.* **2016**, *31*, 5122–5132. [CrossRef]
117. Zhang, J.; Du, M.; Jing, L.; Wei, K.; Hurley, W.G. IGBT Junction Temperature Measurements: Inclusive of Dynamic Thermal Parameters. *IEEE Trans. Device Mater. Reliab.* **2019**, *19*, 333–340. [CrossRef]
118. Luo, H.; Chen, Y.; Li, W.; He, X. Online high-power pin diode junction temperature extraction with reverse recovery fall storage charge. *IEEE Trans. Power Electron.* **2016**, *32*, 2558–2567. [CrossRef]
119. Luo, H.; Li, W.; He, X. Online high power Pin diode chip temperature extraction and prediction method with maximum recovery current di/dt. *IEEE Trans. Power Electron.* **2015**, *30*, 2395–2404. [CrossRef]
120. Xiang, D.; Ran, L.; Tavner, P.; Yang, S. Condition monitoring power module solder fatigue using inverter harmonic identification. *IEEE Trans. Power Electron.* **2012**, *27*, 235–247. [CrossRef]
121. Bryant, A.; Yang, S.; Mawby, P.; Xiang, D. Investigation into IGBT dV/dt during turn-off and its temperature dependence. *IEEE Trans. Power Electron.* **2011**, *26*, 3019–3031. [CrossRef]
122. Chen, Y.; Luo, H.; Li, W.; He, X.; Iannuzzo, F.; Blaabjerg, F. Analytical and Experimental Investigation on a Dynamic Thermo-Sensitive Electrical Parameter with Maximum dI_C/dt during Turn-off for High Power Trench Gate/Field-Stop IGBT Modules. *IEEE Trans. Power Electron.* **2017**, *32*, 6394–6404. [CrossRef]
123. Zheng, R.; Haoge, X.; Chengmin, L.; Wuhua, L.; Xiangning, H.; Luo, H.; Li, D. Online Aging Parameter Extraction with Induced Voltage v_{eE} between Kelvin and Power Emitter in Turn-off Progress for IGBT Modules. In Proceedings of the IEEE Energy Conversion Congress and Exposition (ECCE), Portland, OR, USA, 23–27 September 2018; pp. 362–366.
124. Zeng, Z.; Li, X.; Zhang, X.; Cao, L. Comparative Evaluation of Kelvin Connection for Current Sharing of Multi-Chip Power Modules. In Proceedings of the IEEE Energy Conversion Congress and Exposition (ECCE), Portland, OR, USA, 23–27 September 2018; pp. 4664–4670.
125. Chen, H.L.; Li, W.; He, X. A thermo-sensitive electrical parameter with maximum dIC/dt during turn-off for high power trench/field-stop IGBT modules. In Proceedings of the 31th Annual IEEE Applied Power Electronics Conference and Exposition, Long Beach, CA, USA, 20–24 March 2016; pp. 499–504.
126. Mautry, P.G.; Trager, J. Investigation of self-heating in VLSI and ULSI MOSFETs. In Proceedings of the International Conference on Microelectronic Test Structures, San Diego, CA, USA, 5–7 March 1990; pp. 221–226.
127. Baker, N.; Munk-Nielsen, S.; Liserre, M.; Iannuzzo, F. Online junction temperature measurement via internal gate resistance during turn-on. In Proceedings of the 16th European Conference on Power Electronics and Applications, Lappeenranta, Finland, 26–28 August 2014; pp. 1–10.
128. Lembeye, Y.; Schanen, J.L.; Keradec, J.P. Experimental characterization of insulated gate power components: Capacitive aspects. In Proceedings of the 32nd IAS Annual Meeting on Industry Applications Conference, New Orleans, LA, USA, 5–9 October 1997; pp. 983–988.
129. Zhou, S.; Zhou, L.; Sun, P. Monitoring potential defects in an IGBT module based on dynamic changes of the gate current. *IEEE Trans. Power Electron.* **2013**, *28*, 1479–1487. [CrossRef]
130. Baker, N.; Munk-Nielsen, S.; Iannuzzo, F.; Liserre, M. IGBT Junction Temperature Measurement via Peak Gate Current. *IEEE Trans. Power Electron.* **2016**, *31*, 3784–3793. [CrossRef]
131. Baker, N.; Dupont, L.; Munk-Nielsen, S.; Iannuzzo, F.; Liserre, M. IR Camera Validation of IGBT Junction Temperature Measurement via Peak Gate Current. *IEEE Trans. Power Electron.* **2017**, *32*, 3099–3111. [CrossRef]

electronics

MDPI

Article

Reliability Analysis and Repair Activity for the Components of 350 kW Inverters in a Large Scale Grid-Connected Photovoltaic System

Filippo Spertino *, Angela Amato, Gabriele Casali, Alessandro Ciocia and Gabriele Malgaroli

Dipartimento Energia, Politecnico di Torino, Corso Duca degli Abruzzi 24, 10129 Torino, Italy; angela.amato@polito.it (A.A.); gabriele.casali@studenti.polito.it (G.C.); alessandro.ciocia@polito.it (A.C.); gabriele.malgaroli@polito.it (G.M.)

* Correspondence: filippo.spertino@polito.it

Abstract: The reliability of photovoltaic (PV) generators is strongly affected by the performance of Direct Current/Alternating Current (DC/AC) converters, being the major source of PV under-performance. However, generally, their reliability is not investigated at component level: thus, the present work presents a reliability analysis and the repair activity for the components of full bridge DC/AC converters. In the first part of the paper, a reliability analysis using failure rates from literature is carried out for 132 inverters (AC rated power of 350 kW each) with global AC power of 46 MW in a large scale grid-connected PV plant. Then, in the second part of the work, results from literature are compared with data obtained by analyzing industrial maintenance reports in the years 2015–2017. In conclusion, the yearly energy losses involved in the downtime are quantified, as well as their availability.

Keywords: photovoltaic systems; DC/AC converter; reliability; maintenance; power system faults; availability

Citation: Spertino, F.; Amato, A.; Casali, G.; Ciocia, A.; Malgaroli, G. Reliability Analysis and Repair Activity for the Components of 350 kW Inverters in a Large Scale Grid-Connected Photovoltaic System. *Electronics* **2021**, *10*, 564. https://doi.org/10.3390/electronics10050564

Academic Editors: Elio Chiodo and Deep Jariwala

Received: 28 January 2021
Accepted: 22 February 2021
Published: 27 February 2021

1. Introduction

In the last decade, energy generation using renewable energy sources (RES) has rapidly grown, also at local level in nearly zero-energy buildings [1,2], to fulfil an increasing amount of electricity demand and, thus, their self-sufficiency. Among RES, solar photovoltaic (PV) is the most important technology because it is clean, inexhaustible, worldwide distributed, and has low installation costs [3]. Moreover, PV technology does not include rotating machinery or components likely to fail easily [4]. Indeed, compared to other RES plants, e.g., wind turbines, PV systems are more reliable, requiring lower operation and maintenance costs [5]. The coupling between PV generators and Direct Current/Alternating Current (DC/AC) converters can be investigated using proper energy models. As well known, it is recommended to remain within the voltage range of the maximum power point (MPP) tracker. According to the max/min temperature variations during the lifetime of the system, the range of DC voltage at the MPP can be predicted using advanced PV models, like the single diode model [6]. Moreover, the reliability of PV systems is strongly affected by the DC/AC converters. In fact, the inverters have very high performance (with DC/AC conversion efficiencies higher than 98%, easily simulated by models for energy balances between RES generation and load [7]), but they are the most subject to failures [8,9]. The failure of a single inverter in a PV plant, actually, may determine a significant loss of power production [10].

In this context, a reliability analysis (RA) is fundamental for operators in order to identify the components most likely to fail. In fact, a correct RA permits to schedule in advance the preventive and corrective maintenance, reducing the energy losses and, thus, optimizing the generation of the PV plant. However, the RA commonly presented in literature analyzes the converters at system level, not investigating the performance of their components. In this context, in [11], a model is proposed to evaluate the reliability of a PV system, whose subsystems are the PV modules and the Insulated Gate Bipolar Transistors (IGBTs). The paper [12] performs a reliability analysis of seven grid-connected PV plants with rated power from 100 kW to 2.5 MW: the components of the inverters are not investigated. The article [13] proposes an approach based on reliability block diagrams to evaluate PV system reliability for several inverter configurations. The RA performed in these works cannot identify the converters' components most likely to fail. Only in a few papers [14,15] their performance is partially investigated. In particular, the article [14] proposes a reliability model to analyze several inverter designs. In this context, the paper groups the devices in subcomponents: however, these subsystems correspond to the main failure modes of the converters rather than to their physical components. On the contrary, in [15], a reliability analysis of converters is performed at component level. However, the analyses presented in [14,15] are performed using data from literature: mainly, failure rates from the standards MIL-HDBK-217F [16]. Generally, these values are average quantities from large experimental datasets, which may be outdated or not representative for the case studies under analysis.

This paper performs a reliability analysis and the repair activity of full bridge DC/AC converters at component level. In particular, a fault tree analysis (FTA) is carried out to identify the most critical components of the converters using data from literature. Moreover, a second reliability analysis is performed using data from industrial maintenance reports. The results of the two analyses are compared in order to investigate the agreement of the model from literature with respect to the reports. The analysis is divided in two parts: in the first part of the work, a reliability analysis using failure rates from literature is carried out for 132 inverters (AC rated power = 350 kW each) with total AC power of 46 MW. In the second part of the paper, industrial maintenance reports of the converters in the years 2015–2017 are analyzed. These reports collect the main data regarding the maintenance operations performed for each converter: the start/end date of maintenance activities, their duration and typology (corrective, preventive maintenance, or monitoring), the involved components and a short description of the operation. Starting from these data, the number of failures corresponding to each subcomponent of the inverters and their average repair time are estimated. Finally, the energy losses consequent to the absence of availability are calculated for the components of the converters.

The paper is organized as follows. Section 2 presents the models performing the reliability analysis using data from literature. In Section 3, the main components of full bridge DC/AC converters and the assumed failure rates are presented. Section 4 presents the converters under analysis and the fault tree used to describe their performance. Section 5 presents the results of the reliability analysis and the repair activity, along with the energy losses and the availability of the inverters under study. Finally, Section 6 contains the conclusions.

2. Reliability and Availability Models Applied to PV Plants

The reliability analysis is a method quantifying the probability of success of an object (component, subsystem or system), i.e., its ability to carry out its functions for a time period Δt under specific environmental and operational conditions [17]. However, this analysis requires the knowledge of some parameters, which are shortly described below.

The failure rate λ represents the probability that an object fails in the time unit [18]. Thus, it provides quantitative information regarding the failure frequency of a device, which is expressed in number of failures per time unit [19]. Moreover, the failure rate is not constant with time because it varies according to the life stage of the component [20]. Generally, the life cycle of a device is divided into three periods: the burn-in, the useful life and the wear-out. A typical profile of the failure rate as a function of time, named "bathtub curve", is presented in Figure 1 [21]. In the first stage, the failure is high and rapidly decreasing: after their manufacturing, a component is subject to validation tests but undesirable failure, named "early failures", may occur due to manufacturing defects or design issues, not detected in the test phase. If early failures do not occur, the component operates in its useful life period: in this stage, its failure probability reaches the minimum value, and random failures may occur only. In this stage, the failure rate may be assumed constant. In the wear-out period, the probability of failure rapidly increases due to degradation and usage [22]. In the present work, components are assumed to work in their useful life with a constant failure rate. Under this assumption, the exponential model used to estimate the reliability function R at any time t is the following:

$$R(t) = e^{-\lambda t} \tag{1}$$

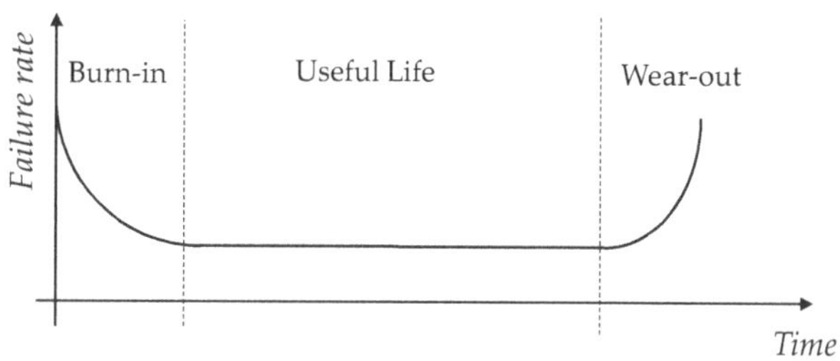

Figure 1. Bathtub curve of a generic component during time.

According to this model, the reliability function is unitary at the beginning of the component life, i.e., $R(t = 0) = 1$.

The mean time to failure (*MTTF*) is a parameter provided for not repairable components, i.e., devices that are not repaired after their failure because their total replacement is more cost-effective and requires less time with respect to their repair. In particular, the *MTTF* is the expected time before a component fails. On the contrary, in case of repairable components, their repair is preferred to their complete replacement. For these devices, the mean time between failure (*MTBF*) quantifies the expected time between two consecutive failures.

These parameters permit to compare the reliability of systems consisting of different components. During their useful life, if repairable components fail with the same failure mode, they can be assumed to recover their total functionality after being repaired (reliability function equal to 1): in this condition, their *MTBF* can be assumed equivalent to their *MTTF*. The *MTTF* can be estimated with the following equation:

$$MTTF = \int_0^{\infty} R(t)dt = \int_0^{\infty} e^{-\lambda t}dt \tag{2}$$

Under the assumption of components operating in their useful life, the *MTTF* can be assumed equivalent to $1/\lambda$. The information provided by the *MTTF* permits to schedule the preventive maintenance of components in order to minimize their risk of failure and, thus, to improve their reliability. In case of a complex system with many identical components, the reliability for each group of identical components R_i can be calculated starting from the reliability of a single device in the following way:

$$R_i(t) = e^{-m_i \cdot \lambda_i \cdot t} \tag{3}$$

where m_i is the number of identical devices for each type of component and λ_i is the corresponding failure rate of a single device.

In case of failures, the mean time to repair (*MTTR*) and the mean down time (*MDT*) provide information regarding the rapidity of maintenance operations. In particular, the *MTTR* is the average time required to restore a component to its full functionality, while the *MDT* is the average time between a failure of a device and its restoration to normal operation [23]. Thus, in addition to *MTTR*, *MDT* includes delays due to failure

4. Description of the Case Study under Analysis

The fault tree analysis (FTA) is a diagrammatic method commonly used to improve the reliability of a system by reducing its risk of failure. This technique selects the top event of the analysis, i.e., the undesired event to avoid during operation [31]. Then, the possible combinations between the failures determining the top event are identified [32]. The top event of the FTA may consist of the failure or the breakage of a component: in the present work, the top event is the loss of output power on the AC side of the converters. Under this assumption, the system under analysis is the group of inverters (132) in the PV plant. In particular, the fault tree has a single level including the failures for the different components, i.e., the fault of any converter component determines the top event (Figure 4).

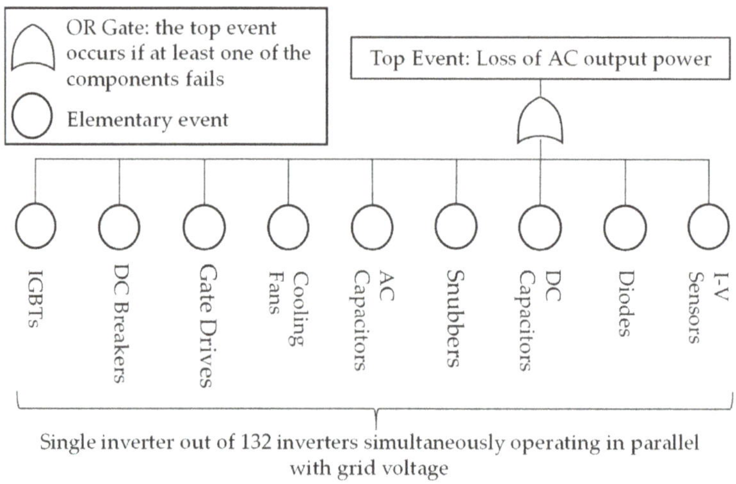

Figure 4. Fault tree for the components of the DC/AC converters under analysis.

In this analysis, each failure is assumed independent of the others and the events are supposed not mutually exclusive, i.e., the failure of a component cannot exclude other faults. The failures are elementary events: the fault of a component does not cause another component to fail. This work does not take into account faults due to design/installation errors: each device is supposed to be properly designed and installed. Furthermore, this paper does not consider losses of PV production due to electrical mismatch, which lowers the performance of the plant and generates harmonic distortion phenomena [33]. Finally, the components of the converters are assumed repairable: in such condition, the *MTTF* coincides with the *MTBF* [34].

In the first part of this work, the reliability for the subcomponents of full bridge DC/AC converters is evaluated. In particular, 132 inverters with a total AC rated power of 46 MW in Northern Italy were analyzed using their technical datasheets and failure rates from literature. The designer of this PV system decided to undersize the converters with respect to their own PV generators (rated power = 48 MW). The ratio between the DC rated power (peak power at standard test conditions) of the PV modules and the total AC nominal power of the converters is 104%. This ratio can affect the reliability of the system. In the second part of the paper, the reliability of the components is calculated using maintenance reports and monitoring data in the years 2015–2017. Finally, the energy losses involved in the downtime and the availability are estimated for each group of components.

The converters belong to centralized configuration [35], having a rated power of 350 kW each, and they are placed in concrete cabins. The list of their specifications is presented in Table 1, and the number of components is reported for a single converter in Table 2. In particular, the number of some components (DC capacitors, diodes, snubbers,

DC and AC inductors, and *I-V* sensors) is unknown. According to maintenance reports, they are assumed to be operating for an average time of 11 h per day.

Table 1. Specifications of the converters.

DC [1] power	360 kW
DC [1] voltage	450–830 V
Maximum DC [1] current	856 A
AC [2] power	350 kW
AC [2] voltage	502 V
AC [2] current	697 A
Efficiency	98%
Total harmonic distortion of current waveform	<3%

[1] Direct Current (DC); [2] Alternating Current (AC).

Table 2. Number of components for a single converter.

AC capacitors	5
AC inductors	Unknown
Cooling fans	7
DC breakers	1
DC capacitors	Unknown
Diodes	Unknown
Gate drivers	1
I-V [1] sensors	Unknown
IGBTs [2]	6
Snubbers	Unknown

[1] Current-Voltage (*I-V*); [2] Insulated Gate Bipolar Transistors (IGBTs).

5. Results

In this section, the results of the reliability analysis and the repair activity, as well as the energy losses and the availability are presented. As previously written, the number of some components (DC capacitors, diodes, snubbers, DC and AC inductors, and *I-V* sensors) is unknown, but the maintenance reports did not detect any failure related to these devices. Moreover, according to data from literature, they are expected to work without failures for a much longer period than the years under analysis. Among these components, according to literature, electrolytic capacitors have the lowest *MTTF*: however, in this case, this quantity may assume values in a wide range, up to ≈15,000 h [36]. Therefore, DC capacitors may work without failing for more than 3 years (as stated in the maintenance reports).

5.1. Results Using Data from Literature

Figure 5 presents the results of the reliability analysis using data from literature: the *MTTF* refers to the global number of identical components in the PV plant. In particular, the IGBTs have the highest *MTTF* (≈1400 h), while the components with the worst performance are the gate driver circuits (*MTTF* ≈ 310 h). Obviously, the *MTTF* of the DC/AC converter (≈140 h) is lower than the single *MTTF* of the components, because the inverter fails if any of its components fail.

In Figure 6, the profiles of the reliability function are presented.

After 1 year, the IGBTs, the cooling fans, the DC breakers, and the AC capacitors have a non-null reliability, i.e., their probability of being healthy after 1 year is not zero. Indeed, their reliability ranges between ≈1% (DC breakers and AC capacitors) and ≈6% (IGBTs). On the contrary, the reliability of the gate drivers is zero after a couple of months, i.e., they are very likely to require maintenance actions after that period. The reliability of the converters is zero after about 1 month.

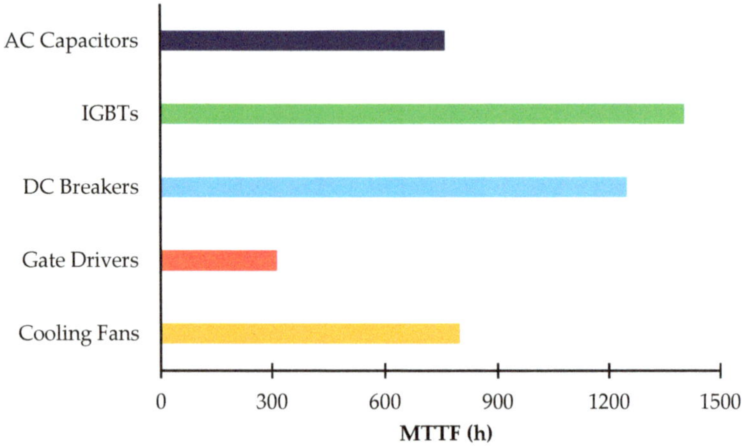

Figure 5. Mean Time To Failure (*MTTF*) of the DC/AC converter components evaluated using data from literature.

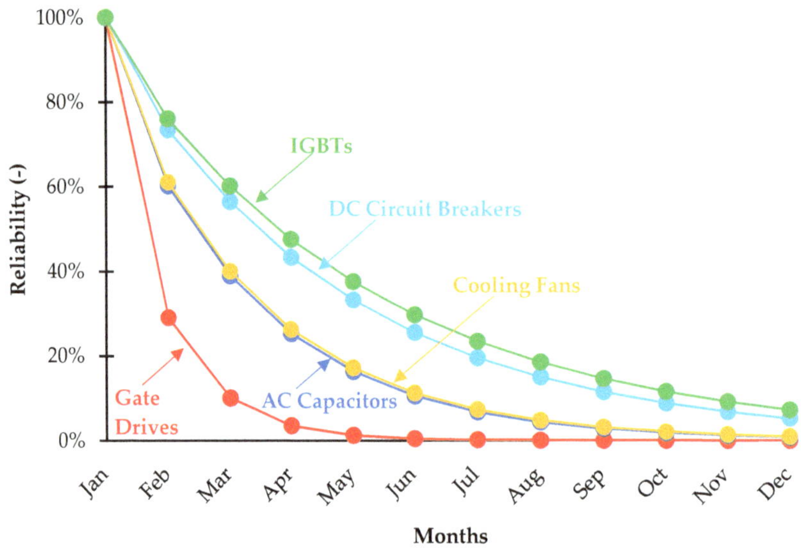

Figure 6. Reliability function of the DC/AC converter components using data from literature (first year of operation).

5.2. Results Using Experimental Data from Industrial Maintenance Reports

In this section, the results of the reliability analysis using data from industrial maintenance reports are presented. Table 3 quantifies the failures of the components according to the reports: the *MTTF* of each group of components is obtained as the ratio between the total operation time and the corresponding number of failures recorded. In Figure 7, the *MTTFs* using data from maintenance reports (red bars) are compared with the results of previous subsection (yellow bars). Regarding AC capacitors, IGBTs and DC breakers, results from literature overestimate the *MTTF*, i.e., these components fail more frequently than expected, having real *MTTF* equal to ≈265, ≈415, and ≈485 h, respectively. On the contrary, the gate drivers and the cooling fans perform better than expected (*MTTF* ≈670 and ≈865 h, respectively); as expected, the inverter has the lowest *MTTF* (≈80 h).

Table 3. Number of failures per year of the components for the 132 inverters in the PhotoVoltaic (PV) plant.

	AC Capacitors	IGBTs	DC Breakers	Gate Drivers	Cooling Fans
2015	9	9	7	5	4
2016	16	9	3	7	6
2017	21	11	15	6	4
2015–2017	46	29	25	18	14

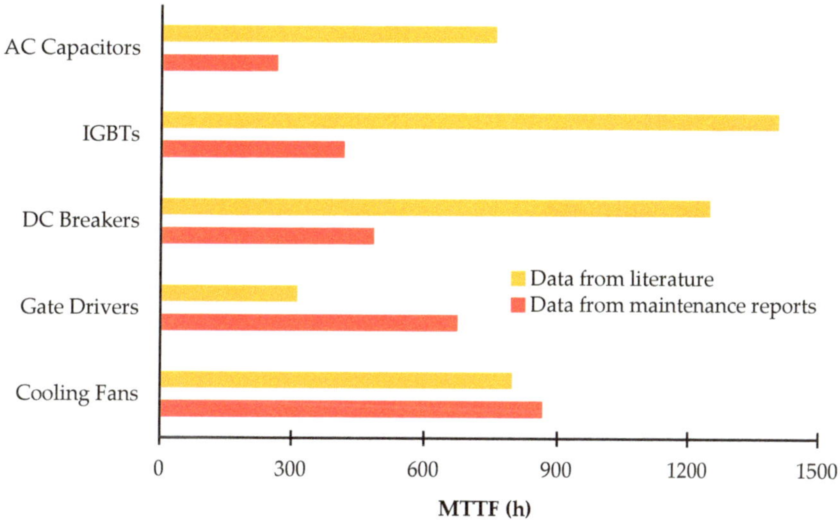

Figure 7. *MTTF* obtained using data from literature (yellow) and from maintenance reports (red).

The deviations between the results using failure rates from literature and data from maintenance reports may be very high, up to several thousand hours. However, the failure rates from literature are average values taken from experimental datasets at international level. Obviously, these data may refer to very different PV plants from the system under analysis. In such condition, high deviations around the mean value may occur, and it is preferred to compare the results using another parameter rather than the *MTTF*. This new quantity can be evaluated for the components and the DC/AC converters as well, being the ratio between two terms. The numerator consists of the *MTTF* deviations (in terms of hours) between values from literature and from maintenance reports, while the denominator is the number of operation hours. In the present paper, this parameter is reasonably low (≈8%) for the components with the lowest *MTTF* from the reports (IGBTs): thus, the agreement of the model used to estimate the reliability of the system starting from literature data with the reports is confirmed. For the other groups of components, this parameter is much lower, being ≈0.6% for the cooling fans and ≈0.5% for the converters.

The *MTTR* of each component group is evaluated from maintenance reports as the ratio between the total repair time and the number of maintenance activities for each group. The components requiring the lowest time to be repaired are the cooling fans (≈2 h) and the DC breakers (≈3 h). On the contrary, the highest *MTTR* is related to the IGBTs (≈95 h), which is more than four times higher than the other components (the highest value is ≈19 h for gate drivers). For this reason, the maintenance operations regarding IGBTs were analyzed in detail, and many delays in the supply of spare components, up to 40 days, were identified. This issue occurred mainly in the first 2 years of the analysis (2015 and 2016), determining a *MTTR* of 175 and 90 h, respectively. On the contrary, in 2017, these problems were solved, and the IGBTs were repaired without delays in a much lower average time (*MTTR*≈16 h). Therefore, the data of the years 2015–2016 regarding the repair of IGBTs were excluded from this analysis; the *MTTR* values are reported in Figure 8. The *MTTR* of the DC/AC converters amounts to ≈9 h.

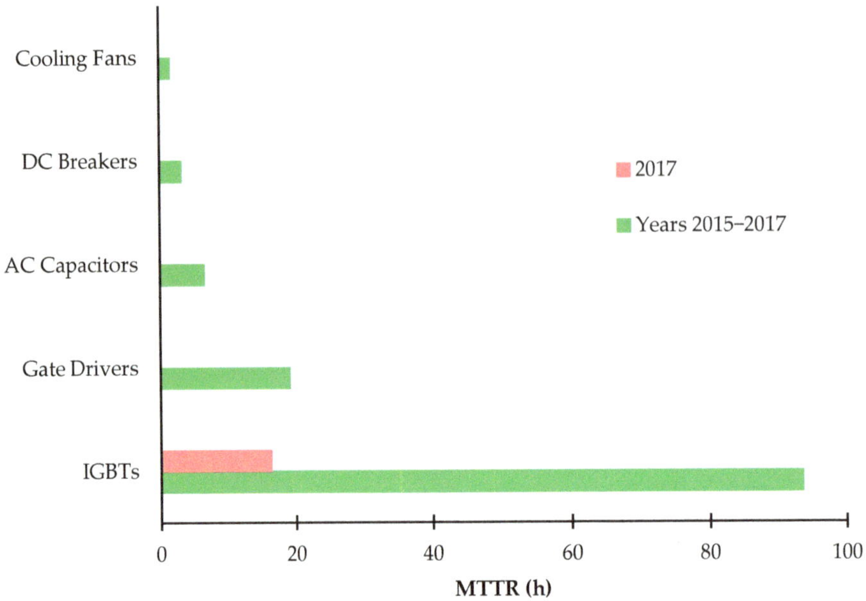

Figure 8. Mean Time To Repair (*MTTR*) of the groups of components.

The energy losses involved in the downtime durations confirm this behavior. In particular, they were estimated starting from the information provided in the reports (the AC energy generated by the single inverters in time intervals of 15 min). In case of failures, the associated energy losses, for each faulty DC/AC converter, were estimated as the difference between the energy produced by another converter in normal operation and the energy generated by the faulty device in the reports. The converters of the 48 MW system are identical, and the connected PV generators have the same orientation and inclination, without shadowing phenomena. Therefore, in case of faults, the first quantity was assumed as the average energy generated by the other inverters, working properly. The total energy losses in the years 2015–2017 are ≈−0.2%.

However, huge differences occur in the 3 years: in fact, the energy losses in 2015 (≈−0.3%) and 2016 (≈−0.2%) are more than four times higher than the corresponding quantity in 2017 (≈−0.06%). This is due to the delayed maintenance that occurred in 2015 and 2016 regarding IGBTs. Finally, the availability ranges between ≈96% (IGBTs) and 100% (cooling fans), determining a converter availability of ≈90% (Table 4).

Table 4. Availability for the components of the DC/AC converters.

Components	Availability
AC capacitors	97.6%
Gate drivers	97.2%
DC breakers	99.3%
IGBTs	96.3%
Cooling fans	99.8%

6. Conclusions

The present work performed a reliability analysis and the repair activity for the components of 132 full bridge inverters with rated power of 350 kW each. Moreover, a second reliability analysis was performed using data from industrial maintenance reports. The results of the two analyses were compared in order to investigate the agreement of the model from literature with respect to the reports. In the first part of the work, reliability results were obtained using failure rates from literature, while in the second part, industrial maintenance reports were analyzed to evaluate the *MTTF* and the *MTTR* for the global number of components in the PV plant. According to the reports, the components with the highest *MTTF* ($\approx$865 h) are the cooling fans, while the AC capacitors have the worst performance (*MTTF*$\approx$265 h). The *MTTF* of the IGBTs is $\approx$415 h, and the overall *MTTF* of DC/AC converters is $\approx$80 h.

According to the results using failure rates from literature, the components with the worst performance are the gate drivers (*MTTF*$\approx$310 h). However, the industrial reports show that these components perform better than expected (*MTTF*$\approx$670 h). As a consequence, operators may schedule a less restrictive maintenance on gate drivers. On the contrary, designers should invest mainly on improving the reliability of other components. Actually, according to the reports, the *MTTF* of AC capacitors, IGBTs and DC breakers is less than 40% of the corresponding *MTTF* estimated using data from literature. Thus, improving the reliability of these components is crucial to increase the reliability and availability of the entire converter.

In addition, the reliability of the converters may be affected by the ratio between the DC and AC rated powers of the PV system. This ratio is, generally, higher than 100% [37] for PV plants actually in operation in order to maximize their power generation at low/intermediate irradiance. In this case, converters may work more frequently close to their rated power: however, this condition increases the thermal stress of their components, which may become more subject to failures. In this work, the ratio between the rated power of PV generators and the AC nominal powers of the 132 converters is $\approx$104%. Thus, the converters are slightly undersized with respect to PV generators, but to improve the reliability of the system, it may be advisable to equalize the two parameters, especially in sunny and windy sites (as in the Mediterranean area). Nevertheless, values lower than 100% are favorable for the reliability but increase the installation costs.

Regarding the *MTTR*, the highest value is for the IGBTs, while the cooling fans have the best performance, being repaired within an average time of 2 h. However, the *MTTR* of IGBTs is more than four times higher than the other components: thus, the maintenance performance of IGBTs was further investigated by analyzing the reports in detail. Indeed, many delays in the supply of spare parts for these components (up to 40 days) are identified for the first 2 years of the analysis (2015 and 2016). These issues are not detected in 2017: the IGBTs are repaired within an average time of 16 h. Thus, the data of the years 2015–2016 regarding the repair of IGBTs were excluded from this analysis, and the *MTTR* of the DC/AC converters is $\approx$9h. The energy losses due to the absence of operation during the downtime confirm this behavior, being more than four times higher in 2015 ($\approx -0.3\%$) and in 2016 ($\approx -0.2\%$) with respect to 2017 ($\approx -0.06\%$). Finally, the availability for the different groups of components is higher than 97%, with an overall value of 90% for the inverters.

Author Contributions: Conceptualization, F.S.; methodology, F.S., G.C., G.M.; writing—original draft preparation, A.A., G.C., G.M.; writing—review and editing, F.S., A.A., G.C., A.C., G.M.; visualization, F.S., A.A., G.C., A.C., G.M.; supervision, F.S.. All authors have read and agreed to the published version of the manuscript.

Funding: This research received no external funding.

Data Availability Statement: Data are not available on a publicly accessible repository and they cannot be shared under request.

Conflicts of Interest: The authors declare no conflict of interest.

References

1. Spertino, F.; Fichera, S.; Ciocia, A.; Malgaroli, G.; Di Leo, P.; Ratclif, A. Toward the complete self-sufficiency of an NZEBS microgrid by photovoltaic generators and heat pumps: Methods and applications. *IEEE Trans. Ind. Appl.* **2019**, *55*, 7028–7040. [CrossRef]
2. Di Leo, P.; Spertino, F.; Fichera, S.; Malgaroli, G.; Ratclif, A. Improvement of self-sufficiency for an innovative nearly zero energy building by photovoltaic generators. In Proceedings of the 2019 IEEE Milan PowerTech, Milan, Italy, 23–27 June 2019; pp. 1–6. [CrossRef]
3. Ciocia, A.; Di Leo, P.; Fichera, S.; Giordano, F.; Malgaroli, G.; Spertino, F. A novel procedure to adjust the equivalent circuit parameters of photovoltaic modules under shading. In Proceedings of the 2020 International Symposium on Power Electronics, Electrical Drives, Automation and Motion (SPEEDAM), Sorrento, Italy, 24–26 June 2020; pp. 711–715. [CrossRef]
4. Spertino, F.; Chiodo, E.; Ciocia, A.; Malgaroli, G.; Ratclif, A. Maintenance Activity, Reliability Analysis and Related Energy Losses in Five Operating Photovoltaic Plants. In Proceedings of the 2019 IEEE International Conference on Environment and Electrical Engineering and 2019 IEEE Industrial and Commercial Power Systems Europe (EEEIC/I&CPS Europe), Genova, Italy, 11–14 June 2019; pp. 1–6. [CrossRef]
5. Spertino, F.; Chiodo, E.; Ciocia, A.; Malgaroli, G.; Ratclif, A. Maintenance Activity, Reliability, Availability, and Related Energy Losses in Ten Operating Photovoltaic Systems up to 1.8 MW. *IEEE Trans. Ind. Appl.* **2021**, *57*, 83–93. [CrossRef]
6. Humada, A.M.; Darweesh, S.Y.; Mohammed, K.G.; Kamil, M.; Mohammed, S.F.; Kasim, N.K.; Tahseen, T.A.; Awad, O.I.; Mekhilef, S. Modeling of PV system and parameter extraction based on experimental data: Review and investigation. *Sol. Energy* **2020**, *199*, 742–760. [CrossRef]
7. Spertino, F.; Ahmad, J.; Chicco, G.; Ciocia, A.; Di Leo, P. Matching between electric generation and load: Hybrid PV-wind system and tertiary-sector users. In Proceedings of the 2015 50th International Universities Power Engineering Conference (UPEC), Stoke on Trent, UK, 1–4 September 2015; pp. 1–6. [CrossRef]
8. El-Metwally, M.; EL-Shimy, M.; Elshahed, M.; Sayed, A. Detailed Analyses of the Failure and Repair Rates of Wind and Solar- PV Systems for RAM Assessment. In Proceedings of the 11th International Conference on Electrical Engineering (ICEENG), Cairo, Egypt, 3–5 April 2018; Volume 11, pp. 1–16. [CrossRef]
9. Yang, Y.; Wang, H.; Sangwongwanich, A.; Blaabjerg, F. 45-Design for Reliability of Power Electronic Systems. In *Power Electronics Handbook*; Elsevier Science & Technology: Amsterdam, The Netherlands, 2018; pp. 1423–1440.
10. Koutroulis, E.; Blaabjerg, F. Design optimization of transformerless grid-connected PV inverters including reliability. *IEEE Trans. Power Electron.* **2013**, *28*, 325–335. [CrossRef]
11. Umarani, D.; Seyezhai, R. Investigation of Reliability Aspects of Photovoltaic Quasi Z-Source Inverter. In Proceedings of the 2018 IEEE International Conference on Power Electronics, Drives and Energy Systems (PEDES), Chennai, India, 18–21 December 2018; pp. 1–5. [CrossRef]
12. Ahadi, A.; Ghadimi, N.; Mirabbasi, D. Reliability assessment for components of large scale photovoltaic systems. *J. Power Sources* **2014**, *264*, 211–219. [CrossRef]
13. Shi, X.; Bazzi, A.M. Solar photovoltaic power electronic systems: Design for reliability approach. In Proceedings of the 2015 17th European Conference on Power Electronics and Applications (EPE'15 ECCE-Europe), Geneva, Switzerland, 8–10 September 2015; pp. 1–8. [CrossRef]
14. Ristow, A.; Begović, M.; Pregelj, A.; Rohatgi, A. Development of a methodology for improving photovoltaic inverter reliability. *IEEE Trans. Ind. Electron.* **2008**, *55*, 2581–2592. [CrossRef]
15. Ma, Z.J.; Thomas, S. Reliability and maintainability in photovoltaic inverter design. In Proceedings of the 2011 Proceedings—Annual Reliability and Maintainability Symposium, Lake Buena Vista, FL, USA, 24–27 January 2011; pp. 1–5. [CrossRef]
16. Department of Defense of the USA. Reliability Prediction of Electronic Equipment. In *Military Handbook MIL-HDBK-217F*; Department of Defense of the USA: Washington, DC, USA, 1991; MIL-HDBK-217F (NOTICE 1).
17. Song, Y.; Wang, B. Survey on reliability of power electronic systems. *IEEE Trans. Power Electron.* **2013**, *28*, 591–604. [CrossRef]
18. Rausand, M.; Hoyland, A. *System Reliability Theory: Models, Statistical Methods, and Applications*; John Wiley & Sons: Hoboken, NJ, USA, 2004.
19. Obeidat, F.; Shuttleworth, R. Reliability prediction of PV inverters based on MIL-HDBK-217F N2. In Proceedings of the 2015 IEEE 42nd Photovoltaic Specialist Conference (PVSC), New Orleans, LA, USA, 14–19 June 2015; pp. 1–6. [CrossRef]
20. Yu, X.; Khambadkone, A.M. Reliability analysis and cost optimization of parallel-inverter system. *IEEE Trans. Ind. Electron.* **2012**, *59*, 3881–3889. [CrossRef]
21. Cooper, M. Observations on component infant mortality and burn-in effectiveness. *IEEE Trans. Components Packag. Technol.* **2008**, *31*, 914–916. [CrossRef]
22. Alvarez-Alvarado, M.S.; Jayaweera, D. Bathtub curve as a Markovian process to describe the reliability of repairable components. *IET Gener. Transm. Distrib.* **2018**, *12*, 5683–5689. [CrossRef]
23. Baschel, S.; Koubli, E.; Roy, J.; Gottschalg, R. Impact of component reliability on large scale photovoltaic systems' performance. *Energies* **2018**, *11*, 1579. [CrossRef]
24. Selvik, J.T.; Ford, E.P. Down Time Terms and Information Used for Assessment of Equipment Reliability and Maintenance Performance. In *System Reliability*; InTech Open: Rijeka, Croatia, 2017. [CrossRef]
25. Elsayed, E.A. *Reliability Engineering*; Addison Wesley, Reading: Boston, MA, USA, 1996.

26. Gao, D.; Sun, K. 16-DC–AC inverters. In *Electric Renewable Energy Systems*; Academic Press: Cambridge, MA, USA, 2016; pp. 354–381.
27. Abedinpour, S.; Shenai, K. 5-Insulated Gate Bipolar Transistor. In *Power Electronics Handbook*, 2nd ed.; Elsevier: Amsterdam, The Netherlands, 2007; pp. 71–88.
28. IGBT Technologies and Applications Overview: How and When to Use an IGBT. Available online: https://www.onsemi.com/pub/Collateral/TND6235-D.PDF (accessed on 28 January 2021).
29. Ciocia, A.; Di Leo, P.; Fichera, S.; Malgaroli, G.; Russo, A.; Spertino, F.; Tzanova, S.; Dalanbayar, B. Innovative Laboratories for Teaching on Photovoltaic Generation in Higher Education. In Proceedings of the 2020 XXIX International Scientific Conference Electronics (ET), Sozopol, Bulgaria, 16–18 September 2020; pp. 1–4. [CrossRef]
30. Vujacic, M.; Hammami, M.; Srndovic, M.; Grandi, G. Analysis of dc-link voltage switching ripple in three-phase PWM inverters. *Energies* **2018**, *11*, 471. [CrossRef]
31. Golnas, A. PV system reliability: An operator's perspective. *IEEE J. Photovolt.* **2013**, *3*, 416–421. [CrossRef]
32. Kritzinger, D. 4-Fault Tree Analysis. In *Aircraft System Safety*; Elsevier Science & Technology: Amsterdam, The Netherlands, 2017; pp. 59–99.
33. Chicco, G.; Corona, F.; Porumb, R.; Spertino, F. Experimental indicators of current unbalance in building-integrated photovoltaic systems. *IEEE J. Photovolt.* **2014**, *4*, 924–934. [CrossRef]
34. Zini, G.; Mangeant, C.; Merten, J. Reliability of large-scale grid-connected photovoltaic systems. *Renew. Energy* **2011**, *36*, 2334–2340. [CrossRef]
35. Spertino, F.; Chicco, G.; Ciocia, A.; Malgaroli, G.; Mazza, A.; Russo, A. Harmonic distortion and unbalance analysis in multi-inverter photovoltaic systems. In Proceedings of the 2018 International Symposium on Power Electronics, Electrical Drives, Automation and Motion (SPEEDAM), Amalfi, Italy, 20–22 June 2018; pp. 1031–1036. [CrossRef]
36. Flicker, J.D. *Capacitor Reliability in Photovoltaic Inverters*; Sandia National Laboratories: Albuquerque, NM, USA, 2015.
37. Oversizing of SolarEdge Inverters, Technical Note—SolarEdge. Available online: https://www.solaredge.com/sites/default/files/inverter_dc_oversizing_guide.pdf (accessed on 16 February 2021).

electronics

MDPI

Article

Forecasting the Reliability of Components Subjected to Harmonics Generated by Power Electronic Converters

Giovanni Mazzanti [1,*], Bassel Diban [1], Elio Chiodo [2], Pasquale De Falco [3] and Luigi Pio Di Noia [4]

1 Department of Electrical, Electronic and Information Engineering "Guglielmo Marconi", University of Bologna, 40136 Bologna, Italy; bassel.diban2@unibo.it
2 Department of Industrial Engineering, University of Napoli "Federico II", 80125 Naples, Italy; chiodo@unina.it
3 Department of Engineering, University of Napoli Parthenope, 80143 Naples, Italy; pasquale.defalco@uniparthenope.it
4 Department of Electrical Engineering and Information Technology, University of Naples Federico II, 80125 Napoli, Italy; luigipio.dinoia@unina.it
* Correspondence: giovanni.mazzanti@unibo.it

Received: 18 May 2020; Accepted: 3 August 2020; Published: 7 August 2020

check for updates

Abstract: This paper aims at refining an experimentally based reliability model for the insulation of power components subjected to the randomly varying harmonics generated by power electronic converters. Compared to previous papers of the same authors and to the existing literature, here the model is re-formulated from the theoretical viewpoint focusing on the foremost role played by low percentiles of time to failure—in particular by the 1st percentile—selected as the rated life in the framework of modern probabilistic design of components. This is not only more correct from the viewpoint of component design, but also on the safe side as for the reliability of devices. Moreover, the application of the model is broadened to treat the whole sequence of odd voltage harmonics from the 5th to the 25th, i.e., those taken as the most significant in power systems according to international standards. The limits to voltage distortion set in Standard EN50160 are the reference for establishing parametrically a series of typical distorted voltage waveshape analyzed in the applicative part, which account for the possible phase-shift angles between voltage harmonics. The effect of current harmonics is also considered, from both the theoretical and applicative viewpoint. As a last, but not least novelty, the reliability model is used here for life and reliability estimates not only of Medium Voltage (MV)/Low Voltage (LV) capacitors and cables—already studied in the previous stages of this investigation—but also of induction motors and transformers in the presence of harmonics from power converters.

Keywords: current harmonics; voltage harmonics; power electronic converters; reliability; cables; capacitors

---

1. Introduction

The worldwide diffusion of electric transportation systems and of smart grid technologies call for better performance of power electronic converters and components. However, in turn power electronic devices are well known to act as distorting loads, which inject voltage and current harmonics into the alternating current (AC) grid. These harmonics may hamper the reliability of power components—such as cables, capacitors, transformers, electrical motors—connected to the grid, because of the potential increase of thermal and electrical stress associated with current and voltage harmonics. Such a situation raises the reliability challenges to power components in electric transportation systems and smart grid installations. For this reason, in the last three decades some international standards have set distortion limits. IEEE 519 in 1993 [1] established limits to voltage and current harmonics, but later on IEC 61000-2-2 [2], IEC 61000-2-4 [3], EN 50160 [4] set limits to voltage distortion only, by fixing the maximum values of low-order voltage harmonics in LV/MV grids in unperturbed conditions.

Forecasting the reliability of the components in the presence of current and voltage harmonics is not an easy task. Traditional approaches rely on accelerated life testing (ALT) and on historical failure databases, but the fast technological development-leading to highly reliable devices-and the difficulty of performing sound testing campaigns under harmonic distortion make these data hardly available. The estimation of reliability in distorted conditions is made more difficult by the random nature of harmonic distortion brought about by power electronic devices. This requires proper statistical models and methods capable of correlating the electric and thermal stress, associated with voltage and current harmonics, to life and the reliability of components.

To overcome all these difficulties, a probabilistic electro-thermal life model—that can also be referred to as "electro-thermal reliability model" [5]—has been developed and proved to be capable of forecasting the life and reliability of components subjected to randomly time-varying harmonics generated by power electronic converters [6–9]. This model is based on a broad and innovative experimental campaign of testing insulating specimens (flat samples, mini-cables, twisted pairs) for power components (cables, capacitors, transformers, rotating machines) subjected to a big deal of combinations of voltage harmonics, as shown in [10–13]. Other tests of this kind are described e.g., in [14–18], but with particular emphasis on water tree growth in cross-linked polyethylene (XLPE) insulation for power cables in a wet environment. However, results of aging tests of insulation under distorted voltage are rare in the literature, due to the experimental difficulties in arranging test set-ups for aging of insulation in the presence of distorted voltage. For this reason, attention in the literature of reliability models under harmonic distortion is mainly concentrated on the thermal effect of current harmonics [19–23] although applications relevant to the electrical aging can also be found e.g., in [24–30].

Following the streamline of this investigation over the years, in the very first application of this electro-thermal reliability model for insulation under harmonic distortion the level of distortion was either set parametrically [6] or derived experimentally for a particular case [7]. Then, the model was used to estimate the reliability of MV/LV components affected by voltage harmonics matching exactly the limits established in [4] (which are numerically the same as those in [2,3]); the study was broadened from the 11th, 13th voltage harmonics treated in [8], to the combination of the 5th, 7th, 11th, 13th treated in [9], showing that—notwithstanding the compliance with [4]—the reliability of components decreased significantly with respect to rated sinusoidal conditions.

Here, as the ultimate stage of the investigation specifically conceived for this Special Issue, the reliability model is refined and better formulated from the theoretical viewpoint. Furthermore, its application is significantly broadened in this paper, since here many new case studies are examined and two more power system components (induction motors and transformers) are treated in addition to those (cables and capacitors) already studied in the previous stages of this investigation. The main novelties in this paper compared to previous papers devoted to the same topic [8,9] are as follows.

From the theoretical viewpoint, in our previous papers and in the existing literature the theory of life and reliability estimation of the insulation of components in distorted conditions relied on the 63.2th percentile of failure time and on the mean time to failure (MTTF). Here, in the theoretical treatment the 63.2th percentile is replaced with design life, L_D, given at design failure probability, P_D. As a consequence, the theory is focused on a conservatively-low percentile of times to failure (in particular on the 1st, see below): this is both more correct from the viewpoint of modern probabilistic design of power components and on the safe side as for power component reliability [5]. Furthermore, the whole theory is formalized more extensively and carefully, with a more detailed treatment of the effect of current harmonics and of a possible increase of thermal and electric stress associated with the sinusoidal components of voltage and current.

From the applicative viewpoint, a first novelty compared to previous papers is the treatment of both the 5th, 7th, 11th, 13th voltage harmonics already studied in [9], and of the 17th, 19th, 23rd, 25th, tackled here for the first time. Such harmonics are characteristic of 6-pulse—some of them of 12-pulse—alternating/direct current (AC/DC) converters; they play a major role, since are not only the highest in the spectra of voltage and current harmonics typically measured at the bus-bars of MV/LV grids [31,32], but also those for which Standards IEC 61000 and EN50160 set limits [2–4].

Another applicative novelty is that in our previous papers and in the existing literature the calculations were concentrated on the MTTF (design life itself was given as the design value of the MTTF) and only a spot estimation of reliability was carried out at a service time equal to design life [8,9]; moreover, the amplitudes of voltage harmonics were selected so as to match the limits in [4] exactly, only. Here, on the contrary—consistently with what is said above—the calculation is focused on the $100 \times P_D$th percentile of times to failure, and for failure probability a conservatively-low value $P_D = 0.01 = 1\%$ is chosen to be on the safe side as for power component reliability. One more novelty is that, pragmatically, design life L_D of power components is the typical service life of power systems affected by harmonics where the components are located. In addition, as for reliability estimation, reliability is evaluated throughout the service life of power components. Regarding the selected amplitude of voltage harmonics, beside the case where the amplitudes of voltage harmonics match the limits after EN 50160 exactly, two more cases are considered where voltage harmonics are 25% above and 25% below the limits in [4], so as to illustrate respectively the problems that may arise if these limits are overcome and the problems which may still remain even if these limits are matched with an apparently-broad safety margin—e.g., resorting to passive and/or active filters [31].

As a last, but not least applicative novelty-as hinted at above—two more components of power systems, i.e., induction motors and MV/LV transformers, are studied here in the presence of harmonics from power electronic converters in addition to cables and capacitors, already studied in the previous stages of this investigation.

On the whole, it is worth emphasizing that the calculations in this paper are completely new compared with those in previous ones [8,9], that have intentionally been cited here to allow a direct comparison. As highlighted hereafter at Section 3 in comprehensive tables for the cable and the capacitor, as well as for the motor and the transformer, such a comparison shows that all results and figures in this paper—although being intentionally homologous to those in previous papers for enabling a straightforward comparison—are not only different as for the values obtained and the curves plotted, but also on the safe side with respect to those in [8,9] This highlights the need for the more accurate analysis performed for the first time in this paper vs. the simpler one carried out in previous papers. This overall comparison is also opportune in a Special Issue paper, which can take the chance for reviewing and completing previous investigations so as to outline the state of the art and the ultimate achievements in this field.

Last but not least, an aspect to mention as a closure of this Introduction is that—as hinted at above—similar models were also used focusing on the effect of either current harmonics on thermal aging as in [33,34], or voltage harmonics on electrical aging as in [35,36], with results which agree with those found in this investigation. However, the full application to electrical and thermal aging of insulation in under current and voltage harmonics has been carried out only in the streamline of the development and refinement of the electro-thermal reliability model used here, which finds in this paper its conclusive application.

2. Theoretical Background of Insulation Aging in the Presence of Voltage and Current Harmonics

The weakest part of a power system component is mostly its insulation [37]. In the presence of the harmonic distortion generated by a power electronic device, a reduction of insulation life at a given failure probability—or, conversely, of insulation reliability at a given service time—may be observed vs. rated sinusoidal life and/or reliability due to a possible rise of [6,7]:

1. temperature, which involves an increase of the thermal stress. Indeed, current harmonics in the conducting parts and voltage harmonics in the dielectrics may warm up the insulation. In the following, only the increase of temperature due to harmonic currents is considered, while the warming-up effect of voltage harmonics is neglected. Indeed, the main applications here are conceived for MV and LV systems (see Section 3), where dielectric losses are negligible;
2. electric stress, associated with the non-sinusoidal voltage waveshape.

Let us now treat cases 1 and 2 separately (Sections 2.1 and 2.2), then combine them (Section 2.3) and later on recast them into a probabilistic time-varying framework (Sections 2.4 and 2.5).

2.1. The Role Played by Current Harmonics

Treating case 1 of Section 2 first, let us assume that a power electronic device generates M current harmonics (of root mean square (rms) value I_h, $h = 1, \ldots, M$), and that such current harmonics lead to a non-negligible increase (current harmonics superimposed to fundamental current can only increase the losses in conducting parts, thus the temperature of the component. Of course, such an increase might be negligible, i.e., non-measurable.) ΔT_{arm} of the temperature (herein, all temperatures T are meant in K, while the corresponding temperatures in degrees Celsius are indicated as θ) of the insulation of a nearby power component. Let us further hypothesize that the temperature of the insulation of the affected component increases from the nominal–design–sinusoidal temperature T_S to a "non-sinusoidal" temperature T_{NS} equal to:

$$T_{NS} = T_s + \Delta T_{arm} \tag{1}$$

Then, the following thermal life model can be used for the estimation of the time to failure of the insulation of the power component in the presence of the M current harmonics generated by the power electronic device and affecting the component [6–9]:

$$L_{NS,I} = L_S exp(-B\Delta T'_{harm}) \tag{2}$$

where:

- $L_{NS,I}$ = insulation time-to-failure (life) in the presence of non-sinusoidal current I_{NS}, thus of non-sinusoidal temperature T_{NS};
- L_S = life at nominal–design–sinusoidal current (of rms value I_S) and voltage (of rms value V_S), thus in the presence of design values of temperature T_S and electric field E_S (a trivial geometrical proportionality factor relates design electric field E_S to design voltage V_S, as well as non-sinusoidal electric field E_{NS} to non-sinusoidal voltage V_{NS} (see Section 2.2.));
- $B = \Delta W/k_B$ is the well-known parameter typical of the Arrhenius thermal life model, being ΔW the activation energy of the main thermal degradation reaction and k_B the Boltzmann constant [38,39];
- $\Delta T'_{harm}$ = a quantity depending on ΔT_{arm} and equal to (see Equation (1)):

$$\Delta T'_{harm} = \frac{1}{T_S} - \frac{1}{T_{NS}} = \frac{1}{T_S} - \frac{1}{(T_S + \Delta T_{arm})} \tag{3}$$

From (3) it is readily seen that the greater the increase ΔT_{arm}—if any—of the temperature of the insulation, the shorter is life in the presence of current harmonics.

Of course, the use of model (2) requires that the relationship between the M current harmonics and the relevant variation ΔT_{arm} of the temperature of the insulation is known.

2.2. The Role Played by Voltage Harmonics

Coming now to case 2 of Section 2, let us assume that a power electronic device generates N voltage harmonics (of rms value V_h, $h = 1, \ldots, N$), and that such voltage harmonics distort non-negligibly the nominal–design–sinusoidal voltage waveshape at power frequency applied to the insulation of a nearby component connected to the AC power grid. The distortion is said to be "non-negligible" if at least one of the three following quantities associated with the distorted voltage waveshape change sensibly (i.e., in a measurable way) from their nominal sinusoidal value:

1. the rms value of voltage, which changes from the nominal sinusoidal rms value $V_S = V_{1,n}$ (rms nominal sinusoidal voltage of harmonic order $h = 1$ or rms fundamental voltage at power frequency (For components connected directly to the AC power grid, the frequency of the fundamental harmonic f_1 coincides with the power frequency, f_0 = 50/60 Hz.)) to V_{NS}, the rms value of distorted voltage. Then, the rms value of distorted voltage in p.u. of the rms value of rated sinusoidal voltage, v_{NS}, can be written as follows:

$$v_{NS} = \frac{V_{NS}}{V_S} = \frac{V_{NS}}{V_{1,n}} \tag{4}$$

2. the peak value of voltage, which changes from the peak value of the nominal sinusoidal voltage $V_{S,p} = V_{1,n,p}$ to $V_{NS,p}$, the rms value of distorted voltage. Then, the peak value of distorted voltage in p.u. of the peak value of rated sinusoidal voltage, $v_{NS,p}$, can be written as follows:

$$v_{NS,p} = \frac{V_{NS,p}}{V_{S,p}} = \frac{V_{NS,p}}{V_{1,n,p}} \tag{5}$$

3. the rms value of the derivative of voltage, which changes from the rms value of the derivative of nominal sinusoidal voltage $hV_{1,n} = V_{1,n}$ to the rms value of the derivative of distorted voltage, where N different voltage harmonics of order h are now present. Then, by defining the ratio "rms value of the hth voltage harmonics" over "rms fundamental voltage at power frequency" $\alpha_h = V_h / V_{1,n}$, the rms value of the derivative of each of these voltage harmonics in p.u. of the rms value of nominal sinusoidal voltage can be written as follows:

$$hV_h / V_{1,n} = h\alpha_h \tag{6}$$

The distortion of the voltage waveshape, in turn, may result in an increase of the electric stress acting on the insulation of the power component. Such increase—if any—is due to the distorted electric field (of rms value E_{NS}) within the insulation caused by the non-sinusoidal voltage (of rms value V_{NS}). The increase of the electric stress in distorted regime is quantified by the so-called "voltage waveshape factors", which can be defined from Equations (4)–(6) as follows [6–10]: peak factor:

$$K_p = v_{NS,p} \tag{7}$$

waveshape factor (or slew rate) (If—differently from here—one wants to study the reliability of the insulation of a component not directly connected to the AC power grid—e.g., a PWM inverter controlling a three-phase motor whose working frequency f_1 is different from power frequency f_0, then the frequency of the fundamental harmonic is f_1 and the ratio f_1/f_0 should appear as a multiplying factor of the right-hand side of Equation (8)):

$$K_w = \sqrt{\sum_{h=1}^{N} h^2 \alpha_h^2} \tag{8}$$

rms factor:

$$K_r = v \tag{9}$$

Equations (7)–(9) clearly demonstrate that, when nominal sinusoidal voltage is applied, the three voltage waveshape factors are equal to unity, namely $K_p = K_w = K_r = 1$. By contrast, when N voltage harmonics are superimposed onto nominal sinusoidal voltage, it is observed that:

(a) K_p can be either greater or lower than 1, depending on the phase-shift angle ϕ_h between the voltage harmonics V_h and the fundamental V_1. Typically, the peak factor affects insulation life at most among the voltage waveshape factors, as made clear in Section 3 [8,9,35].
(b) K_w and K_r are always greater than 1, being definite positive quantities with lower value equal to 1, but the higher are the values of N and V_h, the higher are the values of K_w and K_r, and the more distorted is the voltage waveshape. This observation is consistent with the definition of the total harmonic distortion factor of the voltage, THD_V, related to K_w as follows [2–4]:

$$THD_v = \sqrt{\sum_{h=2}^{N} h^2 \alpha_h^2} = \sqrt{K_w{}^2 - 1} \tag{10}$$

Then, the following electrical life model can be used for the estimation of the time to failure of the insulation of a power component in the presence of the N voltage harmonics generated by the power electronic device [6–10]:

$$L_{NS,V} = L_S K_p^{-n_p} K_w^{-n_w} K_r^{-n_r} \tag{11}$$

where:

- $L_{NS,V}$ = insulation time-to-failure (life) under non-sinusoidal voltage/field, V_{NS}/E_{NS}, but at nominal—design—sinusoidal current I_S, thus in the presence of design temperature T_S;
- L_S = same as in Equation (2), namely life at nominal sinusoidal current I_S and voltage V_S, thus in the presence of design values of temperature T_S and electric field E_S;
- n_p = exponent that accounts for the aging acceleration effect of K_p, if any [35];
- n_w = exponent that accounts for the aging acceleration effect of K_w;
- n_r = exponent that accounts for the aging acceleration effect of K_r.

From Equation (11) it is readily seen that the greater the values of K_p, K_w, K_r with respect to unity, the shorter is life in the presence of voltage harmonics. However, it must be highlighted that—as pointed out above and in [36]—it is not granted that a distorted voltage waveshape necessarily leads to a value of $K_p > 1$, as clearly shown in Section 3.

2.3. The Combination of Current and Voltage Harmonics

When a power system component is subjected to both current and voltage harmonics generated by a power electronic device, then a combination of the life models (2) and (11) is required for a thorough and complete evaluation of the effect of distorted current and voltage on the life of the component. By combining these models, the following electro-thermal life model for distorted conditions is obtained:

$$L_{NS} = L_S exp(-B\Delta T'_{harm}) K_p^{-n_p} K_w^{-n_w} K_r^{-n_r} \tag{12a}$$

where L_{NS} is insulation life in the presence of non-sinusoidal voltage, V_{NS}, and non-sinusoidal temperature T_{NS}.

A careful analysis of Equations (4)–(11) emphasizes that voltage waveshape factors K_p, K_w, K_r are defined with respect to nominal sinusoidal voltage, hence life models (11) and (12) implicitly assume that the fundamental component of the distorted voltage waveshape (of rms value V_1) is equal to the rated sinusoidal voltage (of rms value $V_S = V_{1,n}$). However, it might happen that the fundamental component of distorted voltage has rms value V_H higher-than-rated sinusoidal voltage V_S, for instance because a highly capacitive load—e.g., a capacitor bank—is also supplied by the voltage source (only an increase of sinusoidal voltage is considered, since a lower than rated sinusoidal voltage is not common in unperturbed conditions, even under full inductive load; moreover considering a decrease of sinusoidal voltage is not on the safe side as for the estimation of life and reliability of power components). This situation tends to reduce the life of the insulation of power components with respect to nominal sinusoidal life, too: not because of voltage harmonics, but because a higher-than-rated sinusoidal electric field, E_H, is applied to the insulation. Such an effect has to be accounted for, resorting to the well-known Inverse Power Law electrical life model (IPM), that can be written as follows:

$$L_H = L_S(E_H/E_S)^{-n_S} \tag{12b}$$

where:

- L_H = insulation life in the presence of higher-than-rated non-sinusoidal voltage/field, V_H/E_H, and sinusoidal temperature T_S;
- n_S = the so-called life exponent or voltage endurance coefficient, that rules the life variation due to a change of sinusoidal voltage/field compared to design sinusoidal voltage/field [38,39].

Thereafter, by combining Equations (12a) and (12b), one obtains the following comprehensive life model:

$$L_{NS} = L_S(E_H/E_S)^{-n_S} exp(-B\Delta T'_{harm})K_p^{-n_p}K_w^{-n_w}K_r^{-n_r} \tag{13}$$

In addition, when looking carefully at Equations (1)–(3) it can be argued that non-sinusoidal temperature T_{NS} is defined with respect to nominal sinusoidal temperature T_S, hence life models (2), (12) and (13) implicitly assume that the fundamental component of the distorted current waveshape (of rms value I_1)—let us call it "sinusoidal current"—is equal to the rated sinusoidal current (of rms value $I_S = I_{1,n}$). However, it might happen that the fundamental component of distorted current has rms value I_{HL} higher/lower than rated sinusoidal current I_S, for instance due to a temporary overload/underload—e.g., because linear loads demand a higher/lower-than-rated current. This situation tends to reduce/increase the life of the insulation of the power component with respect to nominal sinusoidal life, too: not because of current harmonics, but because a higher/lower-than-rated "sinusoidal" temperature, T_{HL}, is applied to the insulation. If this is the case, let us write T_{HL} as follows:

$$T_{HL} = T_S + \Delta T_{HL} \tag{14a}$$

where ΔT_{HL} is higher/lower than zero depending on whether linear loads demand a higher/lower-than-rated current. Moreover, let us introduce the overall temperature variation ΔT_{tot} with respect to nominal sinusoidal temperature T_S resulting from:

- the temperature variation caused by harmonic currents, ΔT_{arm};
- the temperature variation caused by a change of sinusoidal current with respect to rated sinusoidal current, ΔT_{HL}.

Thus ΔT_{tot} can be written as follows:

$$\Delta T_{tot} = T_S + \Delta T_{arm} + \Delta T_{HL} \tag{14b}$$

and a quantity $\Delta T'_{tot}$—analogous to $\Delta T'_{trm}$ defined in Equation (3)—can be introduced, as follows:

$$T'_{tot} = 1/T_S - 1/(T_S + \Delta T_{arm} + \Delta T_{HL}) = 1/T_S - 1/(T_S + \Delta T_{tot}) \tag{15}$$

Therefore, by combining Equation (13) with Equations (14)–(15) one obtains the following electro-thermal life model holding in the presence of voltage and current harmonics, as well as of higher-than-rated sinusoidal voltage and current [9]:

$$L_{NS} = L_S(E_H/E_S)^{-n_S} exp(-B\Delta T'_{tot})K_p^{-n_p}K_w^{-n_w}K_r^{-n_r} \quad (16)$$

Model (16) is a comprehensive relationship which accounts for:

- a possible higher-than-rated sinusoidal electric field/voltage, $E_H > E_S/V_H > V_S$;
- a possible higher/lower than rated sinusoidal temperature, $T_{HL} > T_S/T_{HL} < T_S$, caused by a higher/lower than rated sinusoidal current, $I_{HL} > I_S/I_{HL} < I_S$;
- a distorted current due to M voltage harmonics, which causes an increase ΔT_{arm} of temperature with respect to rated sinusoidal temperature;
- a distorted voltage due to N voltage harmonics, which causes a change of voltage waveshape with respect to rated sinusoidal voltage.

2.4. Reliability Model in the Presence of Current and Voltage Harmonics

The electro-thermal breakdown of the insulation of a power component is an inherently random phenomenon. This is due first and foremost to the non-uniform composition of the dielectric material which constitutes the insulation, as well as to the randomly distributed defects within the insulation itself. A further reason for the random behavior of insulation breakdown is the uncertainty of the values of applied electrical and thermal stress [40,41]. For these reasons, all the above models—and in particular the general electro-thermal life model (16) holding in the presence of voltage and current harmonics—have to be recast into a probabilistic framework, whereby life (time to failure) of insulation is a random variable associated with a certain failure probability. This requires the introduction of a proper probability distribution of failure times. As is well known, failure statistics in polymeric insulation fit well the Weibull probability distribution [5,42], whereby insulation time to failure t_F can be represented by means of a 2-parameter Weibull cumulative probability distribution function (cdf) of failure times, given in the following equation:

$$P(t_F) = P = 1 - exp[-(t_F/\alpha_t)]^{\beta_t} \quad (17)$$

where P is cumulative failure probability corresponding to time to failure t_F, α_t is the scale parameter (the 63.2th percentile of the life) and β_t is the shape parameter of the Weibull probability distribution function. This means that insulation time to failure t_F is a random variable associated with a certain value of failure probability P, or—conversely—of reliability $R = 1 - P$. Equation (17) can be easily recast in terms of α_t as follows:

$$\alpha_t = t_F/[-ln(1-P)]^{1/\beta_t} \quad (18)$$

As a consequence, the reliability of a power component subjected to current and voltage harmonics can be estimated from the general model (16) by assuming that component life is Weibull-distributed according to (17). In this respect, it is worth pointing out that here—contrary to previous papers [6–9] based on the 63.2th percentile of time to failure, α_{NS}, and on the expected value of time to failure, μ_{NS}—the theory and the calculations for distorted conditions are focused on noteworthy $100 \times P$th percentiles of the Weibull distribution of times to failure in distorted conditions, $t_{P,NS}$, which are compared in the applicative section with design life in sinusoidal conditions, L_D, given at a certain cumulative failure probability, P_D: this is consistent with the modern probabilistic approach to the design of power component insulation. Therefore, in order to recast model (16) in terms of $t_{P,NS}$, L_D and P_D, the following procedure is required.

Let us first write Equation (18) twice:

1. a first time for $t_{P,NS}$, namely insulation life in non-sinusoidal conditions at a generic failure probability P, as follows:

$$P(t_{P,NS}) = P = 1 - exp[-(t_{P,NS}/\alpha_{NS})]^{\beta_t} \tag{19}$$

where α_{NS} indicates the 63.2th percentile (or scale parameter) of non-sinusoidal life distribution;

2. a second time for L_D, namely insulation life in nominal sinusoidal conditions at design failure probability P_D—indeed life is now a random variable, hence also design life L_D is associated with a certain fixed and known design failure probability, P_D, see above—as follows:

$$P(L_D) = P_D = 1 - exp[-(L_D/\alpha_S)]^{\beta_t} \tag{20}$$

where α_S indicates the 63.2th percentile (or scale parameter) of sinusoidal life distribution.

Thereafter, using Equation (18) let us explain (19) and (20) in terms of the scale parameter, as follows:

$$\alpha_{NS} = t_{P,NS}/[-ln(1-P)]^{1/\beta_t} \tag{21}$$

$$\alpha_S = L_D/[-ln(1-P_D)]^{1/\beta_t} \tag{22}$$

Considering now model (16), it is readily seen that this model has to be written for a given failure probability as well, in such a way that life in the presence of current and voltage harmonics, L_{NS} (at the left-hand side) and sinusoidal life L_S at the right-hand side are relevant to the same failure probability: indeed, L_{NS} at the left-hand side and L_S at the right-hand side are so far the only random variables in model (16) (for the moment, K_p, K_w and K_r are assumed as deterministic and known, since the N voltage harmonics are assumed as deterministic and known so far. In the following, also the typical random variation of harmonics is accounted for). Therefore, model (16) can be written, e.g., for 63.2% failure probability, so that $L_{NS} = \alpha_{NS}$ and $L_S = \alpha_S$. In this way, one obtains:

$$\alpha_{NS} = \alpha_S(E_H/E_S)^{-n_S}exp(-B\Delta T'_{tot})K_p^{-n_p}K_w^{-n_w}K_r^{-n_r} \tag{23}$$

Then, by expressing α_{NS} via (21) and α_S via (22), relationship (23) becomes:

$$\frac{t_{P,NS}}{[-ln(1-P)]^{1/\beta_t}} = \frac{L_D}{[-ln(1-P_D)]^{1/\beta_t}}(E_H/E_S)^{-n_S}exp(-B\Delta T'_{tot})K_p^{-n_p}K_w^{-n_w}K_r^{-n_r} \tag{24}$$

which can be eventually recast in the following form—reported here for the first time:

$$t_{P,NS} = L_D\frac{[-ln(R)]^{1/\beta_t}}{[-ln(1-P_D)]^{1/\beta_t}}(E_H/E_S)^{-n_S}exp(-B\Delta T'_{tot})K_p^{-n_p}K_w^{-n_w}K_r^{-n_r} \tag{25}$$

Relationship (25) is the so-called electro-thermal "probabilistic life model"—or "reliability model"—for the insulation of a power component affected by current and voltage harmonics. It is a quite powerful tool that, for given values of design life L_D and design failure probability P_D, relates the 100 × Pth percentile of insulation time-to-failure (life) in non-sinusoidal regime, $t_{P,NS}$, to reliability $R = 1 - P$ and applied stresses, i.e., temperature associated with linear and no linear loads, sinusoidal voltage and voltage waveshape factors. Therefore, this relationship yields the main functions and parameters needed for a thorough reliability analysis of power system components, namely [8]:

Reliability function

$$R(t_{P,NS}) = exp\left\{-\left\{\frac{[-ln(1-P_D)]^{1/\beta_t}\,t_{P,NS}}{L_D(E_H/E_S)^{-n_S}exp\left(-B\Delta T'_{tot}\right)K_p^{-n_p}K_w^{-n_w}K_r^{-n_r}}\right\}^{\beta_t}\right\} \tag{26}$$

Failure probability

$$P(t_{P,NS}) = 1 - R(t_{P,NS}) \tag{27}$$

Hazard function

$$h(t_{P,NS}) = \frac{[-ln(1-P_D)]^{1/\beta_t}\,\beta_t(t_{P,NS})^{\beta_t-1}}{\left\{L_D(E_H/E_S)^{-n_S}exp\left(-B\Delta T'_{tot}\right)K_p^{-n_p}K_w^{-n_w}K_r^{-n_r}\right\}^{\beta_t}} \tag{28}$$

Mean Time To Failure (MTTF)

$$MTTF = \frac{L_D}{[-ln(1-P_D)]^{1/\beta_t}}(E_H/E_S)^{-n_S}exp\left(-B\Delta T'_{tot}\right)K_p^{-n_p}K_w^{-n_w}K_r^{-n_r}\Gamma(1+1/\beta_t) \tag{29}$$

where Γ is the Euler Gamma function.

2.5. Reliability Model in the Case of Randomly Time-Varying Distortion

Voltage and current harmonics have amplitude and phase-angle that are stochastically varying with time. Therefore, they have to be regarded as random variables (RVs), in line with [4] and as done hereafter from both the theoretical and the applicative viewpoint.

If the probability density functions (pdfs) of voltage and current harmonics at a certain node of the grid are known, life and reliability of power components at that node can be inferred by applying the cumulative damage law of Miner [43] to the reliability model (25). Then, the following probabilistic electrothermal life model for randomly time-varying distortion is obtained:

$$t_{P,NS} = \frac{[-ln(R)]^{1/\beta_t}}{\int_0^{I_{1,max}}\cdots\int_0^{I_{M,max}}\int_0^{V_{1,max}}\cdots\int_0^{V_{N,max}}\cdots\frac{f(I_1,\ldots,I_N;V_1,\ldots,V_N)\prod_{h=1}^{N}dV_h\prod_{h=1}^{M}dI_h}{\frac{L_D}{[-ln(1-P_D)]^{1/\beta_t}}\left(\frac{E_H}{E_S}\right)^{-n_S}exp(-B\Delta T'_{tot})K_p^{-n_p}K_w^{-n_w}K_r^{-n_r}}} \tag{30}$$

where: $f(I_1, \ldots, I_M; V_1, \ldots, V_N)$ is the multivariate pdf of rms current harmonics ($I_1,..,I_M$) and rms voltage harmonics ($V_1, \ldots, V_N$); $I_{1,max},..I_{M,max}$ and $V_{1,max},..V_{N,max}$ are, respectively, the maximum values reached with time by rms harmonic currents and voltages. It must be emphasized that in LV and MV systems ΔT_{arm}—thus $\Delta T'_{tot}$, see Equation (15)—depends directly on the amplitudes of current harmonics only (Dielectric losses are practically negligible in LV and MV systems.), while K_p, K_w and K_r depend directly on the amplitude and phase of voltage harmonics (the dependence on the phases of voltage harmonic is omitted in (30) for the sake of simplicity).

Let us now assume that the temperature variation caused by harmonic currents, ΔT_{arm}, and the temperature variation caused by a change of sinusoidal current with respect to rated sinusoidal current, ΔT_{HL}, can be regarded as constant with time, since temperature exhibits moderate fluctuations of a few K around a mean value $\langle T_{tot}\rangle$ which is essentially constant in the rms sense [2,3]. Then $\langle T_{tot}\rangle$ can be essentially regarded as a deterministic quantity, constant with time in the rms sense as well; this means that the thermal effects of currents can be also essentially regarded as deterministic. As a consequence, (30) can be made much simpler, as follows:

$$t_{P,NS} = \frac{\frac{L_D[-ln(R)]^{1/\beta_t}}{[-ln(1-P_D)]^{1/\beta_t}}\left(\frac{E_H}{E_S}\right)^{-n_S}exp(-B\langle\Delta T'_{tot}\rangle)}{\int_0^{K_{p,max}}\int_0^{K_{w,max}}\int_0^{K_{r,max}}\frac{f\left(K_p,K_w,K_r\right)dK_p dK_w dK_r}{K_p^{-n_p}K_w^{-n_w}K_r^{-n_r}}} \tag{31}$$

where:

- $f(K_p, K_w, K_r)$ is the multivariate pdf of K_p, K_w and K_r, correlated in turn with $f(V_1,..,V_N;\ \phi_1, \ldots, \phi_N)$, the multivariate pdf of rms values, $V_1,..,V_N$, and phase-shift angles $\phi_1, \ldots, \phi_N$ of voltage harmonics;

- $\langle \Delta T'_{tot} \rangle$ is defined as follows (see Equation (15)):

$$\langle \Delta T'_{tot} \rangle = 1/T_S - 1/(T_S + \langle \Delta T_{arm} + \Delta T_{HL} \rangle) = 1/T_S - 1/(T_S + \langle \Delta T_{tot} \rangle) \tag{32}$$

Equation (31) can be rewritten in a more compact form as follows:

$$t_{P,NS} = \frac{L_D[-ln(R)]^{1/\beta_t}(E_H/E_S)^{-n_S} exp(-B\langle \Delta T'_{tot} \rangle)}{[-ln(1-P_D)]^{1/\beta_t}\, E\left[K_p^{n_p} K_w^{n_w} K_r^{n_r}\right]} \tag{33}$$

where $E\left[K_p^{n_p} K_w^{n_w} K_r^{n_r}\right]$ indicates the expected value of $K_p^{n_p} K_w^{n_w} K_r^{n_r}$.

The above hypothesis that ΔT_{arm} and ΔT_{HL} can be regarded as constant with time is meaningful, particularly as far as ΔT_{arm} is concerned [2,3]. Indeed, load current does vary with time, but the associated variation of Joule losses is more or less compensated by the thermal inertia of the insulation of the component and of the surrounding environment, which have a much longer thermal time constant than the typical variation period of harmonic currents—and sometimes even of sinusoidal current. Therefore, the overall thermal effect of load current can be well approximated as a steady variation of the temperature of the component, proportional to rms load current (the use of rms current is due to the dependence of Joule losses in conducting parts on the rms value of current).

Then, from Equation (33) it follows that life and reliability of the insulation of the component affected by harmonic distortion can be estimated in two steps:

1. evaluation of the 100 × *P*th failure time percentile based on the effect of voltage harmonics only, $t_{P,NS,V}$, resorting to the following equation (derived from Equation (33) by setting $\langle \Delta T'_{tot} \rangle = 0$, which implies $exp(-B\langle \Delta T'_{tot} \rangle) = 1$):

$$t_{P,NS,V} = \frac{L_D[-ln(R)]^{1/\beta_t}(E_H/E_S)^{-n_S}}{[-ln(1-P_D)]^{1/\beta_t} E\left[K_p^{n_p} K_w^{n_w} K_r^{n_r}\right]} \tag{34}$$

2. evaluation of the further possible variation of the 100 × *P*th failure time percentile due to the effect of the mean value of $\Delta T'_{tot}$, $\langle \Delta T'_{tot} \rangle$—guessed in turn through rms load current—resorting to the following equation, derived comparing Equation (33) with Equation (34):

$$t_{P,NS} = t_{P,NS,V}\, exp(-B\langle \Delta T'_{tot} \rangle) \tag{35}$$

Focusing on the 1st step, a proper application of Equation (34) requires an appropriate description of the time variation of voltage harmonics. This variation is inherently stochastic, but two typical situations can occur at a certain node of a grid where a component of interest is located:

(I) a monitoring survey has provided m values of the amplitudes and phases of voltage harmonics measured throughout a time interval t_m, which on the one hand should be as long as possible to well describe component duty service, but on the other hand is restricted by practical constraints, e.g., 1 h, 1 day, 1 week [4]. Monitored data like these are rarely found [7,32,35,36], but if they are available they enable the straightforward computation of the values of K_p, K_w, K_r through Equations (7)–(9) and derivation of the relevant sampling distribution, which is the numerical reproduction of the multivariate pdf $f(K_p,K_w,K_r)$—see [7].

(II) the pdfs of voltage harmonics can be guessed analytically according to a parametric approach. Then, $f(V_1,..,V_N;\, \phi_1, \ldots, \phi_N)$ is readily available, whereby in turn $f(K_p,K_w,K_r)$ can be derived.

In both cases, once the multivariate pdf $f(K_p,K_w,K_r)$ has been obtained, the life and reliability of the power component subjected to randomly time-varying voltage harmonics can be evaluated by means of Equation (34) by computing the expected value $E\left[K_p^{n_p} K_w^{n_w} K_r^{n_r}\right]$.

Coming to the 2nd step, the further effect of time-varying current harmonics—involving in turn a constant (in the rms sense) variation of insulation temperature $\langle \Delta T'_{tot} \rangle$—can be assessed by means of Equation (35).

Since the focus in this paper is on the 1st percentile of times to failure in the presence of harmonic distortion, $t_{1\%,NS}$, its calculation requires rewriting Equation (33) for $R = R_D = 0.99$ (i.e., $P_D = 0.01$), namely:

$$t_{1\%,NS} = \frac{L_D exp(-B\langle \Delta T'_{tot} \rangle)(E_H/E_S)^{-n_S}}{E\left[K_p^{n_p} K_w^{n_w} K_r^{n_r}\right]} \quad (36)$$

Moreover, reliability is calculated as a function of time elapsed in service, t_E, by explaining Equation (33) as a function of R with R_D set to 0.99, as follows:

$$R(t_E) = exp\left\{-\left\{\frac{[-ln(0.99)]^{1/\beta_t}\, t_E\, E\left[K_p^{n_p} K_w^{n_w} K_r^{n_r}\right]}{L_D(E_H/E_S)^{-n_S} exp\left(-B\langle \Delta T'_{tot} \rangle\right)}\right\}^{\beta_t}\right\} \quad (37)$$

By setting $\langle \Delta T'_{tot} \rangle = 0$ in (36), (37), homologous relationships valid in the presence of voltage harmonics only are obtained, i.e.,:

$$t_{1\%,NS} = L_D(E_H/E_S)^{-n_S} / E\left[K_p^{n_p} K_w^{n_w} K_r^{n_r}\right] \quad (38)$$

$$R(t_E) = exp\left\{-\left\{[-ln(0.99)]^{1/\beta_t}\, t_E\, E\left[K_p^{n_p} K_w^{n_w} K_r^{n_r}\right] / \left[L_D(E_H/E_S)^{-n_S}\right]\right\}^{\beta_t}\right\} \quad (39)$$

The reliability model for time-varying distortion as newly formulated here relies on Equations (38) and (39) or (36) and (37)—depending on whether $\langle \Delta T'_{tot} \rangle = 0$ or $\neq 0$, respectively—which differ from those used in [8,9], since these latter are based on the 63.2th percentile of failure time. For the sake of simplicity, hereafter the sinusoidal component of electric field at power frequency is taken equal to rated sinusoidal electric field, namely $E_H = E_S$.

3. Application of the Reliability Model for Randomly Time-Varying Distortion

3.1. Selected Case Studies

In previous papers [8,9] the reliability model for randomly time-varying distortion was applied to power system components affected by voltage harmonics whose amplitudes matched the limits after EN 50160 [4] exactly. The calculations were concentrated on the 63.2th percentile of time to failure, α_{NS}, and on the mean time to failure (MTTF), μ_{NS}; design life itself was given as the design value of the MTTF in sinusoidal conditions set to 40 and 30 years for the cable and the capacitor, the two treated components. Reliability was estimated at a service time equal to design life only.

Here, the calculations are focused on the $100 \times P_D$th percentile of times to failure, which is consistently compared with design life in sinusoidal conditions, L_D, given at design failure probability $P_D = 0.01 = 1\%$. This is in line with the modern probabilistic design of the insulation of power components, which requires high reliability [5]. Moreover, the design life of power components is set pragmatically to $L_D = 20$ years, i.e., the typical order of magnitude of the service life (based on the duration of the mortgages and the amortization times) of industrial plants, traction systems for railways or subways and renewable power plants for photovoltaic (PV) and wind energy generation [44–46]. Grid-connected three-phase power electronic converters are found to connect these plants with the AC power grid and they generate voltage and current harmonics which affect nearby components. This has the first and foremost effect that all estimates obtained here for failure time percentiles and reliability are different from those in previous papers.

In addition, in place of the "spot" reliability estimate at design life in [8,9], reliability is evaluated here in terms of design life of power components—and even beyond it, although results are omitted for brevity. Indeed, as service time t_E goes by, maintenance and/or repair and/or replacement actions might be required on some components of the plant/grid, if their residual reliability is unsatisfactory; conversely, other components may remain in service without any action even after a service time t_E equal to—or even longer than—design life L_D, if their residual reliability is unsatisfactory. For this reason, beside proper diagnostic techniques, a sound reliability model can be extremely useful to provide indications about the residual reliability of components all along their time on duty.

Furthermore, the case studies treated in this paper have been changed and broadened considerably compared to [8,9], as shown in the next section. First of all, the voltage harmonics treated in [8] were the 11th, 13th, to which the 5th, 7th were added in [9], while here for the first time the 17th, 19th, 23rd, 25th are included. It is worth emphasizing that the implementation of the new calculations including the whole set of odd harmonics from the 5th to 25th has required a non-negligible effort for restructuring and modifying the code: analyzing the whole sequence of odd voltage harmonics from the 5th to the 25th, as done here for the first time, is all but trivial and inexpensive from the viewpoint of theoretical and computational efforts. It is also worth outlining that the reliability model (33),(36),(37) works for all harmonic orders [10], but harmonics of order $h > 25$ are neglected here since in [4] no limits are given for voltage harmonics above the 25th, as they are said to be *"usually small, but largely unpredictable due to resonance effects"*.

Moreover, three cases differing as for the magnitude (In agreement with [4] let us consider for each harmonic—in place of the rms value—the *"10 min mean rms value"*, called "magnitude" from now on, and indicate it simply as V_h like the rms value.) of voltage harmonics are treated here:

(i) a case similar (remember that only here the 5th, 7th, 11th, 13th, 17th, 19th, 23rd, 25th harmonics are considered all together) to that in [8,9], where the magnitudes V_h of the treated voltage harmonics are random variables (see Equation (30)) correlated (this hypothesis is acceptable, as shown in [7] from a data set of measured voltage harmonics with phase-shift angles [32], used here as the experimental case. Amplitudes were above the limits after [4]) to each other and following a Gaussian pdf, with such mean value $\mu_h = \mu(V_h)$ and standard deviation $\sigma_h = \sigma(V_h)$ that the 95th percentile of V_h of each harmonic matches the limits set in Tables 1 and 4 of Standard EN 50160 [4] exactly. As pointed out in [8,9], the Gaussian pdf was chosen mainly as it is a good approximation of several other pdfs [47] and it is the most practical to reproduce measured values of randomly-varying quantities such as harmonic voltages [32].

(ii) a new case, where the Gaussian pdfs of the magnitudes V_h of voltage harmonics have such values of μ_h and σ_h that the limits in EN 50160 are exceeded by 25%, so as to illustrate the problems that may arise if these limits are overcome (e.g., in a plant without passive or active filters);

(iii) another new case, where the Gaussian pdfs of V_h have such values of μ_h and σ_h that the limits in EN 50160 are matched with a 25%-safety margin, so as to show that problems may still remain even if limits are matched with a seemingly-broad safety margin.

Three more applicative novelties compared to previous papers [8,9] of this investigation are:

- in addition to $<\Delta\theta_{tot}> = 0$, the case $<\Delta\theta_{tot}> = -15$ °C—a value in between those treated in [8,9], i.e., −10 and −20 °C—is considered in order to account for the effect of current harmonics;
- the chosen value of the shape parameter of Weibull distribution is 2 instead of 3 used in [8,9], as explained hereafter;
- as hinted at in the Abstract and Introduction, two more components of power systems, i.e., induction motors and MV/LV transformers, are treated here in addition to cables and capacitors, already studied in the previous stages of this investigation.

As for the magnitude of voltage harmonics, it can be argued that, anyway, the reference case for setting the correct values of μ_h and σ_h in cases (i)–(iii) above is case (i). Since measurements of harmonic voltages that match the probabilistic limits set in [4] exactly are rare in the literature, the random magnitudes of harmonic voltages matching the limits set in [4] exactly are derived here via a heuristic parametric approach [8,9] based on a priori hypotheses about the Coefficient of Variation $CV_h = \sigma_h/V_h$

of harmonic voltages. Namely, two different values of CV_h are hypothesized so as to reproduce a narrower and a broader spread of V_h:

(1) $CV_{h,1} = 0.1$, i.e., 10% of the mean of V_h;
(2) $CV_{h,2} = 0.3$, i.e., 30% of the mean of V_h.

These values of CV_h were already chosen in [8,9] for the 5th, 7th, 11th, 13th voltage harmonics, being consistent with those measured in the power supply grid of a subway traction system, where CV_h ranged from ≈10% to ≈27% [32]. They are kept here and extended to the 17th, 19th, 23rd, 25th voltage harmonics, for the sake of comparison.

As a result of this heuristic parametric approach, the pdfs (the pdfs describing the randomness of the parameters V_h in Figure 1, as well as the one of other parameters, e.g., K_p and K_w, may be also considered as the "a priori" pdfs in a Bayesian inference framework, which is outside the scope of the present paper. The details of this approach are discussed for the problem under study in [48]) of the 5th, 7th, 11th, 13th, 17th, 19th, 23rd, 25th voltage harmonic magnitudes are obtained, which are Gaussian with μ_h and σ_h such that their magnitudes match exactly the limits set in [4], namely: 6%, 5%, 3.5%, 3%, 2.0%, 1.5%, 1.5%, 1.5% of rms fundamental voltage V_1 for the 95th percentiles of V_5, V_7, V_{11}, V_{13}, V_{17}, V_{19}, V_{23}, V_{25}, respectively [48]. The cdfs of all voltage harmonics, $F(V_h)$, are shown for the sake of illustration in Figure 1a,b for the cases $CV_{h,1} = 0.1$ and $CV_{h,2} = 0.3$, respectively: is readily seen that the limits after [4] are exactly matched for all voltage harmonics.

Now, life and reliability estimation according to model (36) and (37) requires knowledge of the expected value $E\left[K_p^{n_p} K_w^{n_w} K_r^{n_r}\right]$. As it can be argued from Equations (30), (31), (33), this in turn requires the knowledge of the multivariate pdf of K_p, K_w, K_r, $f(K_p, K_w, K_r)$, correlated with $f(V_5, V_7, V_{11}, V_{13}, V_{17}, V_{19}, V_{23}, V_{25}; \phi_5, \phi_7, \phi_{11}, \phi_{13}, \phi_{17}, \phi_{19}, \phi_{23}, \phi_{25})$, the multivariate pdf of magnitudes and phase-shift angles (with respect to the fundamental V_1) of the treated voltage harmonics. The derivation of the pdfs/cdfs of the magnitude of voltage harmonics (Figure 1) has solved the problem of deriving $f(V_5, V_7, V_{11}, V_{13}, V_{17}, V_{19}, V_{23}, V_{25})$, but the phase-shift angles between voltage harmonics and the fundamental is also needed to estimate the part of multivariate pdf relevant to phase-shift angles. This problem is solved—as in [8,9] for the 5th, 7th, 11th, 13th voltage harmonics—by taking phase-shift angles ϕ_5, ϕ_7, ϕ_{11}, ϕ_{13}, ϕ_{17}, ϕ_{19}, ϕ_{23}, ϕ_{25} as known deterministic quantities, both to make the treatment easier and to carry out a parametric sensitivity analysis; thereafter, $f(K_p, K_w, K_r)$ can be readily derived.

Such analysis focuses on three different cases (a), (b), (c) of phase-shift angles, which have quite different (or even extreme and opposite) effects on the distorted voltage waveshape. These cases are illustrated in the reference Figure 2, which shows the mean waveform of harmonic voltage pdfs for phase-shift cases (a), (b), (c) obtained using $CV_{h,1} = 0.1$ and μ_h as the magnitude of the hth harmonic in the case of exact matching of the limits after [4]. Cases (a), (b), (c) of phase-shift angles are as follows.

(a) The sum of voltage harmonic peaks, shown in Figure 2a [49]. It is called "worst case", since it leads to the highest peak of the distorted voltage waveshape, thus to the maximum value of the peak factor K_p. The worst-case features the following set of phase-shift angles: $\phi_5 = 0$, $\phi_7 = 180°$, $\phi_{11} = 180°$, $\phi_{13} = 0$, $\phi_{17} = 0$, $\phi_{19} = 180°$, $\phi_{23} = 180°$, $\phi_{25} = 0$. In [35] the "worst case" is regarded as uncommon in a self-judged "*indicative but not exhaustive*" experimental survey. Hence it is treated here, as in [8,9], since it might happen and it is on the safe side as for the reliability analysis.
(b) The so-called "best case", shown in Figure 2b. Indeed, it leads to the flattest distorted voltage with the lowest peak, thus to the minimum value of the peak factor K_p, namely the less prone to speed-up the degradation of the insulation. The best case features the following set of phase-shift angles: $\phi_5 = 180°$, $\phi_7 = 180°$, $\phi_{11} = 0$, $\phi_{13} = 0$, $\phi_{17} = 180°$, $\phi_{19} = 180°$, $\phi_{23} = 0$, $\phi_{25} = 0$.
(c) The so-called "experimental case", shown in Figure 2c. This case is taken from an experimental data set—hardly found in the literature—of phase-shift angles measured on the supply system of a subway with large size (3.5 MW) 12-pulse AC/DC converters as main distorting loads, plus lower size converters and other loads generating 5th and 7th voltage harmonics [32]. The experimental case features the following set of phase-shift angles: $\phi_5 = 109°$, $\phi_7 = 300°$, $\phi_{11} = 317°$, $\phi_{13} = -15°$,

$\phi_{13} = 0$, $\phi_{17} = -16.6°$, $\phi_{19} = -17.4°$, $\phi_{23} = -19°$, $\phi_{25} = 11°$. The values of ϕ_5, ϕ_7, ϕ_{11}, ϕ_{13}, ϕ_{23}, ϕ_{25} are the mean of the measured values for these harmonics, while the values of ϕ_{17}, ϕ_{19} have been attained through a linear interpolation of the mean values of ϕ_{13} and ϕ_{23}—the 17th and 19th voltage harmonics are not found in [32]. The experimental case leads to a distorted voltage waveshape fairly similar to those observed in the field, as obvious when considering its experimental origin.

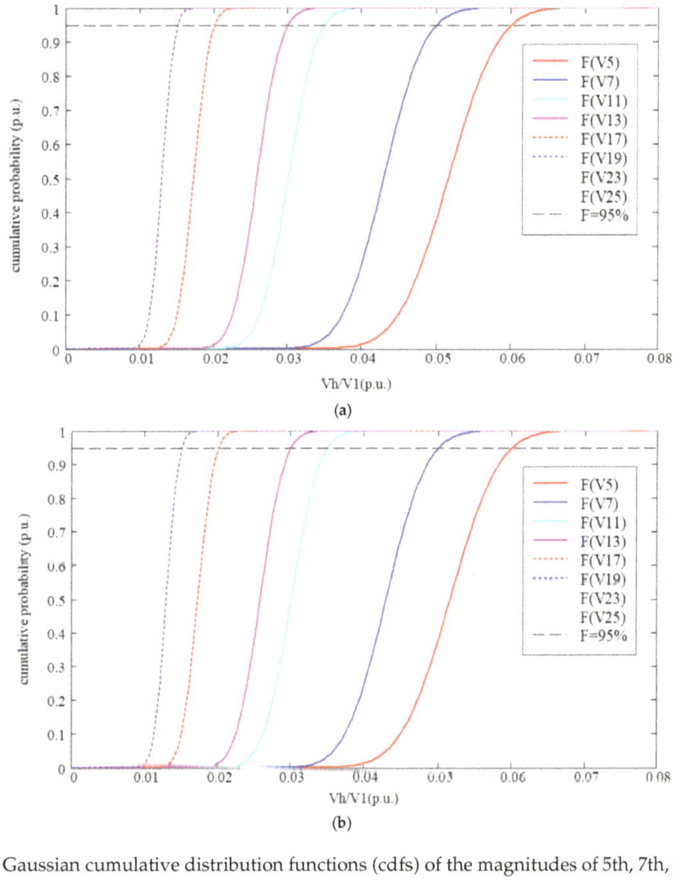

Figure 1. Gaussian cumulative distribution functions (cdfs) of the magnitudes of 5th, 7th, 11th, 13th, 17th, 19th, 23rd, 25th voltage harmonics, with μ_h and σ_h such as to match the limits in [4] exactly (6%, 5%, 3.5%, 3%, 2%, 1.5%, 1.5%, 1.5% of V_1 for the 95th percentiles): (**a**) $CV_h = 0.1$; (**b**) $CV_h = 0.3$.

Coming to the role played by current harmonics, as outlined in Section 2.5 it is assumed that ΔT_{tot} can be replaced by its mean value (in the rms sense) $<\Delta T_{tot}>$, constant with time and deterministic value. Two values of $<\Delta T_{tot}>$ are considered here, namely:

A. $<\Delta T_{tot}> = 0$ K, i.e., $<\Delta\theta_{tot}> = 0$ °C, like in [8,9], so as to emphasize the role played by voltage harmonics. Setting $<\Delta T_{tot}> = 0$ corresponds to step 1 of the procedure outlined in Section 2.5; this implies $\langle \Delta T'_{tot} \rangle = 0$, thus the reliability model (38) and (39) shall be used, focusing on the role played by voltage harmonics only. In fact, setting $<\Delta T_{tot}> = 0$ means that in practice the temperature of component oscillates, with negligible fluctuations, around design temperature;

B. $<\Delta T_{tot}> = -15$ K, i.e., $<\Delta\theta_{tot}> = -15$ °C, a value in between those treated in [8,9], i.e., −10 and −20 °C. Setting $<\Delta T_{tot}> \neq 0$ corresponds to step 2 of the procedure outlined in Section 2.5; this implies $\langle \Delta T'_{tot} \rangle \neq 0$, thus reliability model (36) and (37) shall be used, including the role played by current harmonics and sinusoidal current. Setting a value of $<\Delta T_{tot}> < 0$ implies acknowledging that the component—notwithstanding the presence of current harmonics—works typically at a lower- than-design temperature. This holds in particular for MV distribution cables—whose design temperature oscillates between 75 °C for "vintage" paper-insulated lead-covered cables to 90 °C for extruded cables—which have usually an overall load well below the rated one [50].

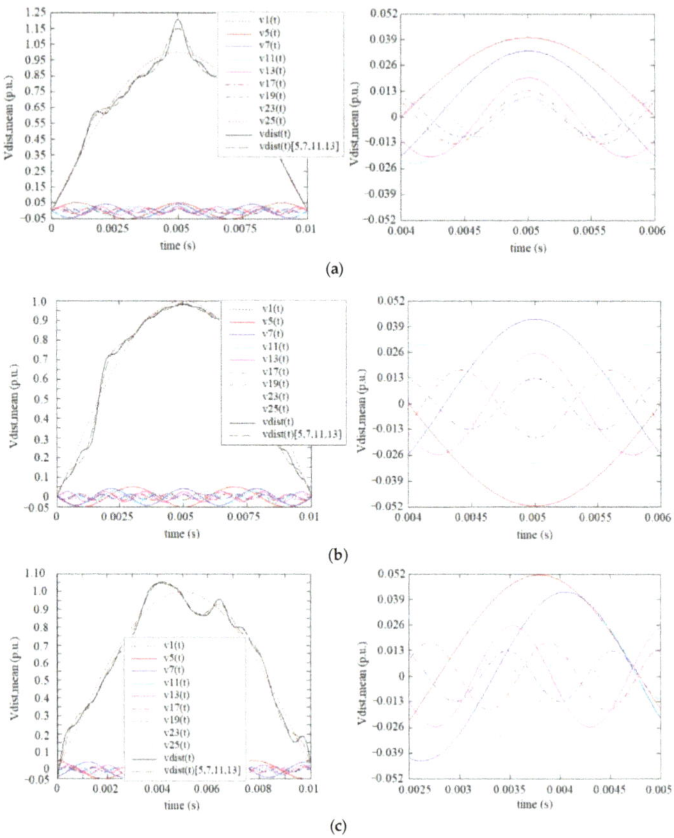

Figure 2. Mean voltage waveshapes for case studies (**a**–**c**) obtained using $CV_{h,1} = 0.1$ and μ_h as the magnitude of the hth harmonic in the case of exact matching of the limits after [4]. Left: positive halfwaves; right: zoomed-in view showing the phase-shift between harmonics. *vdist*(t) is total distorted voltage with all harmonics, while *vdist*(t) [5,7,32,38] with the 5th, 7th, 11th, 13th only as in [9]. THDv = 8.3%.

By combining above points (i)–(iii), (1)–(2), (a)–(c), A.–B., the case studies of distorted voltage waveshapes treated here are 36, as summarized in Table 1, where the relevant value of $THDv$ is also reported; for making the table more compact, the cases corresponding to points (i)–(iii) above (i.e., pdfs of voltage harmonic magnitude V_h that either match the limits in [4] exactly, or overcome them by 25%, or are below them by 25%) are quoted on one single row: this implies that any one of case studies (a)–(l) degenerates in turn into three more case studies depending on the pdfs of V_h. For the sake of illustration, the mean voltage waveshapes of case studies (a), (b), (c) for the exact matching of the limits in [4] (obtained by using μ_h as the magnitude of the hth harmonics) are plotted in Figure 2a–c.

Other distorted voltage waveshapes than these may exits, but they tend to be in between those considered here, thus they tend to have intermediate effects on insulation life and reliability.

Table 1. Summary of cases of distorted voltage waveshapes studied here (interestingly, with the whole set of odd harmonics non-multiple of 3 from the 5th to the 25th, the 8% limit for THD_v after [4] is slightly overcome when CV_h = 0.1 even when the amplitudes of each single harmonic voltage strictly match the limits in [4]. This does not happen when 5th, 7th, 11th, 13th harmonics only are treated [9]. This is another new outcome of this study, which would deserve a deeper investigation).

Case Study	Scatter CV_h	Mean Magnitude μ_h Matching of EN 50160 Limits			Phase-Shift	$\langle THD_v \rangle$ [%]			$\langle \Delta\theta_{tot} \rangle$ [°C]
						Exact	+25%	−25%	
(a)	0.1	Exact	25%-above	25%-below	worst case	8.3	10.0	6.2	0
(b)	0.1	Exact	25%-above	25%-below	best case	8.3	10.0	6.2	0
(c)	0.1	Exact	25%-above	25%-below	experimental	8.3	10.0	6.2	0
(d)	0.3	Exact	25%-above	25%-below	worst case	6.5	8.1	4.8	0
(e)	0.3	Exact	25%-above	25%-below	best case	6.5	8.1	4.8	0
(f)	0.3	Exact	25%-above	25%-below	experimental	6.5	8.1	4.8	0
(g)	0.1	Exact	25%-above	25%-below	worst case	8.3	10.0	6.2	−15
(h)	0.1	Exact	25%-above	25%-below	best case	8.3	10.0	6.2	−15
(i)	0.1	Exact	25%-above	25%-below	experimental	8.3	10.0	6.2	−15
(j)	0.3	Exact	25%-above	25%-below	worst case	6.5	8.1	4.8	−15
(k)	0.3	Exact	25%-above	25%-below	best case	6.5	8.1	4.8	−15
(l)	0.3	Exact	25%-above	25%-below	experimental	6.5	8.1	4.8	−15

The reliability model for distorted current and voltage (36)–(39) is applied to these 36 case studies for estimating the life and reliability of:

- cross-linked polyethylene (XLPE)-insulated cables and all-film polypropylene (PP) insulated capacitors for MV/LV power systems, already examined in [8,9];
- induction motors and MV/LV transformers, tackled here for the first time in the streamline of this investigation.

The values of the parameters of Equations (36)–(39) used in the application are quoted in Table 2. As far as the cable and the capacitor are concerned, the values of n_p, n_w, n_r—derived by processing the results of ALTs performed on insulating samples subjected to various combinations of voltage harmonics [11,12]—and of B in Table 2 are the same as in [8,9] for the sake of comparison, but here the analysis is broader and different with respect to that carried out in [8,9]. Indeed here, as partly explained above:

- the case studies treated here are many more. In particular, the 24 cases with voltage harmonics 25% above and 25% below the limits in [4] are tackled for the first time, as outlined in Table 1;
- the chosen values of design life and probability are now set to 20 years and 1%, respectively, to be closer to the modern probabilistic design of power plant components on the one side, and to focus on the typical useful service life of the power plant;
- the chosen value of the shape parameter of Weibull distribution is 2 instead of 3 [8,9], in order to account for a milder degradation of component insulation during the life of the power plant.

Table 2. Values of the parameters of model (36)–(39) used in the application.

Component	β_t [adim.]	L_D [y]	P_D [%]	n_p [adim.]	n_w [adim.]	n_r [adim.]	B [K]
XLPE cable	2	20	1	14.80	4.90	1.20	12,430
All-film capacitor	2	20	1	6.20	0.56	1.80	12,500
induction motors	2	20	1	9.00	0.88	0.36	12,600
MV/LV transformers	2	20	1	11.00	0.99	0.44	12,600

As far as the induction motors and MV/LV transformers are concerned, the values of n_p, n_w, n_r and B in Table 2 have been derived by reprocessing the results in [10,51,52]. In fact, experimental results relevant to twisted pairs (i.e., twin varnished copper conductors wrapped on each other which reproduce on a smaller scale the full-size insulation of induction motors) indicate for n_p a value of 8–10, that can be used also for induction motor insulation. In the case of transformers, the dielectric performances of both epoxy resin and oil-paper insulation for MV/LV transformers suggest a slightly higher value, namely $n_p = 10 - 12$. For parameter B, a value of the order of those reported in Table 2 for the cable and the capacitor seems reasonable for both the induction motor and the transformer (from [52] a value of B = 12,600 can be derived for a bisphenolic epoxy resin), even if the thermal endurance of the epoxy insulation is usually larger than that of XLPE and PP.

A first observation drawn from Table 2 is that for all components the value of n_p overwhelms that of n_w and n_r; hence, K_p should have a foremost effect on the life and reliability of components subjected to voltage harmonics, if K_p, K_w and K_r have the same order of magnitude.

3.2. Results

For the XLPE cable and the all-film capacitor working in the distorted conditions of case studies (a)–(f) of Table 1, i.e., those relevant to $\Delta\theta_{tot} = 0$, as well as of case studies (g)–(l) of Table 1, i.e., those relevant to $\Delta\theta_{tot} = -15$ °C, Table 3 reports the estimates of the following quantities:

- the 1st percentile of failure time, $t_{1\%,NS}$, in p.u. of design life L_D, computed via Equations (38) and (36);
- the reliability after a service time t_E equal to design life, $R(L_D)$, computed via Equations (39) and (37);

Table 3 also reports the 1st percentile of failure time, $t_{1\%,NS,OLD}$, in p.u. of design life L_D, computed via Equation (38) for $\Delta\theta_{tot} = 0$ with the voltage harmonics treated in [9] only, i.e., the 5th, 7th, 11th, 13th.

It is worth recalling that the cases 25% above and 25% below the limits after EN 50160 are totally new and treated here for the first time. Moreover, also the values of $t_{1\%,NS,OLD}$ in Table 3 differ from the values of the 1st percentile of failure time in [9]; indeed, the design hypothesis in [9] was to set the MTTF in sinusoidal conditions to 40 and 30 years for the cable and the capacitor, while here design life in sinusoidal conditions is set for both components to L_D = 20 years at design failure probability $P_D = 0.01 = 1\%$.

Analogously to Table 3, the homologous Table 4 reports for the induction motor and MV/LV transformer, working in the distorted conditions of case studies (a)–(f) ($\Delta\theta_{tot} = 0$) and (g)–(l) ($\Delta\theta_{tot} = -15$ °C) of Table 1, the estimates of the following quantities:

- the 1st percentile of failure time, $t_{1\%,NS}$, in p.u. of design life L_D, computed via Equations (38) and (36);
- the reliability after a service time t_E equal to design life, $R(L_D)$, computed via Equations (39) and (37);

Table 4 also reports the 1st percentile of failure time, $t_{1\%,NS,OLD}$, in p.u. of design life L_D, computed via Equation (38) for $\Delta\theta_{tot} = 0$ with the voltage harmonics treated in [9] only, i.e., the 5th, 7th, 11th, 13th.

It is worth recalling that all cases in Table 4 are totally new and treated here for the first time, since they refer to the induction motor and the MV/LV transformer, tackled here for the first time in the streamline of this investigation.

Table 3. Values of 1st percentile of time to failure, $t_{1\%,NS}$, and reliability at rated life L_D, $R(L_D)$, estimated with reliability models (36)–(39) for the XLPE cable and the all-film capacitor subjected to 5th, 7th, 11th, 13th, 17th, 19th, 23rd, 25th voltage harmonics in case studies (a)–(f) ($\Delta\theta_{tot}$ = 0), and (g)–(l) ($\Delta\theta_{tot}$ = −15 °C), for the various matching of the limits in [4], see Table 1. The values of 1st percentile of time to failure, $t_{1\%,NS,OLD}$, calculated with the 5th, 7th, 11th, 13th voltage harmonics only as in [9], are also reported.

Component ↓	Estimated Quantity ↓	$\Delta\theta_{tot}$[°C]	Case Study → Limit Match ↓	(a)	(b)	(c)	(d)	(e)	(f)
cable	$t_{1\%,NS}$ [p.u. of L_D]	0	Strict	0.015	0.337	0.125	0.026	0.456	0.201
			25%-above	4.6×10^{-3}	0.213	0.052	7.9×10^{-3}	0.311	0.09
			25%-below	0.049	0.523	0.283	0.079	0.642	0.409
cable	$t_{1\%,NS,OLD}$ [p.u. of L_D]	0	Strict	0.054	0.510	0.203	0.087	0.628	0.200
			25%-above	0.023	0.368	0.103	0.039	0.487	0.101
			25%-below	0.124	0.681	0.378	0.183	0.775	0.381
cable	$R(L_D)$ [%]	0	Strict	3×10^{-18}	91.54	52.28	2.7×10^{-5}	95.28	77.90
			25%-above	$<10^{-18}$	80.08	2.526	$<10^{-18}$	90.10	29.21
			25%-below	1.505	96.39	88.24	20.21	97.59	94.18
capacitor	$t_{1\%,NS}$ [p.u. of L_D]	0	Strict	0.265	0.951	0.635	0.342	0.976	0.739
			25%-above	0.192	0.918	0.521	0.259	0.952	0.633
			25%-below	0.370	0.982	0.765	0.451	0.996	0.846
capacitor	$t_{1\%,NS,OLD}$ [p.u. of L_D]	0	Strict	0.380	0.959	0.658	0.461	0.978	0.654
			25%-above	0.297	0.930	0.555	0.376	0.958	0.550
			25%-below	0.486	0.983	0.771	0.564	0.993	0.773
capacitor	$R(L_D)$ [%]	0	Strict	86.69	98.90	97.54	91.75	98.95	98.17
			25%-above	76.05	98.82	96.36	86.07	98.90	97.52
			25%-below	92.92	98.96	98.30	95.19	98.99	98.61

Component↓	Estimated Quantity↓	ΔT_{tot}[°C]	Case study→Limit match↓	(g)	(h)	(i)	(j)	(k)	(l)
cable	$t_{1\%,NS}$ [p.u. of L_D]	−15	Strict	**0.065**	1.476	**0.545**	**0.113**	1.995	**0.876**
			25%-above	**0.020**	**0.931**	**0.229**	**0.035**	1.359	**0.395**
			25%-below	**0.214**	2.287	1.240	**0.347**	2.809	1.792
cable	$R(L_D)$ [%]	−15	Strict	**9.54**	99.54	**96.67**	**45.35**	99.75	**98.70**
			25%-above	**1.1×10^{-9}**	**98.85**	**82.52**	**2.5×10^{-2}**	99.46	**93.77**
			25%-below	**80.32**	99.81	99.35	**91.99**	99.87	99.69
capacitor	$t_{1\%,NS}$ [p.u. of L_D]	−15	Strict	1.170	4.196	2.801	1.508	4.305	3.259
			25%-above	**0.838**	4.018	2.280	1.133	4.165	2.771
			25%-below	1.619	4.297	3.346	1.975	4.360	3.703
capacitor	$R(L_D)$ [%]	−15	Strict	99.27	99.943	99.87	99.56	99.946	99.90
			25%-above	**98.58**	99.938	99.81	99.22	99.942	99.87
			25%-below	99.62	99.946	99.91	99.74	99.947	99.93

Table 4. Values of 1st percentile of time to failure, $t_{1\%,NS}$, and reliability at rated life L_D, $R(L_D)$, estimated with reliability models (36)–(39) for the induction motor and MV/LV transformer subjected to 5th, 7th, 11th, 13th, 17th, 19th, 23rd, 25th voltage harmonics in case studies (a)–(f) ($\Delta\theta_{tot}$ = 0), and (g)–(l) ($\Delta\theta_{tot}$ = −15 °C), for the various matching of the limits in [4], see Table 1. The values of 1st percentile of time to failure, $t_{1\%,NS,OLD}$, calculated with the 5th, 7th, 11th, 13th voltage harmonics only as in [9], are also reported.

Component ↓	Estimated Quantity ↓	$\Delta\theta_{tot}$ [°C]	Case Study → Limit Match ↓	(a)	(b)	(c)	(d)	(e)	(f)
motor	$t_{1\%,NS}$ [p.u. of L_D]	0	Strict	0.143	0.920	0.511	0.202	0.957	0.633
			25%-above	0.089	0.872	0.382	0.133	0.921	0.502
			25%-below	0.234	0.967	0.672	0.308	0.990	0.778
motor	$t_{1\%,NS,OLD}$ [p.u. of L_D]	0	Strict	0.244	0.937	0.542	0.319	0.965	0.537
			25%-above	0.170	0.896	0.422	0.235	0.936	0.417
			25%-below	0.349	0.973	0.684	0.430	0.988	0.686

Table 4. *Cont.*

Component ↓	Estimated Quantity ↓	$\Delta\theta_{tot}$ [°C]	Case Study → Limit Match ↓	(a)	(b)	(c)	(d)	(e)	(f)
motor	$R(L_D)$ [%]	0	Strict	61.31	98.82	96.22	78.22	98.91	97.52
			25%-above	28.06	98.69	93.34	56.51	98.82	96.08
			25%-below	83.20	98.93	97.80	89.94	98.98	98.35
transformer	$t_{1\%,NS}$ [p.u. of L_D]	0	Strict	0.095	0.924	0.449	0.141	0.963	0.577
			25%-above	0.053	0.872	0.317	0.084	0.924	0.435
			25%-below	0.171	0.974	0.624	0.236	0.997	0.741
transformer	$t_{1\%,NS,OLD}$ [p.u. of L_D]	0	Strict	0.180	0.936	0.478	0.246	0.966	0.473
			25%-above	0.117	0.891	0.354	0.169	0.934	0.349
			25%-below	0.278	0.974	0.633	0.356	0.991	0.636
transformer	$R(L_D)$ [%]	0	Strict	32.67	98.83	95.14	60.17	98.92	97.03
			25%-above	2.886	98.69	90.49	23.80	98.83	94.83
			25%-below	70.96	98.95	97.45	83.47	98.99	98.19
Component↓	**Estimated Quantity↓**	ΔT_{tot}[°C]	**Case study→Limit match↓**	**(g)**	**(h)**	**(i)**	**(j)**	**(k)**	**(l)**
motor	$t_{1\%,NS}$ [p.u. of L_D]	−15	Strict	**0.639**	4.103	2.277	**0.902**	4.269	2.823
			25%-above	**0.397**	3.889	1.703	**0.592**	4.107	2.236
			25%-below	1.042	4.313	2.997	1.373	4.413	3.469
motor	$R(L_D)$ [%]	−15	Strict	**97.57**	99.940	99.81	**98.77**	99.945	99.87
			25%-above	**93.81**	99.934	99.65	**97.17**	99.940	99.80
			25%-below	99.08	99.946	99.89	99.47	99.948	99.92
transformer	$t_{1\%,NS}$ [p.u. of L_D]	−15	Strict	**0.423**	4.118	2.003	**0.627**	4.293	2.573
			25%-above	**0.237**	3.890	4.122	**0.373**	4.122	1.940
			25%-below	**0.763**	4.341	2.781	1.052	4.444	3.305
transformer	$R(L_D)$ [%]	−15	Strict	**94.53**	99.941	99.75	**97.48**	99.945	99.85
			25%-above	**83.67**	99.934	99.50	**93.03**	99.941	99.73
			25%-below	**98.29**	99.947	99.87	99.10	99.949	99.91

Figure 3 illustrates the reliability vs. service time (in p.u. of design life) for the XLPE cable with $\Delta\theta_{tot} = 0$ and voltage harmonics that: (a) exactly match, (b) are 25% above, (c) are 25% below the limits in [4]. In the legend, $R_S(t)$ indicates reliability vs. time in the rated sinusoidal case, while $R_a(t)$, $R_b(t)$, $R_c(t)$, $R_d(t)$, $R_e(t)$, $R_f(t)$, denotes reliability vs. time in distorted case studies (a)–(f).

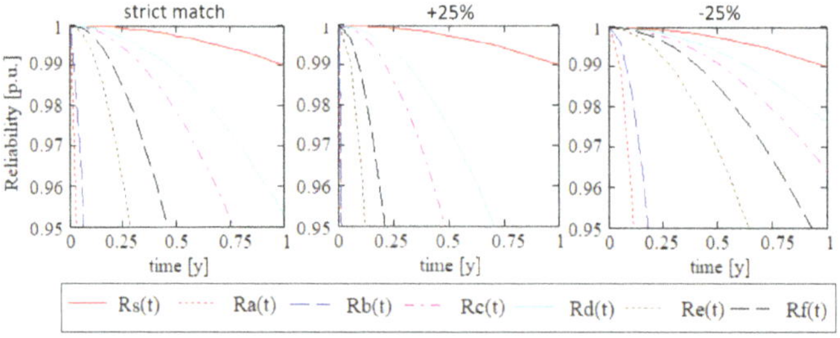

(a) (b) (c)

Figure 3. Reliability vs. service time (in p.u. of design life) for the XLPE cable with voltage harmonics that: (**a**) exactly match, (**b**) are 25% above, (**c**) are 25% below the limits in [4]. $\Delta\theta_{tot} = 0$.

Figure 4 shows the same quantities in the same conditions as Figure 3, but for the capacitor. Similar figures could have been drawn for the induction motor and MV/LV transformer, but they are omitted here for brevity.

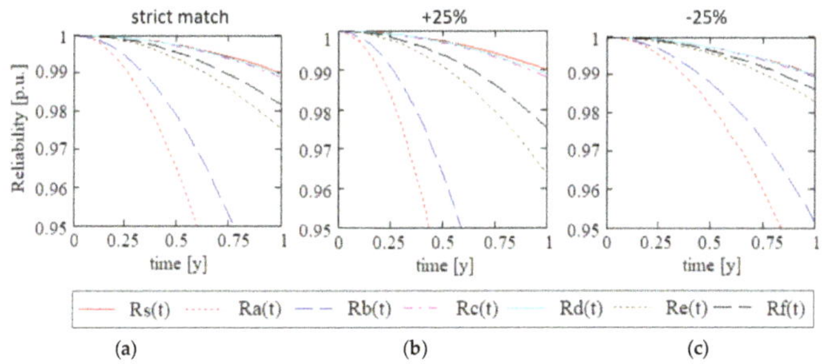

Figure 4. Reliability vs. service time (in p.u. of design life) for the all-film capacitor with voltage harmonics that: (**a**) exactly match, (**b**) are 25% above, (**c**) are 25% below the limits in [4]. $\Delta\theta_{tot} = 0$.

Figure 5a, reports the pdfs of K_p for case studies (a), (b), (c) with the strict match of the limits in [4] in the presence of the 5th, 7th, 11th, 13th, 17th, 19th, 23rd, 25th voltage harmonics, as well as of the 5th, 7th, 11th, 13th only as in [9]. Figure 5b is the same as Figure 5a, but for K_r instead of K_w. Figure 6 displays the pdfs of K_r for case studies (a), (b), (c), (d) with the 5th, 7th, 11th, 13th, 17th, 19th, 23rd, 25th voltage harmonics, as well as for case studies (a), (b) with 5th, 7th, 11th, 13th only.

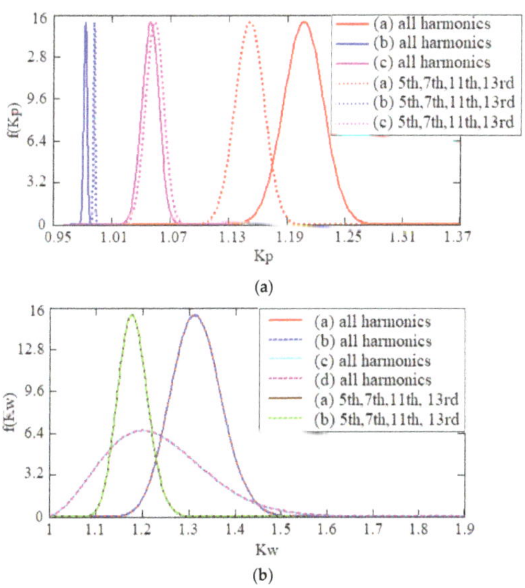

Figure 5. Probability density functions of: (**a**) K_p for case studies (a), (b), (c) with the strict match of the limits in [4] in the presence of the 5th, 7th, 11th, 13th, 17th, 19th, 23rd, 25th voltage harmonics, as well as of the 5th, 7th, 11th, 13th only; (**b**) K_w for case studies (a), (b), (c), (d) with the strict match of the limits in [4] in the presence of the 5th, 7th, 11th, 13th, 17th, 19th, 23rd, 25th voltage harmonics, as well as for case studies (a), (b) with 5th, 7th, 11th, 13th only.

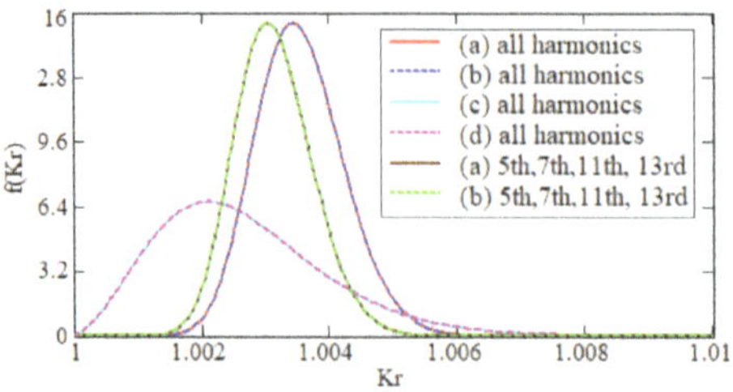

Figure 6. Probability density functions of K_r for case studies (a), (b), (c), (d) with the strict match of the limits in [4] in the presence of the 5th, 7th, 11th, 13th, 17th, 19th, 23rd, 25th voltage harmonics, as well as for case studies (a), (b) in the presence of the 5th, 7th, 11th, 13th only.

Figure 7 illustrates the reliability vs. service time (in p.u. of design life) for the XLPE cable with $\Delta\theta_{tot} = -15$ °C and voltage harmonics that: (a) exactly match, (b) are 25% above, (c) are 25% below the limits in [4].

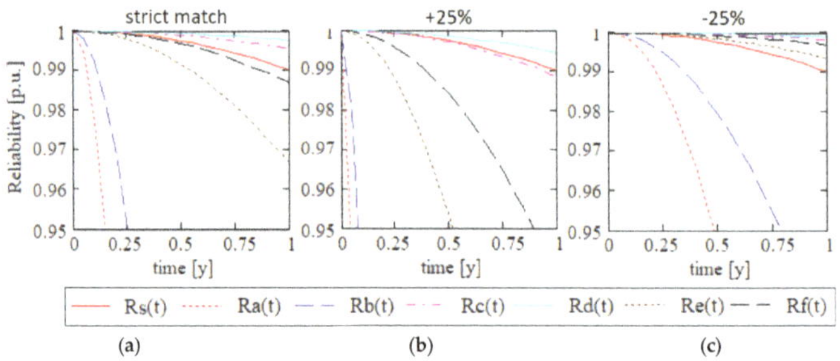

Figure 7. Reliability vs. service time (in p.u. of design life) for the XLPE cable with voltage harmonics that: (**a**) exactly match, (**b**) are 25% above, (**c**) are 25% below the limits in [4]. $\Delta\theta_{tot} = -15$ °C.

3.3. Discussion

3.3.1. The Effect of Voltage Harmonics Only

Let us discuss the results of case studies (a)–(f) in Table 3 first, so as to deal with the effect of voltage harmonics only. The 1st percentile of failure time is lower—in most cases far lower—than design life for all case studies (a)–(f) and for both the cable and the capacitor; correspondingly, the reliability at design life L_D is always lower than design reliability $R_D = 1 - P_D = 0.99 = 99\%$. Hence, for such devices a combination of the 5th, 7th, 11th, 13th, 17th, 19th, 23rd, 25th voltage harmonics seems to be detrimental irrespective of their phase-shift angle.

This is also confirmed by the plots of reliability vs. service time for the cable and the capacitor reported in Figures 3 and 4, respectively, which show that the reliability in distorted cases (a)–(f) is always below that in rated sinusoidal conditions (solid red curve). Surprisingly, this holds not only when the harmonics comply with the limits in [4] exactly—as already found in [8] for the 11th, 13th harmonics and in [9] for the 5th, 7th, 11th, 13th—but also when they are 25% below these limits: hence, this analysis suggests that the distorted voltage waveshapes of case studies (a)–(f) can reduce the life and reliability of the insulation of the cable and the capacitor with respect to design life and reliability even if harmonics have a 25% safety margin compared to the limits in standards [2–4].

Obviously, the lowest values of $t_{1\%,NS}$ are found when harmonics are 25% above the limits: thus the limits in [4] are still a key reference for designing effective filters to limit voltage harmonics [31].

The effect on insulation life and reliability is different among cases (a)–(f): indeed, Table 3 and Figures 3 and 4 indicate that the reduction of component life and reliability is:

(1) the greatest for worst cases (a), (d) leading to the highest peak of distorted voltage, see Figure 2a;
(2) the smallest for best cases (b), (e) leading to the lowest peak of distorted voltage, see Figure 2b;
(3) intermediate for experimental cases (c), (f) leading to an intermediate peak of distorted voltage, see Figure 2c.

From Table 3, case studies (a)–(f), it is also clear that the 1st percentile of failure time and reliability at design life increase as CV_h rises from 0.1 for cases (a), (b), (c) to 0.3 for cases (d), (e), (f)—i.e., with the scatter of the magnitude of voltage harmonics. This is because the 95th percentiles of the magnitude of voltage harmonics are fixed—i.e., either equal to, or 25% above, or 25% below the limits in [4]—and greater dispersion means that the cdfs $F(V_h)$/pdfs $f(V_h)$ encompass lower values of V_h, as deduced by comparing Figure 1b with Figure 1a.

Overall, it is worth emphasizing again that the calculations in this paper for the cable and the capacitor—summarized in Table 3—are completely new compared to those in previous ones [8,9], that have intentionally been cited here to allow a direct comparison. Such comparison shows that all results and figures in this paper—although being intentionally homologous to those in previous papers for enabling a straightforward comparison—are different as for the values obtained and the curves plotted from those reported in the previous papers. However, although being different, the results reported in this paper for the cable and the capacitor are in line with those previously obtained, but is also interesting to observe that—for both the cable and the capacitor—the whole sequence of odd voltage harmonics from the 5th to the 25th analysed here is more challenging than the simple combination of the 5th, 7th, 11th, 13th voltage harmonics treated in [9]. Indeed, in Table 3 the values of $t_{1\%,NS,}$ are all lower than the homologous values of $t_{1\%,NS,O}$: thus, it is worth performing the much more cumbersome analysis carried out here, since it is not only more accurate, but also on the safe side from the viewpoint of life and reliability estimation of power components.

Another essential comment stemming from the results in Table 3 and Figures 3 and 4 for case studies (a)–(f) is that the main outcomes hold for both the cable and the capacitor. Therefore, these outcomes seem to be general and independent of the values of n_p, n_w and n_r and of the uncertainty in their evaluation: indeed, in Table 2 the values of exponents n_p, n_w and n_r for the cable are far apart from those for the capacitor, thus their effect on a same set of values of K_p, K_w, K_r will be different as well.

This comment gives a chance of a deeper understanding of cases (a)–(f), which requires dealing precisely with the values of K_p, K_w,K_r. In fact, it should be understood that the values of K_p, K_w, K_r are not the same, either in a given case study, or among different case studies, as outlined in Figures 5 and 6.

Starting from Figure 5a, it is readily seen that the pdfs of K_p in case studies (a)–(c) fully agree with what previously observed at points (1)–(3) above for the same case studies, i.e., the pdf of K_p is located in correspondence of:

(i) the highest values of K_p—always >1—for worst case (a). Thus, in this case study the peak value of distorted voltage fosters a strong reduction of life and reliability, as n_p>>1 (see Table 2);
(ii) the smallest values of K_p—almost always slightly <1—for best case (b). Thus, in this case study the peak value of distorted voltage fosters a slight, but non-negligible increase of life;
(iii) intermediate values of K_p—almost always >1—for experimental case (c). Thus, in this case study the peak value of distorted voltage fosters a reduction of life, but less than in the worst case (a).

Furthermore, Figure 5a shows that the pdfs of cases (a), (b), (c) are located differently when the 5th, 7th, 11th, 13th voltage harmonics only are dealt with: the consideration of all harmonics moves the pdf of K_p towards higher values in case (a), slightly lower values in cases (b), (c). This agrees well with Figure 2a–c, since Figure 2a exhibits a higher peak of distorted voltage, while Figure 2b,c a slightly lower peak of distorted voltage when all harmonics are considered rather than the 5th, 7th, 11th, 13th only. This is due to the more favourable combination of phase-shift angles of cases (b),(c) when all harmonics than when the 5th, 7th, 11th, 13th only are included, since a better compensation of the peaks of the various harmonics is achieved in the former than in the latter case.

Analogous comments hold for the pdfs of K_p in the worst case (d), best case (e) and experimental case (f)—omitted in Figure 5a for brevity. However, in these case studies the pdfs of K_p are displaced towards lower values, for the same reasons explained above for the pdfs of V_h: namely cases (d), (e), (f) have a higher CV_h—0.3 vs. 0.1—than cases (a), (b), (c).

Anyway, as pointed out above when commenting on Table 3, the life and reliability of components is lower when all harmonics are considered. This is mainly because of the effect of the shape factor K_w, as straightforwardly understood from Figure 5b. Figure 5b—when compared to Figure 5a—demonstrates that, while K_p depends strongly on phase-shift angles ϕ_h, K_w does not, since it is affected only by the order h and rms value V_h of voltage harmonics, see Equations (7)–(9). Thus the pdf of K_w is the same for worst case (a) as for best case (b) and experimental case (c) (this latter omitted in Figure 5b for the sake of graphical clearness)—since these cases differ only as for the values of ϕ_h. Similarly, the pdf of K_w is the same for worst case (d) as for best case (e) and experimental case (f) (this latter omitted, too). However, when all voltage harmonics are considered the pdf of K_w moves towards much higher values than with the 5th, 7th, 11th, 13th harmonics only, and since exponent n_w is significantly >1—although being much lower than n_p—this involves that life and reliability of the component with all voltage harmonics decrease with respect to the 5th, 7th, 11th, 13th harmonics only. On the other hand, such a decrease is much more remarkable for the cable than for the capacitor, as n_w is much higher for the former than for the latter.

One could guess the rms factor K_r plays a more significant role than K_w for the capacitor. Indeed, as shown in Table 2, the capacitor features $n_r = 1.8 > n_w = 0.56$, while the cable features $n_r = 1.2 << n_w = 4.9$. This is not true, as seen in Figure 6—the same as Figure 5b, but for K_r instead of K_w. In fact, Figure 6 shows that $K_r \approx 1$ in all case studies treated here (Experimental cases (c) and (f) are omitted in Figure 6 like in Figure 5b for the sake of graphical clearness, since they yield the same pdfs as cases (a),(b) and (d),(e), respectively.), hence its effect on life and reliability of power components is negligible in these cases. This conclusion can be essentially generalized to the most applicable cases, unless the total harmonic distortion of voltage THD_V is really severe, say, above 20% (see Equation (10) for the relationship between THD_V and K_r): here $THD_V \leq 10\%$, see Table 1.

Figures 5 and 6 also confirm that—as stated in Section 2.1 when commenting on Equations (7)–(9)—K_p can be either <1 or >1, while K_w and K_r are always ≥1. However, K_w can attain much higher values than:

- K_p, as straightforwardly understood when comparing Figure 5b with Figure 5a;
- K_r in particular, as readily seen by comparing Figure 5b with Figure 6.

Like the pdf of K_p, also the pdfs of K_w and K_r are displaced towards lower values of K_w and K_r for a higher dispersion: the reason is same as for the pdfs of V_h and K_p, see above.

As for the differences between the cable and the capacitor, all quantities in Table 3 are much greater for the all-film capacitor than for the XLPE cable in all case studies. This is because the cable features much greater values of the exponents n_p and n_w (see Table 2), which weigh the role played by K_p and K_w, the most dominant ageing factors not only in the case studies analysed here, but under distorted voltage in general [35,36]. On the other hand, unforeseen parallel resonances between the capacitor and the equivalent impedance of the grid—skipped here for the sake of brevity—might increase a few voltage harmonics strongly, thereby causing a premature failure of the capacitor.

Another difference between capacitor and cable is that the former is truly placed at a certain node in the network, where voltage distortion changes only with time, while cable lines actually stretch from one node to another, experiencing a voltage distortion that typically changes both with time and along the line. Hence, voltage distortion at a given time is more or less deterministically distributed along cable lines. Here cables are implicitly regarded as a discrete number of series-connected cable lengths with a constant—in space—voltage distortion level within each length.

Coming now to the induction motor and the transformer, as done in the case of Table 3 for the cable and the capacitor, let us discuss the results of case studies (a)–(f) in Table 4 first, so as to deal with the effect of voltage harmonics only. Table 4 confirms the following observations already made when commenting on the homologous case studies (a)–(f) in Table 3 for the cable and the capacitor.

(1) The 1st percentile of failure time is lower—in most cases far lower—than design life for all case studies (a)–(f) and for both the induction motor and the transformer; correspondingly, the reliability at design life L_D is always lower than design reliability $R_D = 1 - P_D = 0.99 = 99\%$. Hence, for such devices a combination of the 5th, 7th, 11th, 13th, 17th, 19th, 23rd, 25th voltage harmonics seems to be detrimental irrespective of their phase-shift angle.

(2) Surprisingly, this holds not only when the harmonics comply with the limits in [4] exactly, but also when they are 25% below these limits. Hence, this analysis suggests that the distorted voltage waveshapes of case studies (a)–(f) can reduce the life and reliability of the insulation of the induction motor and the transformer with respect to design life and reliability even if harmonics have a 25% safety margin compared to the limits in standards [2–4].

(3) Obviously, the lowest values of $t_{1\%,NS}$ are found when harmonics are 25% above the limits. Thus also for the induction motor and the transformer the limits in [4] are a key reference for designing effective filters to limit voltage harmonics.

(4) The effect on insulation life and reliability is different among cases (a)–(f). Indeed, Table 4 indicates that the reduction of component life and reliability is:

    (i) the greatest for worst cases (a), (d) leading to the highest peak of distorted voltage, see Figure 2a;
    (ii) the smallest for best cases (b), (e) leading to the lowest peak of distorted voltage, see Figure 2b;
    (iii) intermediate for experimental cases (c), (f) leading to an intermediate peak of distorted voltage, see Figure 2c.

(5) The 1st percentile of failure time and reliability at design life increase as CV_h rises from 0.1 for cases (a), (b), (c) to 0.3 for cases (d), (e), (f)—i.e., with the scatter of the magnitude of voltage harmonics.

(6) For both the induction motor and the transformer the whole sequence of odd voltage harmonics from the 5th to the 25th analysed here is more challenging than the simple combination of the 5th, 7th, 11th, 13th voltage harmonics treated in [9]. Indeed, in Table 4 the values of $t_{1\%,NS,}$ are all lower than the homologous values of $t_{1\%,NS,O}$: thus, it is worth performing the much more cumbersome analysis carried out here, since it is not only more accurate, but also on the safe side from the viewpoint of life and the reliability estimation of power components subjected to harmonics generated by power electronic converters (cables, capacitors, induction motors, transformers).

(7) The main outcomes hold for both the induction motor and the transformer, as well as for the cable and the capacitor before, thus for all treated components. Therefore, these outcomes seem to be general and independent of the values of n_p, n_w and n_r and of the uncertainty in their evaluation. Indeed, in Table 2 the values of exponents n_p, n_w and n_r differ more or less remarkably among the various components, thus their effect on a same set of values of K_p, K_w, K_r will be different as well.

Other comments stemming from a comparison between Table 3 for the cable and the capacitor and Table 4 for the transformer and induction motor—as well as from the inspection of Figure 5—is that the role played by K_w vs. K_p is more noteworthy for the cable, less significant for the other components, essentially because the capacitor, motor and transformer have all a lower value of n_w, as readily seen in Table 2. In this respect it can be said that the motor and transformer have a closer behavior to the capacitor than to the cable; this is not surprising, when considering that typically the composite/lapped structure of the insulation of motors and MV/LV transformers is more similar to the lapped insulation of the PP capacitor than to the quasi-homogeneous insulation of the XLPE cable.

On the other hand, it is also interesting to point out that the motor and the transformer exhibit lower values of time to failure and reliability than the capacitor in the same case studies. This is essentially due to the higher value of n_p featured by the motor ($n_p = 9$) and the transformer ($n_p = 11$)

compared to the capacitor (n_p = 6.2). For the same reason, in turn, the transformer exhibits lower values of time to failure and reliability than the motor in the same case studies.

In summary, the hierarchy of the values of failure times and reliability is as follows: capacitor < motor < transformer < cable. However, it should be pointed out that this hierarchy has to be taken with great care, since it has been obtained from the values of reliability model parameters reported in Table 2, which are subjected to all uncertainties related to the relevant experimental campaigns and failure of data processing. Moreover, these values hold for the particular insulating specimens considered in those campaigns, which of course cannot be considered as fully representative of all insulations of existing capacitors, motors, transformers and cables found in MV and LV systems.

3.3.2. The Effect of Current Harmonics and Sinusoidal Current

Coming now to the effect of current harmonics, one could guess that the relevant Joule losses might cause a further reduction of lifetime and reliability of power components compared to those quoted for case studies (a)–(f) in Table 3 and in Table 4, as outlined in [6,7]. However, load demand and common service practices imply that distribution grids and the relevant power components mostly work at a load substantially lower than the design "sinusoidal" load. In this way, the thermal effect of harmonic currents is insufficient to let the maximum temperature of components reach the rated sinusoidal temperature. This is also because some design practices recommended to oversize conductors to be on the safe side as for possible harmonic currents. This situation, rather common especially for MV distribution cables [50], might hide the decrease of life and reliability under distorted voltage found for case studies (a)–(f) in Table 3. To check this effect, Table 3 includes case studies (g)–(l), identical to case studies (a)–(f) apart that $<\Delta\theta_{tot}> = <\Delta\theta_{arm}> + <\Delta\theta_{HL}> = -15$ °C (see Table 1).

Case studies (g)–(l) in Tables 3 and 4 show that this effect can really occur. Indeed, for the treated components all values of failure times and reliability increase significantly in case studies (g)–(l) with respect to their homologous values in case studies (a)–(f). In particular, the 1st percentile of time to failure rises by a factor $exp(-B\langle\Delta T'_{tot}\rangle)$—compare Equation (36) with Equation (38)—which is ≈4.4 for the cable and the capacitor, ≈4.5 for the motor and the transformer: thus this factor is quite similar for all treated components, since the relevant values of B are quite close (see Table 2). However, being times to failure at $<\Delta\theta_{tot}> = 0$ °C the highest for the capacitor, the least for the cable, intermediate for the motor and the transformer, the increase involved by $<\Delta\theta_{tot}> = -15$ °C leads to:

- values of the 1st percentile of time to failure more or less broadly above design life (thus values of reliability at design life more or less broadly >99%) in all case studies for the capacitor—apart case (a) with voltage harmonics 25% above the limits in [4] (see bold numbers in Table 3);
- values of the 1st percentile of time to failure below design life (thus values of reliability at design life <99%) in worst cases (g) and (j) with voltage harmonics matching exactly or 25% above the limits in [4] for the motor (see bold numbers in Table 4);
- values of the 1st percentile of time to failure below design life (thus values of reliability at design life <99%) in the following case studies for the transformer (see bold numbers in Table 4):
    - worst case (g) with all matching types of the limits in [4];
    - worst case (j) with voltage harmonics matching exactly or 25% above the limits in [4];
- values of the 1st percentile of time to failure below design life (thus values of reliability at design life <99%) in the following case studies for the cable (see bold numbers in Table 3):
    - worst cases (g) and (j) with all matching types of the limits in [4];
    - experimental cases (i) and (l) with voltage harmonics matching exactly or 25% above the limits in [4];
- values of the 1st percentile of time to failure more or less broadly above design life (thus reliability at design life more or less broadly >99%) in best cases (h) and (k) for the cable.

Focusing on the cable, these potentially harmful cases are also confirmed by the plots of reliability vs. service time for the cable with $<\Delta\theta_{tot}> = -15$ °C reported in Figure 7, which show the many previously listed cases where cable reliability in distorted conditions falls below that in rated sinusoidal conditions (solid red curve). Therefore, as such "pessimistic" cases cannot be fully ruled out for a cable subjected to voltage and current harmonics generated by power electronic devices, such a cable faces a risk of premature failure although its average temperature is 15 °C below design temperature. This holds also for potentially harmful cases for the motor and the transformer emphasized with bold characters in Table 4.

3.3.3. Further Remarks

As a closure of this discussion, it must be emphasized that the analysis carried out here in the streamline of the present investigation does not have a particular type of converter as a reference—e.g., the 6-pulse or the 12-pulse, although the 6-pulse and 12-pulse SCR-based converters are quite often encountered in LV and MV grids. Rather, this analysis aims at reproducing the typical situation of harmonic distortion level which can be typically found in LV and MV grids, based on the typical distorting loads found in such grids and on the typical filtering strategies to limit the effects of such distorting loads. This is the reason why international standards EN50160, IEC 61000-2-2, IEC 61000-2-4 are taken as a reference here; based on the limitations established by these authoritative international standards to voltage harmonics—and in particular by the probabilistically-established limits after EN50160—we have chosen as case-study examples three typical sets of possible voltage distortion levels, i.e.,:

- exact match of the limits, in order to reproduce situations where the plant is designed so as to comply with the limits, but with no safety margins (to spare money);
- 25% below the limits, in order to reproduce situations where the plant is designed so as to comply with the limits with a heuristically found and apparently broad enough safety margin, thereby spending more money than strictly needed to increase the power quality of the grid;
- 25% above the limits, in order to reproduce situations where the plant is designed carelessly with respect to the limits (this might happen from time to time in huge industrial plants with their own MV and LV grids to spare further money).

However, each of these three criteria is only one among infinite possible ways to set the voltage distortion level. Once the distortion level is set parametrically or known from measurements in a particular point of common coupling (PCC) in the grid or internal point of coupling (IPC) in a plant, the method proposed here—namely Equations (31)–(39)—can be applied anyway.

4. Conclusions

This paper has refined the theory and broadened the applications of an electro-thermal probabilistic life model—also referred to as electro-thermal reliability model—developed in previous studies for the insulation of power components affected by current and voltage harmonics. In line with previous papers, the outcomes of this article confirm that the limits to voltage harmonics set in international standards may be not totally conservative as for the reliability of components subjected to the harmonics generated by power electronic converters. This holds even if a fairly broad safety margin of 25% is kept from such limits. This effect is due not only to the peak of the distorted voltage, accounted for by the peak factor K_p, but also to the shape of the distorted voltage—thus to its time derivative—accounted for by the shape factor K_w. The effect of voltage harmonics may be compensated for by a lower-than-rated load current of components. However, only a strong reduction of the load can hide the effect of distorted voltage totally.

Another important conclusion of this study is that for all treated components (cables, capacitors, induction motors, transformers) and for all case studies examined, the whole sequence of odd voltage harmonics from the 5th to the 25th analysed here is more challenging that the simple combination of the 5th, 7th, 11th, 13th voltage harmonics treated previously. Thus, it is worth performing the much more cumbersome analysis carried out here, since it is not only more accurate, but also on the safe side from the viewpoint of life and reliability estimation of power components.

The results also point out that a wide variety of distorted regimes can be encountered in practice, each having different effects on component reliability. Therefore, a reliability model like that presented here is a quite useful tool to be used in each case of interest beside other methods, like careful experimental measurements and sound diagnostic techniques. All these techniques employed together may provide a fairly exhaustive picture of the effect of voltage and current harmonics generated by power electronic devices on the life and reliability of power components.

It is also worth emphasizing that the reliability model is used here for life and reliability estimates of the MV/LV capacitor and cable, but it can be extended to other insulations, e.g., those of induction motors and transformers. Indeed, the treated components feature quite different values of the parameters of the reliability model: since the main results obtained here for all case studies hold for both the cable and the capacitor, these results appear to be general and independent of the values of the parameters, as well as of the uncertainty in their evaluation.

Furthermore, the investigations presented in this paper for the international standards EN50160, IEC 61000-2-2, and IEC 61000-2-4 could be conducted in future works considering also other standards from the IEC 61000 series that are related to specific equipment and converters.

As a final remark, it has to be pointed out that—due to the simplifying assumptions made and the "quasi-parametric" approach followed—the results obtained should be always generalized with care, being indicative but not exhaustive. Indeed, all results reported here have been obtained from particular values of reliability model parameters, which are subjected to all uncertainties related to the relevant experimental campaigns and failure of data processing. Moreover, these values hold for the particular insulating specimens considered in those campaigns, which of course cannot be considered as fully representative of all insulations of existing capacitors, motors, transformers and cables found in MV an LV system.

Author Contributions: Conceptualization, G.M.; methodology G.M.; validation, G.M. and B.D.; formal analysis, G.M., B.D., E.C., P.D.F., L.P.D.N.; investigation, G.M.; writing—original draft preparation, G.M.; writing—review and editing, G.M. and B.D. All authors have read and agreed to the published version of the manuscript.

Funding: This research received no external funding

Conflicts of Interest: The authors declare no conflict of interest.

References

1. IEEE Standard 519-1992. *IEEE Recommended Practices and Requirements for Harmonic Control. in Electric Power Systems*; IEEE: Piscataway, NJ, USA, 1993.
2. IEC Standard 61000-2-2. *Environment—Compatibility Levels for Low-Frequency Conducted Disturbances and Signalling in Public Low-Voltage Power Supply Systems*; IEC: Geneva, Switzerland, 2002.
3. IEC Standard 61000-2-4. *Environment—Compatibility Levels in Industrial Plants for Low-Frequency Conducted Disturbances*; IEC: Geneva, Switzerland, 2002.
4. European Standard EN Standard 50160. *Voltage Characteristics of Electricity Supplied by Public Distribution Networks*, 3rd ed.; European Committee for Standardization: Brussels, Belgium, 2010.
5. Chiodo, E.; Mazzanti, G. Theoretical and practical aids for the proper selection of reliability models for power system components. *Int. J. Reliab. Saf.* **2008**, *2*, 99–128. [CrossRef]
6. Mazzanti, G.; Passarelli, G. A probabilistic life model for reliability analysis of power cables feeding electric traction systems. In Proceedings of the 2006 IEEE International Symposium on Power Electronics, Electrical Drives, Automation and Motion IEEE SPEEDAM 2006, Taormina, Italy, 23–26 May 2006; pp. S30_17–S30_22.
7. Mazzanti, G.; Passarelli, G.; Russo, A.; Verde, P. The effects of voltage waveform factors on cable life estimation using measured distorted voltages. In Proceedings of the 2006 IEEE Power Engineering Society General Meeting IEEE PES GM 2006, Montreal, QC, Canada, 18–22 June 2006.

8. Mazzanti, G. Are the Limits to Harmonic Voltages given by International Standards on the Safe Side from the Viewpoint of Power Component Reliability? *Int. Rev. Electr. Eng. (IREE)* **2011**, *6*, 2762–2772.
9. Mazzanti, G. Reliability Evaluation of Insulation subjected to Harmonic Voltages within the Limits set by International Standards. *IEEE Trans. Dielectr. Electr Insul.* **2014**, *21*, 2037–2046. [CrossRef]
10. Cavallini, A.; Fabiani, D.; Mazzanti, G.; Montanari, G.C.; Contin, A. Voltage endurance of electrical components supplied by distorted voltage waveforms. In Proceedings of the International Symposium on Electrical Insulation (IEEE/ISEI), Anaheim, CA, USA, 2–5 April 2000; pp. 73–76.
11. Cavallini, A.; Fabiani, D.; Mazzanti, G.; Montanari, G.C. Models for degradation of self-healing capacitors operating under voltage distortion and temperature. In Proceedings of the 6th International Conference on Properties and Applications of Dielectric Materials (Cat. No.00CH36347), Xi'an, China, 6 August 2002; pp. 108–111.
12. Cavallini, A.; Mazzanti, G.; Montanari, G.C.; Fabiani, D. The effect of power system harmonics on cable endurance and reliability. In Proceedings of the IEEE IAS 35th Annual Meeting, Rome, Italy, 8–12 October 2000; pp. 240–247.
13. Montanari, G.C.; Fabiani, D. The effect of voltage distortion on ageing acceleration of insulation systems under partial discharge activity. *IEEE Ind. Electron. Mag.* **2001**, *17*, 24–33.
14. Sletbak, J.; Botne, A. A Study on Inception and Growth of Water Trees and Electrochemical Trees in Polyehtylene and Cross Linked Polyehtylene Insulations. *IEEE Trans. Electron. Insul.* **1977**, *12*, 383–389. [CrossRef]
15. Ildstad, E. *High. Voltage AC and DC Aging of an XLPE Distribution Cable in Wet Enviroment*; Nord-IS: Trondheim, Norway, 1988.
16. Ildstad, E.; Bardsen, H.; Faremo, H.; Knutsen, B. Influence of Mechanical and Frequency on Water Treeing in XLPE Cable Insulation. In Proceedings of the IEEE International Symposium on Electrical Insulation, Toronto, ON, Canada, 3–6 June 1990.
17. Ross, R. Inception and propagation mechanisms of water treeing. *IEEE Trans. Dielectr. Electr. Insul.* **1998**, *5*, 660–680. [CrossRef]
18. Mauseth, F.; Amundsen, M.; Lind, A.; Faremo, H. Water Tree Growth of Wet XLPE Insulation Stressed with DC and High Frequency AC. In Proceedings of the 2012 Annual Report Conference on Electrical Insulation and Dielectric Phenomena, Montreal, QC, Canada, 14–17 October 2012; pp. 692–695.
19. Patil, K.D.; Gandhare, W.Z. Threat of harmonics to underground cables. In Proceedings of the 2012 Students Conference on Engineering and Systems, Allahabad, Uttar Pradesh, 16–18 March 2012.
20. Sadati, S.B.; Yazdani-Asrami, M.; Taghipour, M. Effects of harmonic current content and ambient temperature on load ability and life time of distribution transformers. *Int. Rev. Electr. Eng.* **2010**, *3*, 1444–1451.
21. Bihari, S.; Goswami, U.; Goswami, A.; Sadhu, P.K. Fast Estimating Method of Effective Life of Power Cable in Industrial Environment. In Proceedings of the 2019 23rd International Conference Electronics, Palanga, Lithuania, 17–19 June 2019.
22. Yaghoobi, J.; Alduraibi, A.; Martin, D.; Zare, F.; Eghbal, D.; Memisevic, R. Impact of high-frequency harmonics (0–9 kHz) generated by grid-connected inverters on distribution transformers. *Int. J. Electr. Power Energy Syst.* **2020**, *122*, 106177. [CrossRef]
23. Thango, B.A.; Jordaan, J.A.; Nnachi, A.F. Effects of Current Harmonics on Maximum Loading Capability for Solar Power Plant Transformers. In Proceedings of the 2020 International SAUPEC/RobMech/PRASA Conference, Cape Town, South Africa, 29–31 January 2020.
24. Montanari, G.C. *Power Electronics and Electrical Apparatus: A Threat?* NORD-IS: Lingby, Denmark, 2007.
25. Cavallini, A.; Fabiani, D.; Montanari, G.C. Power electronics and electrical insulation systems—part 2: Life modeling for insulation design. *IEEE Electr. Insul. Mag.* **2010**, *26*, 33–39. [CrossRef]
26. Lebey, T.; Castelan, P.; Montanari, G.C.; Ghinello, I. Influence of PWM-type voltage waveforms on reliability of machine insulation system. In Proceedings of the 8th International Conference on Harmonics and Quality of Power. Proceedings, Athens, Greece, 14–16 October 1998.
27. Montanari, G.C.; Hebner, R.; Morshuis, P.; Seri, P. An Approach to Insulation Condition Monitoring and Life Assessment in Emerging Electrical Environments. *IEEE Trans. Power Deliv.* **2019**, *34*, 1357–1364. [CrossRef]
28. Carboni, A.; ElShawarby, K.; Foglia, G.M.; Perini, R.; Gerlando, A.D.; Ragaini, E. Electric Stress in Power Electronics Applications. In Proceedings of the 2019 IEEE Milan PowerTech, Milan, Italy, 23–27 June 2019.

29. Lv, C.; Liu, J.; Zhang, Y.; Lei, W.; Cao, R.; Lv, G. Reliability Modeling for Metallized Film Capacitors Based on Time-Varying Stress Mission Profile and Aging of ESR. *IEEE J. Emerg. Sel. Top. Power Electron.* **2020**. [CrossRef]
30. Borthakur, D.P.; Das, S. Life Estimation of XLPE Cable Under Electrical and Thermal Stress. In Proceedings of the International Conference on High Voltage Engineering and Technology (ICHVET), Hyderabad, India; 7–8 February 2019. [CrossRef]
31. Arillaga, J.; Bradley, D.A.; Bodge, P.S. *Power System Harmonics*; Wiley: New York, NY, USA, 1988.
32. Lamedica, R.; Marzinotto, M.; Prudenzi, A. Harmonic amplitude and harmonic phase angle monitored in an electrified subway system during rush hours traffic. In Proceedings of the 4th IASTED, Rhodes, Greece, 28–30 June 2004.
33. Caramia, P.; Carpinelli, G.; Verde, P.; Vitali, F. Probabilistic Evaluation of the Cable Thermal Useful Life in MV/LV Energy Systems. In Proceedings of the 5′ International Conference on Probabilistic Methods applied to Power Systems, Vancouver, BC, Canada, 21–25 September 1997.
34. Caramia, P.; Carpinelli, G.; Russo, A.; Verde, P. Estimation of Thermal Useful Life of MV/LV Cables in Presence of Harmonics and Moisture Migration. In Proceedings of the IEEE Bologna Power Tech Conference, Bologna, Italy, 23–26 June 2003.
35. Gallo, D.; Langella, R.; Testa, A. Predicting voltage stress effects on MV/LV components. In Proceedings of the IEEE Bologna Power Tech Conference, Bologna, Italy, 23–26 June 2003.
36. Gallo, D.; Langella, R.; Testa, A. Is it always possible to separately analyze different power quality phenomena? The case of the voltage peak. *L'Energia Elettr.* **2004**, *81*, 162–167.
37. Montanari, G.C.; Mazzanti, G. From Thermodynamic to Phenomenological Multistress Models for Insulating Materials without or with evidence of Threshold. *J. Phys. D Appl. Phys.* **1994**, *27*, 1691–1702. [CrossRef]
38. Montanari, G.C.; Mazzanti, G.; Simoni, L. Progress in electrothermal life modeling of electrical insulation over the last decades. *IEEE Trans. Dielectr. Electr. Insul.* **2002**, *9*, 730–745. [CrossRef]
39. Mazzanti, G.; Montanari, G.C.; Simoni, L. Insulation Characterization in Multistress Conditions by Accelerated Life Tests: An Application to XLPE and EPR for High-Voltage Cables. *IEEE Electr. Insul. Mag.* **1997**, *13*, 24–33. [CrossRef]
40. Dissado, L.A.; Fothergill, J.C. *Electrical Degradation and Breakdown in Polymers*; Peregrinus: London, UK, 1992.
41. Mazzanti, G.; Marzinotto, M. *Extruded Cables for High. Voltage Direct Current Transmission: Advances in Research and Development*; Wiley-IEEE Press: Hoboken, NJ, USA, 2013.
42. Cacciari, M.; Mazzanti, G.; Montanari, G.C. Comparison of Maximum Likelihood Unbiasing Methods for the Estimation of the Weibull Parameters. *IEEE Trans. Dielectr. Electr. Insul.* **1996**, *3*, 18–27. [CrossRef]
43. Miner, M.A. Cumulative damage in fatigue. *J. Appl. Mech.* **1945**, A159–A163.
44. Ministerial Decree 6 July 2012—Incentives for Energy from Non-Photovoltaic Renewable Electricity Sources. Available online: https://www.mise.gov.it/index.php/it/normativa/decreti-ministeriali/2023799-decreto-ministeriale-6-luglio-2012-ed-allegati-incentivi-per-energia-da-fonti-rinnovabili-elettriche-non-fotovoltaiche (accessed on 13 May 2020).
45. Ministerial Decree 4 July 2019—Methods for Calculating Storage Obligations. Available online: https://www.gse.it/documenti_site/Documenti%20GSE/Servizi%20per%20te/FER%20ELETTRICHE/NORMATIVE/DM%204%20luglio%202019%20-%20Incentivazione%20dell%e2%80%99energia%20elettrica%20prodotta%20dagli%20impianti.pdf (accessed on 13 May 2020).
46. The National Renewable Energy Laboratory of the U.S. Department of Energy. Available online: https://www.nrel.gov/analysis/tech-footprint.html (accessed on 13 May 2020).
47. Chiodo, E.; Mazzanti, G. Mathematical and physical properties of reliability models in view of their application to modern power system components. In *Innovations in Power Systems Reliability*; Anders, G.J., Vaccaro, A., Eds.; Springer: London, UK, 2011; pp. 59–140.
48. Chiodo, E.; Mazzanti, G. Bayesian inference of power system insulation reliability in the presence of voltage harmonics. *Int. Rev. Electr. Eng.* **2016**, *11*, 266–276. [CrossRef]
49. Wang, Y.J.; Pierrat, L.; Wang, L. Summation of Harmonic Currents produced by AC/DC Static Converters with Randomly Fluctuating Loads. *IEEE Trans. Power Deliv.* **1994**, *9*, 1129–1135. [CrossRef]
50. Katz, C.; Seman, G.W.; Bernstein, B.S. Low temperature aging of XLPE and EP insulated cables with voltage transients. *IEEE Trans. Power Deliv.* **1995**, *10*, 34–42. [CrossRef]

51. Caramia, P.; Carpinelli, G.; Verde, P.; Mazzanti, G.; Cavallini, A.; Montanari, G.C. An approach to life estimation of electrical plant components in the presence of harmonic distortion. In Proceedings of the 9th IEEE International Conference on Harmonics and Quality of Power (IEEE ICHQP), Orlando, FL, USA, 1–4 October 2000.
52. Montanari, G.C.; Pattini, G. Thermal endurance of insulating materials. *IEEE Trans. Electr. Insul.* **1986**, *21*, 66–75.

electronics

MDPI

Article

The Study of the Single Event Effect in AlGaN/GaN HEMT Based on a Cascode Structure

Yanan Liang [1], Rui Chen [1,2,*], Jianwei Han [1,2], Xuan Wang [1], Qian Chen [1] and Han Yang [1]

[1] National Space Science Center, Chinese Academy of Sciences, Beijing 100000, China; liangyanan@nssc.ac.cn (Y.L.); hanjw@nssc.ac.cn (J.H.); wangxuan171@mails.ucas.ac.cn (X.W.); chenqian16@mails.ucas.ac.cn (Q.C.); yanghan18@mails.ucas.ac.cn (H.Y.)

[2] Institute of astronomy and space, University of Chinese Academy of Sciences, Beijing 100000, China

* Correspondence: chenrui2010@nssc.ac.cn

Abstract: An attractive candidate for space and aeronautic applications is the high-power and miniaturizing electric propulsion technology device, the gallium nitride high electron mobility transistor (GaN HEMT), which is representative of wide bandgap power electronic devices. The cascode AlGaN/GaN HEMT is a common structure typically composed of a high-voltage depletion-mode AlGaN/GaN HEMT and low-voltage enhancement-mode silicon (Si) MOSFET connected by a cascode structure to realize its enhancement mode. It is well known that low-voltage Si MOSFET is insensitive to single event burnout (SEB). Therefore, this paper mainly focuses on the single event effects of the cascode AlGaN/GaN HEMT using technical computer-aided design (TCAD) simulation and heavy-ion experiments. The influences of heavy-ion energy, track length, and track position on the single event effects for the depletion-mode AlGaN/GaN HEMT were studied using TCAD simulation. The results showed that a leakage channel between the gate electrode and drain electrode in depletion-mode AlGaN/GaN HEMT was formed after heavy-ion striking. The enhancement of the ionization mechanism at the edge of the gate might be an important factor for the leakage channel. To further study the SEB effect in AlGaN/GaN HEMT, the heavy-ion test of a cascode AlGaN/GaN HEMT was carried out. SEB was observed in the heavy-ion irradiation experiment and the leakage channel was found between the gate and drain region in the depletion-mode AlGaN/GaN HEMT. The heavy-ion irradiation experimental results proved reasonable for the SEB simulation for AlGaN/GaN HEMT with a cascode structure.

Keywords: AlGaN/GaN HEMT; cascode structure; single event effects; technology computer-aided design simulation; heavy-ion irradiation experiment

check for updates

Citation: Liang, Y.; Chen, R.; Han, J.; Wang, X.; Chen, Q.; Yang, H. The Study of the Single Event Effect in AlGaN/GaN HEMT Based on a Cascode Structure. *Electronics* **2021**, *10*, 440. https://doi.org/10.3390/electronics10040440

Academic Editor: Elio Chiodo

Received: 23 December 2020
Accepted: 20 January 2021
Published: 10 February 2021

1. Introduction

As high-power and miniaturizing electric propulsion technology devises continue to develop, high-performance and high-reliability wide bandgap power electronic devices begin to play an important role in air and space vehicles. Gallium nitride high electron mobility transistors (GaN HEMT) are representative of wide bandgap power electronic devices and could be an attractive candidate for space and aeronautic applications due to their excellent electrical characteristics, such as high electron mobility, high breakdown voltage, and high thermal conductivity. To realize space and aeronautic applications, the radiation effect should be considered, including the total ionizing dose (TID) effect, single event effect (SEE), and the displacement damage dose (DDD) effect. Due to the existence of two-dimensional electron gas (2DEG), AlGaN/GaN HEMT, fabricated on AlGaN/GaN heterojunction, is in depletion mode. To ensure the reliability of the device in space applications, enhancement-mode devices can be adopted. The enhanced mode is mainly realized in AlGaN/GaN HEMT by the following: p-GaN [1,2], F ion implantation [3,4], MIS HEMT [5,6], and cascode structure [7,8]. The total ionizing dose and displacement damage effect on GaN HEMT have been studied by many researchers [9–14]. Because there is no gate dielectric layer in AlGaN/GaN HEMT, GaN HEMT is less sensitive to the TID effect. Additionally, some references [12,13] have shown that the DDD effect impacts the direct-current (DC) characteristics of the AlGaN/GaN HEMTs and the failure mechanisms have been basically studied. However, as one of the most important effects of space environment radiation, many studies focused on the enhancement-mode AlGaN/GaN HEMT based on the p-GaN structure [15–17]. The SEE characteristics and AlGaN/GaN HEMT failure mechanisms based on the cascode structure are not clear. Therefore, in this paper, we focused on studying the SEE of the AlGaN/GaN HEMTs and analyzing the failure mechanism of the single event burnout (SEB) effect.

To study the radiation response and failure mechanism of the SEB effect for GaN HEMTs, TCAD simulation and heavy-ion irradiation experiments were carried out. The remainder of the paper is organized as follows.

Firstly, according to the circuit structure shown in Figure 1, the cascode structure of AlGaN/GaN HEMT, including the depletion-mode AlGaN/GaN HEMT and Si MOSFET, were modeled. Because low-voltage Si MOSFET is not sensitive to the SEB effect, we focused on studying the SEE in the depletion-mode AlGaN/GaN HEMT. The simulation results showed that (1) the SEB effect of the AlGaN/GaN HEMT was influenced by heavy-ion energy and track position. Further, (2) a leakage channel between the gate electrode and drain electrode was observed after heavy-ion striking in depletion-mode AlGaN/GaN HEMT.

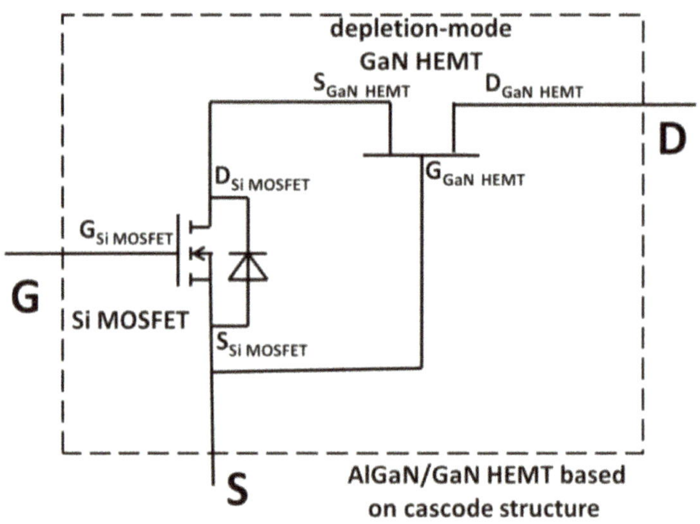

Figure 1. The schematic of aluminium gallium nitride /gallium nitride high electron mobility transistor (AlGaN/GaN HEMT), based on the cascode structure.

Heavy-ion irradiation experiments of the commercial cascode AlGaN/GaN HEMT were conducted in an off-mode condition. The experimental results showed that (1) the depletion-mode AlGaN/GaN HEMT was burnt out under V_{ds} = 50 V for Ge ions at LET = 28.5 MeV·cm^2/mg (about 0.31 pC/μm). Moreover, it showed (2) a leakage channel between the gate electrode and the drain electrode in the depletion-mode, where AlGaN/GaN HEMT was formed after the heavy-ion striking.

Finally, the possible SEE failure mechanism was proposed. We proved that the enhancement of the impact ionization mechanism at the edge of the gate might be an important factor in the formation of the leakage channel.

2. Simulation Results and Discussion

As shown in Figure 1, a high voltage depletion-mode AlGaN/GaN HEMT and a low voltage enhancement-mode silicon (Si) MOSFET were connected to form a high voltage enhancement-mode AlGaN/GaN HEMT in the cascode structure. For the cascode AlGaN/GaN HEMT circuit, the drain electrode of GaN HEMT served as the drain port and the gate electrode of Si MOSFET acted as the gate port. The source port of the device was formed by connecting the source electrode of Si MOSFET and the gate electrode of GaN HEMT. It is well known that the low voltage power MOSFETs are robust to the SEB effect [18–20]. Therefore, to investigate the mechanism of the SEB induced by heavy ion irradiation on cascode AlGaN/GaN HEMT, we studied the SEB effect of the depletion-mode AlGaN/GaN HEMT. TCAD simulation was carried out with the Sentaurus TCAD simulator.

2.1. Modeling

In the simulation, the generic model of the depletion-mode AlGaN/GaN HEMT supported by the foundry was adopted. According to previous reports [21–23], the architecture considered in the simulation is shown in Figure 2. The structure consisted of a Si substrate, a GaN buffer layer, an unintentional doped GaN channel layer, an unintentional doped AlGaN barrier layer, and a SiN passivation layer. The breakdown voltage of the simulated device was higher than 900 V. The breakdown, transfer, and output characteristics of the simulated structure are shown in Figure 3. The threshold voltage (V_{th}) of the AlGaN/GaN HEMT was approximately −2.5 V.

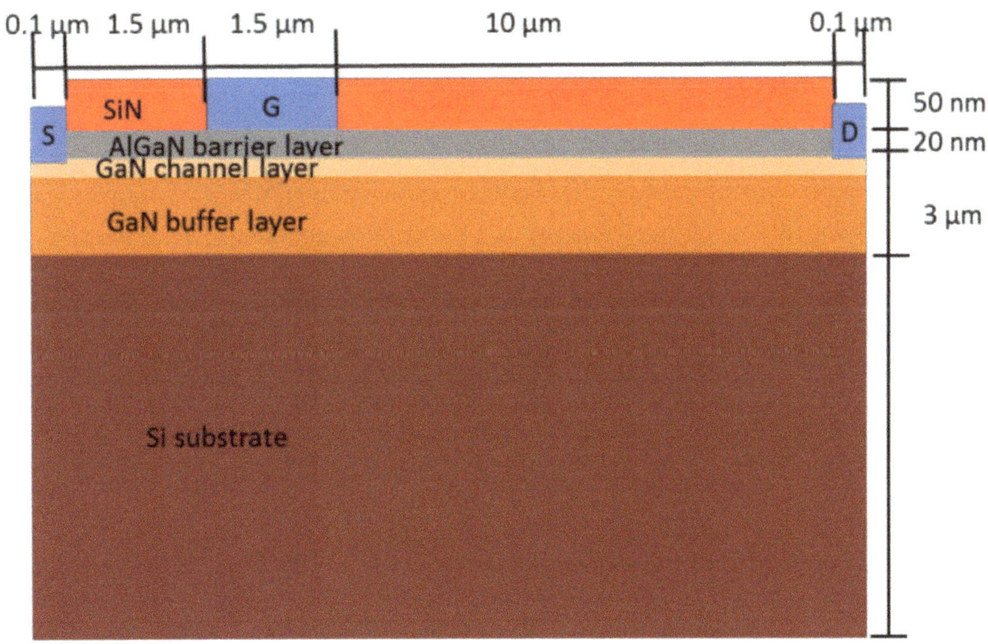

Figure 2. The schematic of the depletion-mode AlGaN/GaN HEMT.

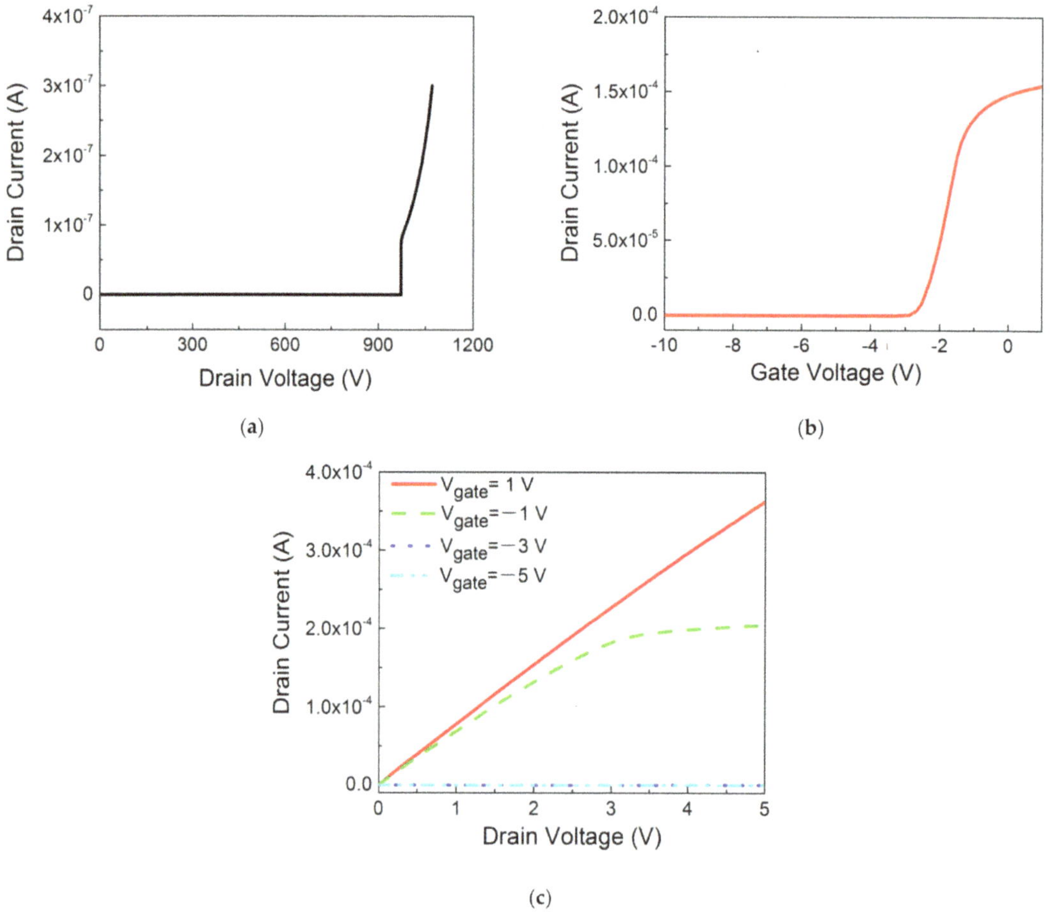

Figure 3. The characteristics of the depletion-mode AlGaN/GaN HEMT: (**a**) the breakdown characteristic, (**b**) the transfer characteristic, and (**c**) the output characteristics.

In the simulation, we used a heavy-ion model in Sentaurus software. The impact ionization model and drift-diffusion model were adopted. To simplify the simulation process, thermal equations (lattice heating) were not considered in this paper. The heavy-ion was vertical incidence on the device from the front side. The Gaussian track radius was 20 nm spatially.

2.2. Simulation Results and Discussion

2.2.1. SEB Characteristics of HMET

To trigger the device's SEE effect, we adopted a track length of 10 μm and linear energy transfers (LET) of 1 pC/μm. The drain, source, and gate port of the depletion-mode AlGaN/GaN HEMT were biased at 400 V, 10 V, and 0 V, respectively ($V_{gs} = -10$ V, $V_{ds} = 390$ V), which ensured that the device was in the off state. The transient drain, source, and gate currents are presented in Figure 4. The currents increased immediately after a heavy-ion struck the device. The drain current increased to about 2.5 mA, the gate current increased to about 1.5 mA, and the source current increased to about 1 mA. Approximately 1 ns after striking, the source current returned to 0.1 μA and the drain current and gate current reached 0.2 mA. The drain current and gate current were about 0.1 mA at 50 ns after striking. This indicated that a leakage channel between the gate electrode and the drain electrode in depletion-mode AlGaN/GaN HEMT was formed.

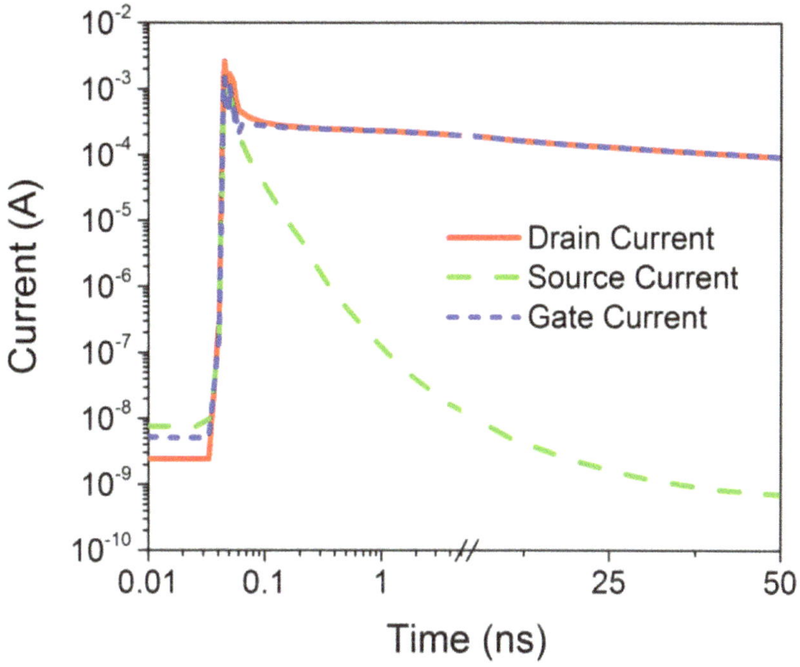

Figure 4. The transient drain, source, and gate currents after heavy-ion striking.

2.2.2. Influence Factors of SEB

- Heavy-ion energy

Heavy-ion energy, heavy-ion track length, and heavy-ion track position are important parameters that affect the single event effect of the depletion-mode AlGaN/GaN HEMT. Heavy-ion energy is related to the electron-hole pairs produced in the device. To study the influence of heavy-ion energy on the SEE of the depletion-mode AlGaN/GaN HEMT, simulation experiments with different LETs (10, 5, 1, 0.1, 0.01, 0.005, 0.001 pC/μm) were carried out. The track length (10 μm) and track position (close to the drain electrode) were adopted in the simulation. The transient drain currents after heavy ion impacts for different LETs were shown in Figure 5. The transient drain current increased with the LET increasing and reached a saturation value until the LET was higher than 0.1 pC/μm. It indicated that the electron-hole pairs generated in the AlGaN and GaN layers increased with the LET grew and reached the maximum at 0.1 pC/μm [16].

- Heavy-ion track length

In addition to heavy-ion energy, the heavy-ion track length is also an important parameter. The impacts of heavy-ion with different track lengths were also studied. In the simulation, the heavy-ion energy (1 pC/μm) and track position (close to the drain electrode) were adopted and the track lengths were 2, 5, 10, and 15 μm. The transient drain currents after heavy ion impacts for different track lengths were shown in Figure 6. The amplitude of transient drain current almost kept constant under these simulation conditions. In these simulation conditions, the heavy-ion had passed through the GaN channel layer and reached the GaN buffer layer or the Si substrate layer. In other words, the heavy-ion had passed through the active region of the depletion-mode AlGaN/GaN HEMT. Therefore, the transient drain currents did not change with heavy-ion track length under these simulation conditions. These results indicated when the heavy-ion passed through the GaN channel layer and reached the GaN buffer layer, the heavy-ion track length had little effect on the SEE of the depletion-mode AlGaN/GaN HEMT.

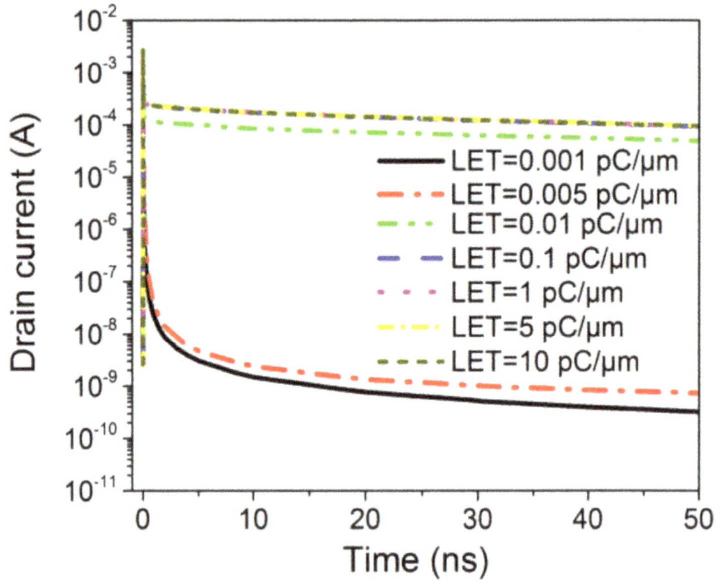

Figure 5. The transient drain currents after heavy-ion striking with different linear energy transfers (LETs).

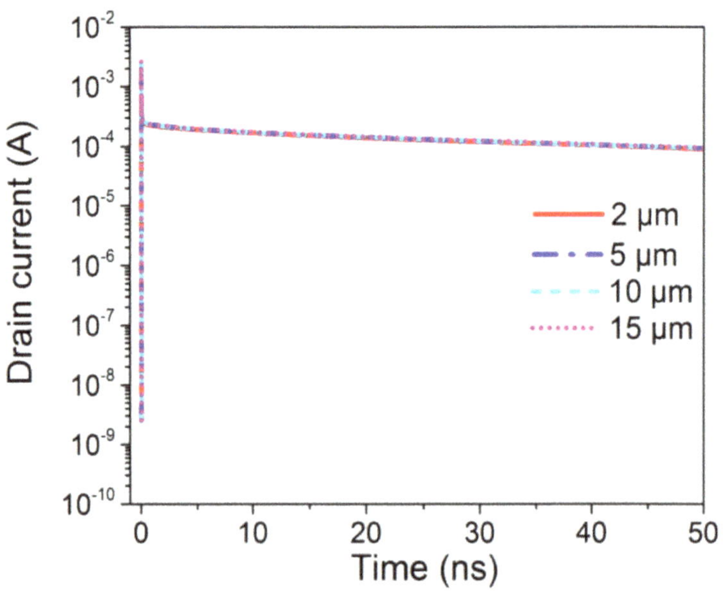

Figure 6. The transient drain currents after heavy-ion striking with different track lengths.

- Sensitive region of SEB

To localize the sensitive region of the depletion-mode AlGaN/GaN HEMT, different track positions of heavy-ion were studied. The heavy-ion energy (1 pC/μm) and track length (10 μm) were adopted in the simulation. The schematic of heavy-ion tracks on the depletion-mode AlGaN/GaN HEMT is revealed in Figure 7. The transient drain currents after heavy-ion impact on different track positions are shown in Figure 8. The drain currents were about 0.1 mA at 50 ns after the heavy-ions striking at P7, P8, and P9, while the drain currents were lower than 10^{-8} A at 50 ns after the heavy-ions striking at P1 to P6. These indicated that a SEB current was triggered when the heavy-ion impact on the location of P7, P8, and P9. Moreover, the closer the track position is to the drain region, the more sensitive it is to the SEB device.

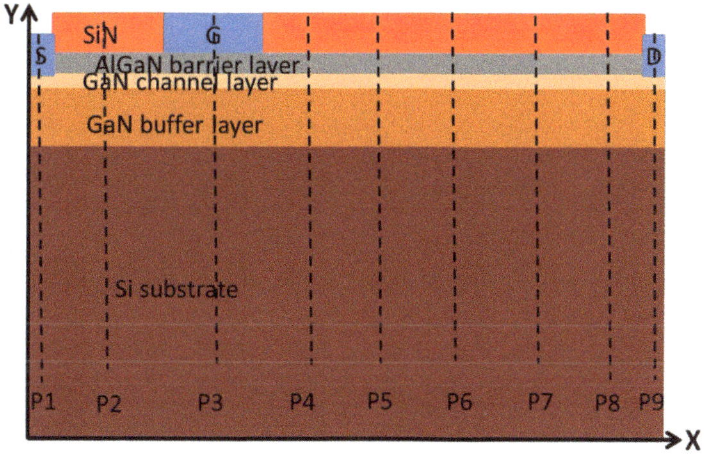

Figure 7. The cross-section of AlGaN/GaN HEMT with different heavy-ion track positions.

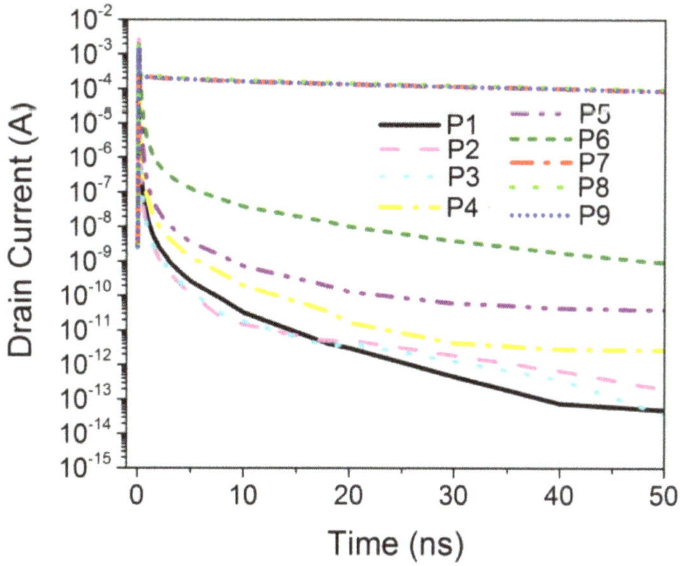

Figure 8. The transient drain currents after heavy-ion striking with different positions.

2.2.3. The Failure Mechanism of SEB

To understand the SEB mechanism, the cross-sections of the total current density after the heavy-ion impact on location P5 and P8 are shown in Figures 9 and 10, respectively. Figures 9a and 10a show the initial states of the total current density before the heavy-ions striking. Figures 9b and 10b show the total current density after the heavy-ions striking at P5 and P8 for 0.05 ns. After the heavy-ions striking, the total current density in the GaN buffer layer increased (from green to orange), indicating that the GaN buffer layer forming a leakage channel. Figures 9c and 10c show the total current density after the heavy-ions strike at P5 and P8 for 50 ns. These showed that the leakage channel disappeared at 50 ns after the heavy-ions striking at P5, while it was still a leakage channel at 50 ns after the heavy-ions striking at P8.

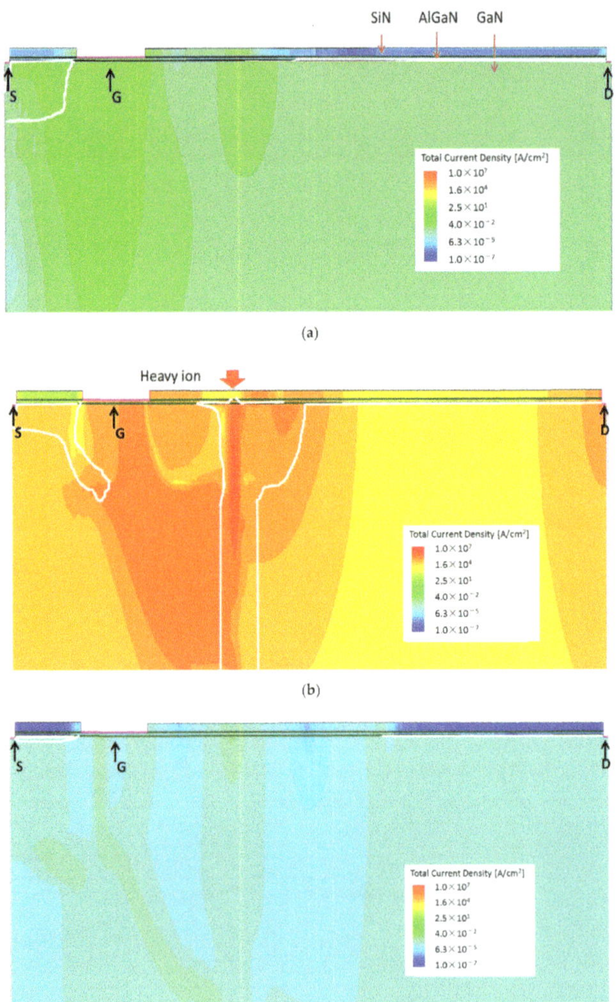

Figure 9. The cross-sections of the total current density with time after the heavy-ion striking at P5: (**a**) before heavy-ion striking, (**b**) 0.05 ns after heavy-ion striking, and (**c**) 50 ns after heavy-ion striking. The white line represented the depletion region.

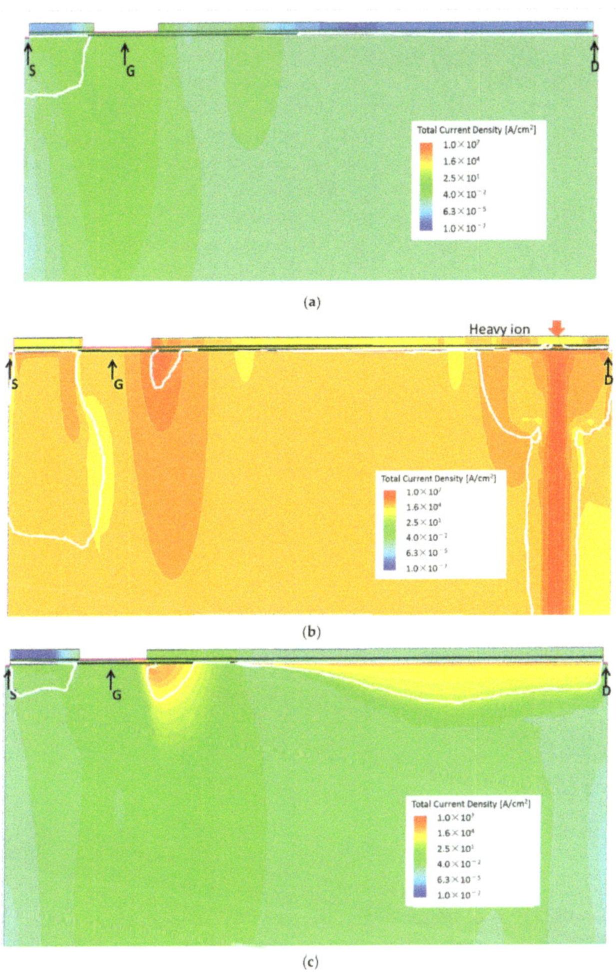

Figure 10. The cross-sections of the total current density with time after heavy-ion striking at P8: (**a**) before heavy-ion striking, (**b**) 0.05 ns after heavy-ion striking, and (**c**) 50 ns after heavy-ion striking. The white line represented the depletion region.

After the heavy-ions strike the device, the electron-hole pairs were generated. Electrons flowed toward the drain under the high drain voltage and collected by the drain electrode. Holes were left in the GaN buffer layer, leading to the electrons injection of the source and gate electrode [16,24]. Thus, the large transient current was generated in the source, gate, and drain regions. The leakage channel of the GaN buffer layer was formed. The electron-hole pairs generated by heavy-ions decreased with time, leading to the leakage current reduced.

In order to further understand the SEB mechanism, the cross-sections of the impact ionization after the heavy-ion striking at P5 and P8 were studied. Figures 11a and 12a were the initial value of the impact ionization before the heavy-ions struck. Figure 11b,c and Figure 12b,c show the impact ionization at 0.05 ns and 50 ns after the heavy-ions striking at P5 and P8. As shown in Figure 11, after the heavy-ions striking at P5, the impact ionization at the edge of the gate region decreased at 0.05 ns and the value was still low at 50 ns. As shown in Figure 12, after the heavy-ions striking at P8, the impact ionization at the edge of the gate region and the drain region increased at 0.05 ns and the value was still high at 50 ns. The electron-hole pairs were generated after the heavy-ion injected to the AlGaN/GaN HEMT. Then the generated electron-hole pairs drifted under the applied voltage. Electrons flowed toward the drain under the high drain voltage and collected by the drain electrode. Holes were left in the GaN buffer layer, leading to the electrons injection of the source, and gate electrode. The movement of the electron-hole pairs led to the changing of the potential distribution. When the heavy-ions struck at P5, which was close to the gate electrode, the electrons were injected from the source and gate electrode quickly. This process decreased the potential near the drain, which decreased the impact ionization at the edge of the gate region. When the heavy-ions struck at P8, which was close to the drain electrode, electrons were quickly collected by the drain and holes left in the GaN buffer layer accumulated. This process increased the potential near the drain, which enhanced the impact ionization at the edge of the gate region. The enhancing impact ionization increased the number of generated carriers, which increased the probability of electron tunneling of the gate region. This might be attributed to the burnout of the gate electron of the AlGaN/GaN HEMT after heavy-ion striking at P8.

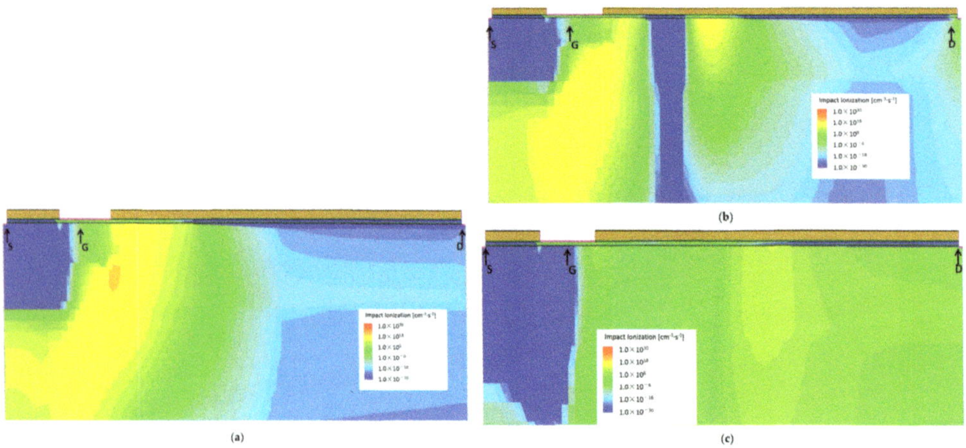

Figure 11. The cross-sections of the impact ionization after the heavy-ion striking at P5: (**a**) before heavy-ion striking, (**b**) 0.05 ns after heavy-ion striking, and (**c**) 50 ns after heavy-ion striking.

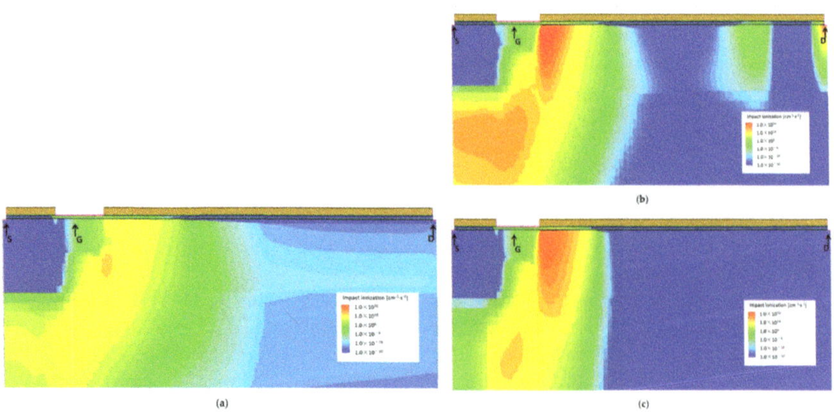

Figure 12. The cross-sections of the impact ionization after the heavy-ion striking at P8: (**a**) before heavy-ion striking, (**b**) 0.05 ns after heavy-ion striking, and (**c**) 50 ns after heavy-ion striking.

3. Heavy-Ion Experimental Results and Discussion

3.1. Experiment Samples and Setup

To further study the phenomenon of the drain current increase obtained in the simulation section, a commercial GaN HEMT with a breakdown voltage of 900 V was chosen. The devices under test (DUTS) were commercial devices employing a cascode structure from transform (TP90H180PS). The radiation experiments were carried out on the heavy-ion accelerator of China Atomic Energy Research Institute, shown in Figure 13. In the heavy-ion radiation experiments, the device was biased at the off state. The source port and gate port were grounded (V_{gs} = 0 V). The drain port was biased at 50 V (V_{ds} = 50 V). The heavy-ion parameters used in the irradiation experiment are shown in Table 1. For the heavy-ion irradiation with the flux of about 10^4 particle/cm^2/s, the devices were irradiated to the fluence of 5×10^6 particle/cm^2. The drain currents were monitored by Keithley 2470 during the heavy-ion irradiation. The drain current and gate current were measured by Keithley 2470 and 2450 after heavy-ion irradiation.

Figure 13. The heavy-ion accelerator of China Atomic Energy Research Institute.

Table 1. Heavy ions used in the experiment.

Heavy-Ion	Energy (MeV)	LET(GaN) (MeV·cm^2/mg)	Strike in GaN (μm)
$^{74}Ge^{11,20+}$	210	28.5	16.21

3.2. Experiment results and discussion

The transient drain currents during the heavy-ion irradiation are shown in Figure 14. When the drain port was biased at 50 V, the drain current quickly increased to 10 mA (current limitation), which took about 45 s at LET = 28.5 MeV·cm^2/mg (about 0.31 pC/μm).

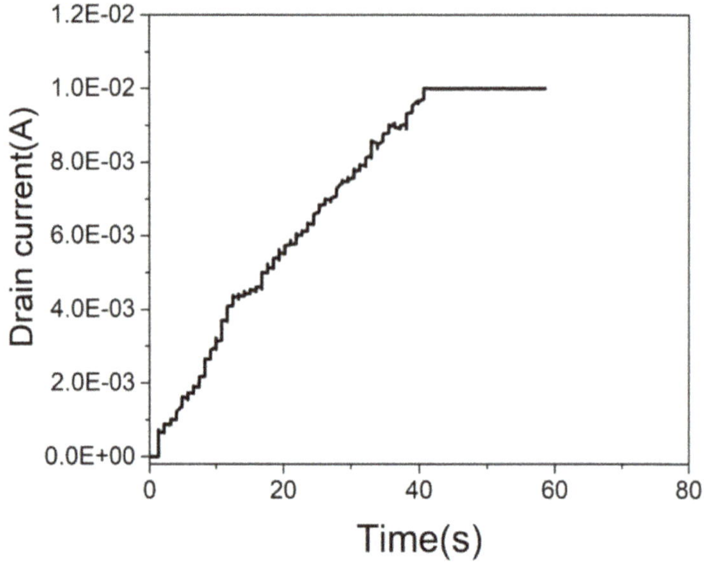

Figure 14. The drain currents of the device at V_{ds} = 50 V for Ge ions at LET = 28.5 MeV·cm^2/mg (about 0.31 pC/μm).

After heavy-ion irradiation, the drain current and gate current were measured (V_{gs} = 0 V), as shown in Figure 15. The drain leakage current increased to approximately 5 mA and the gate leakage current was less than 5 nA at V_{ds} = 20 V, V_{gs} = 0 V. It indicated that a leakage channel was formed between the source port and drain port in the Cascode AlGaN/GaN HEMTs. Analyzing from Figure 1, the source port and drain port of the Cascode AlGaN/GaN HEMTs corresponded to the gate electrode and the drain electrode of the depletion-mode AlGaN/GaN HEMT. Therefore, a leakage channel was formed between the gate electrode and the drain electrode in the depletion-mode AlGaN/GaN HEMT.

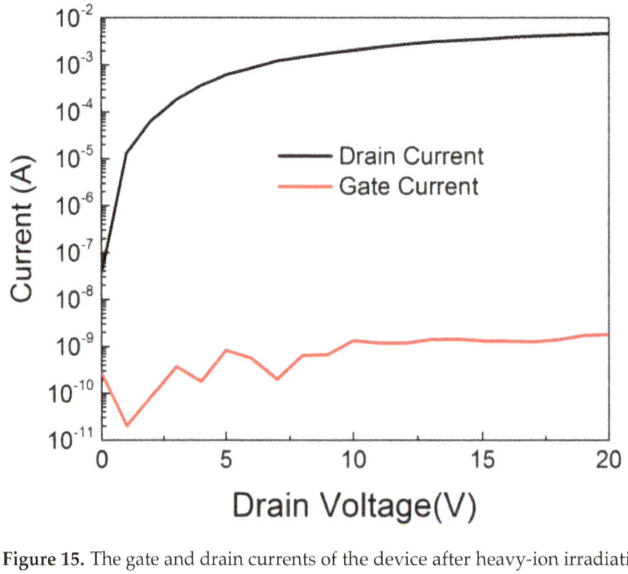

Figure 15. The gate and drain currents of the device after heavy-ion irradiation.

The picture of the DUT before and after heavy-ion irradiation is shown in Figure 16. Figure 16a shows the structure of the cascode AlGaN/GaN HEMT under test and Figure 16b shows a picture of the device after heavy-ion irradiation. The left device was the low voltage enhancement-mode Si MOSFET and the right device was the high voltage depletion-mode AlGaN/GaN HEMT. As shown in Figure 16b, the depletion-mode AlGaN/GaN HEMT was burnt out under V_{ds} = 50 V for Ge ions at LET = 28.5 MeV·cm^2/mg (about 0.31 pC/μm). The Si MOSFET was not sensitive to SEB. Moreover, it was found that the burned area of the depletion-mode AlGaN/GaN HEMT was located at the gate electrode of the device, using layout photographing technology. A burnout area between the drain and gate was found on the surface of depletion-mode AlGaN/GaN HEMT from the heavy-ion experimental results, as shown in Figure 16b, which indicated the sensitive region was between the drain electrode and gate electrode.

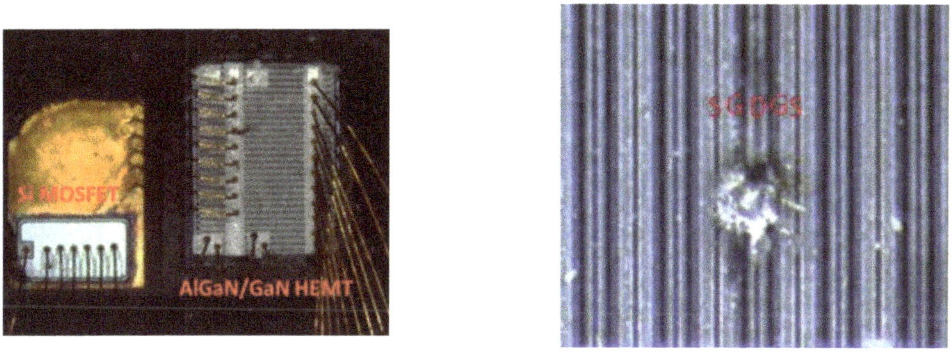

(a) (b)

Figure 16. (**a**) The cascode AlGaN/GaN HEMT under test and (**b**) the picture of the device after heavy-ion irradiation.

Analyzing from the simulation results, the sensitive region of the depletion-mode AlGaN/GaN HEMT was located near the drain. As shown in Figure 16b, the sensitive region was between the drain electrode and gate electrode. The results of the heavy-ion experiment proved reasonable of the SEB simulation for AlGaN/GaN HEMT with the cascode structure. In a word, the simulation and heavy-ion experimental results both indicated that a leakage channel between the gate electrode and the drain electrode in depletion-mode AlGaN/GaN HEMT was formed after the heavy-ion striking. This may be the main failure mechanism leading to the SEB effect.

4. Conclusions

Single event effects on cascode AlGaN/GaN HEMT were studied in this paper. To figure out the mechanism of single event effects, the simulation and experiment were carried out. The simulation results showed that the heavy-ion energy and track position influenced the SEB effect of the depletion-mode AlGaN/GaN HEMT. When the heavy-ion passed through the GaN channel layer and reached the GaN buffer layer, the heavy-ion track length had little effect on the SEE of the depletion-mode AlGaN/GaN HEMT. The sensitive region of the depletion-mode AlGaN/GaN HEMT was located near the drain. When the depletion-mode AlGaN/GaN HEMT was biased at $V_{gs} = -10$ V and V_{ds} = 390 V, heavy-ion (LET = 0.1 pC/μm) struck near the drain and caused a leakage current between the gate electrode and drain electrode in the depletion-mode AlGaN/GaN HEMT. Moreover, the heavy-ion irradiation experiments of a commercial cascode AlGaN/GaN HEMT (TP90H180PS) were carried out. Experimental results showed that the drain current of the device increased to 10 mA (limiting by the instrument), which was about 45 s at LET = 28.5 MeV·cm^2/mg (about 0.31 pC/μm) with V_{ds} = 50 V. Analyzing from the cascode structure, the leakage channel was formed between the gate electrode and the drain electrode in the depletion-mode AlGaN/GaN HEMT. The picture of the DUT after heavy-ion irradiation indicated that the burning area was between the gate electrode and the drain electrode of the depletion-mode AlGaN/GaN HEMT. The results of the heavy-ion experiment proved reasonable for the SEB simulation of AlGaN/GaN HEMT with cascode structure. The enhancement of the impact ionization at the edge of the gate was revealed using simulation and it might be an important factor for SEB effect. It was very useful for the radiation-hardened technique for GaN HEMT design.

Author Contributions: Conceptualization, Y.L.; data curation, Y.L.; formal analysis, X.W.; investigation, Q.C. and H.Y.; methodology, R.C.; project administration, R.C. and J.H.; writing—original draft, Y.L.; writing—review and editing, Y.L. All authors have read and agreed to the published version of the manuscript.

Funding: This work was supported in part by the Fast Support Project (Grant No. E11Z360101) and the Beijing Municipal Commission of Science and Technology (Grant No. E039360101).

Conflicts of Interest: The authors declare no conflict of interest.

References

1. Efthymiou, L.; Longobardi, G.; Camuso, G.; Chien, T.; Chen, M.; Udrea, F. On the physical operation and optimization of the p-GaN gate in normally-off GaN HEMT devices. *Appl. Phys. Lett.* **2017**, *110*, 123502. [CrossRef]
2. Hwang, I.; Choi, H.; Lee, J.; Choi, H.; Kim, J.; Ha, J.; Um, C.-H.; Hwang, S.-K.; Oh, J.; Kim, J.-Y.; et al. 1.6 kV, 2.9 mΩ cm2 normally-off p-GaN HEMT device. In Proceedings of the International Symposium on Power Semiconductor Devices and ICs, Bruges, Belgium, 3–7 June 2012; pp. 41–44.
3. Sun, W.-W.; Zheng, X.-F.; Fan, S.; Wang, C.; Du, M.; Zhang, K.; Chen, W.-W.; Cao, Y.-R.; Mao, W.; Ma, X.-H. Degradation mechanism of enhancement-mode AlGaN/GaN HEMTs using fluorine ion implantation under the on-state gate overdrive stress. *Chin. Phys. B* **2015**, *24*, 017303. [CrossRef]
4. Chen, H.; Wang, M.; Chen, K.J. Self-aligned enhancement-mode AlGaN/GaN HEMTs using 25 keV fluorine ion implantation. In Proceedings of the 68th Device Research Conference, South Bend, IN, USA, 21–23 June 2010.
5. Yatabe, Z.; Hori, Y.; Ma, W.-C.; Asubar, J.T.; Akazawa, M.; Sato, T.; Hashizume, T. Characterization of electronic states at insulator/(Al)GaN interfaces for improved insulated gate and surface passivation structures of GaN-based transistors. *Jpn. J. Appl. Phys.* **2014**, *53*, 100213. [CrossRef]
6. Zhao, Q.-T. Fabrication and Characterization of Enhancement-Mode High-κ LaLuO3-AlGaN/GaN MIS-HEMTs. *IEEE Trans. Electron Devices* **2013**, *60*, 3040–3046.
7. Saito, W.; Saito, Y.; Fujimoto, H.; Yoshioka, A.; Ohno, T.; Naka, T.; Sugiyama, T. Switching controllability of high voltage GaN-HEMTs and the cascode connection. In Proceedings of the 24th International Symposium on Power Semiconductor Devices and ICs, Bruges, Belgium, 3–7 June 2012; pp. 229–232.
8. Liu, Z.; Huang, X.; Lee, F.C.; Li, Q. Simulation model development and verification for high voltage GaN HEMT in cascode structure. In Proceedings of the 2013 IEEE Energy Conversion Congress and Exposition, Denver, CO, USA, 15–19 September 2013.
9. Sun, X.; Saadat, O.I.; Chen, J.; Zhang, E.X.; Cui, S.; Palacios, T.; Fleetwood, D.M.; Ma, T.P. Total-Ionizing-Dose Radiation Effects in AlGaN/GaN HEMTs and MOS-HEMTs. *IEEE Trans. Nucl. Sci.* **2013**, *60*, 4074–4079. [CrossRef]
10. Jiang, R.; Zhang, E.X.; McCurdy, M.W.; Wang, P.; Gong, H.; Yan, D.; Schrimpf, R.D.; Fleetwood, D.M. Dose-Rate Dependence of the Total-Ionizing-Dose Response of GaN-Based HEMTs. *IEEE Trans. Nucl. Sci.* **2019**, *66*, 170–176. [CrossRef]
11. Aktas, O.; Kuliev, A.; Kumar, V.; Schwindt, R.; Toshkov, S.; Costescu, D.; Stubbins, J.; Adesida, I. 60Co gamma radiation effects on DC, RF, and pulsed I–V characteristics of AlGaN/GaN HEMTs. *Solid State Electron.* **2004**, *48*, 471–475. [CrossRef]
12. Stocco, A.; Gerardin, S.; Bisi, D.; Dalcanale, S.; Rampazzo, F.; Meneghini, M.; Meneghesso, G.; Grünenpütt, J.; Lambert, B.; Blanck, H.; et al. Proton induced trapping effect on space compatible GaN HEMTs. *Microelectron. Reliab.* **2014**, *54*, 2213–2216. [CrossRef]
13. Kim, B.-J.; Ahn, S.; Ren, F.; Pearton, S.; Yang, G. Effects of proton irradiation and thermal annealing on off-state step-stressed AlGaN/GaN high electron mobility transistors. *J. Vac. Sci. Technol. B* **2016**, *34*, 041231. [CrossRef]
14. Ives, N.E.; Chen, J.; Witulski, A.F.; Schrimpf, R.D.; Fleetwood, D.M.; Bruce, R.W.; McCurdy, M.W.; Zhang, E.X.; Massengill, L.W. Effects of Proton-Induced Displacement Damage on Gallium Nitride HEMTs in RF Power Amplifier Applications. *IEEE Trans. Nucl. Sci.* **2015**, *62*, 2417–2422. [CrossRef]
15. Koehler, A.D.; Anderson, T.J.; Khachatrian, A.; Roche, N.J.-H.; Buchner, S.; Weaver, B.D.; Hobart, K.D.; Kub, F.J. (Invited) Displacement Damage and Single Event Effects in AlGaN/GaN HEMTs. *ECS Trans.* **2017**, *75*, 13–20. [CrossRef]
16. Zerarka, M.; Austin, P.; Bensoussan, A.; Morancho, F.; Durier, A. TCAD Simulation of the Single Event Effects in Normally-OFF GaN Transistors After Heavy Ion Radiation. *IEEE Trans. Nucl. Sci.* **2017**, *64*, 2242–2249. [CrossRef]
17. Zerarka, M.; Crepel, O. Radiation robustness of normally-off GaN/HEMT power transistors (COTS). *Microelectron. Reliab.* **2018**, *90*, 984–991. [CrossRef]

18. Salame, C.-T.; Hoffmann, A.; Mialhe, P.; Charles, J.P.; Kerns, D.V.; Kerns, S.E. Size effect on seb cross-section of VDMOSFETs. *Radiat. Eff. Defects Solids* **2000**, *152*, 191–200. [CrossRef]
19. Tang, Z.; Li, X.; Tan, K.; Liu, C.; Chen, X.; Fu, X. The Progress of SEB and SEGR Irradiation Hardening Technology for Power MOSFET. In Proceedings of the International Conference on Radiation Effects of Electronic Devices (ICREED), Beijing, China, 16–18 May 2018; pp. 1–4.
20. Li, Z.C.; Li, S.Y.; Liu, J.C.; Cao, Z.; Yang, S.Y. Measurements of the Cross Sections of the Single Event Burnout (SEB) for the Power MOSFET. *At. Energy Sci. Technol.* **2004**, *38*, 395–398.
21. Bulutay, C. Electron initiated impact ionization in AlGaN alloys. *Semicond. Sci. Technol.* **2002**, *17*, L59. [CrossRef]
22. Rowena, I.B.; Selvaraj, S.L.; Egawa, T. Buffer Thickness Contribution to Suppress Vertical Leakage Current With High Breakdown Field (2.3 MV/cm) for GaN on Si. *IEEE Electron Device Lett.* **2011**, *32*, 1534–1536. [CrossRef]
23. Visalli, D.; Van Hove, M.; Srivastava, P.; Derluyn, J.; Das, J.; Leys, M.; DeGroote, S.; Cheng, K.; Germain, M.; Borghs, G. Experimental and simulation study of breakdown voltage enhancement of AlGaN/GaN heterostructures by Si substrate removal. *Appl. Phys. Lett.* **2010**, *97*, 113501. [CrossRef]
24. Onoda, S.; Hasuike, A.; Nabeshima, Y.; Sasaki, H.; Yajima, K.; Sato, S.-I.; Ohshima, T. Enhanced Charge Collection by Single Ion Strike in AlGaN/GaN HEMTs. *IEEE Trans. Nucl. Sci.* **2013**, *60*, 4446–4450. [CrossRef]

electronics

MDPI

Article

On the Lifetime Estimation of SiC Power MOSFETs for Motor Drive Applications

Carmelo Barbagallo, Santi Agatino Rizzo *, Giacomo Scelba, Giuseppe Scarcella and Mario Cacciato

Department of Electrical, Electronics and Computer Engineering, University of Catania, 95129 Catania, Italy; carmelo.barbagallo@unict.it (C.B.); giacomo.scelba@unict.it (G.S.); giuseppe.scarcella@unict.it (G.S.); mario.cacciato@unict.it (M.C.)

* Correspondence: santi.rizzo@unict.it; Tel.: +39-095-738-2308

Abstract: This work presents a step-by-step procedure to estimate the lifetime of discrete SiC power MOSFETs equipping three-phase inverters of electric drives. The stress of each power device when it is subjected to thermal jumps from a few degrees up to about 80 °C was analyzed, starting from the computation of the average power losses and the commitment of the electric drive. A customizable mission profile was considered where, by accounting the working conditions of the drive, the corresponding average power losses and junction temperatures of the SiC MOSFETs composing the inverter can be computed. The tool exploits the Coffin–Manson theory, rainflow counting, and Miner's rule for the lifetime estimation of the semiconductor power devices. Different operating scenarios were investigated, underlying their impact on the lifetime of SiC MOSFETs devices. The lifetime estimation procedure was realized with the main goal of keeping limited computational efforts, while providing an effective evaluation of the thermal effects. The method enables us to set up any generic mission profile from the electric drive model. This gives us the possibility to compare several operating scenario of the drive and predict the worse operating conditions for power devices. Finally, although the lifetime estimation tool was applied to SiC power MOSFET devices for a general-purpose application, it can be extended to any type of power switch technology.

Keywords: AC motor drive; junction temperature; lifetime prediction; power MOSFET; loss modeling; reliability; SiC MOSFET

Citation: Barbagallo, C.; Rizzo, S.A.; Scelba, G.; Scarcella, G.; Cacciato, M. On the Lifetime Estimation of SiC Power MOSFETs for Motor Drive Applications. *Electronics* **2021**, *10*, 324. https://doi.org/10.3390/electronics10030324

Academic Editor: Elio Chiodo

Received: 31 December 2020

Accepted: 25 January 2021

Published: 30 January 2021

1. Introduction

The remarkable properties of silicon carbide (SiC) have made it a perfect candidate for replacing silicon-based power electronic devices in high power, high-temperature applications. In fact, SiC MOSFET technology provides excellent performance to MOSFET power devices in terms of low on-state resistance, high switching frequency, high breakdown voltage, high current capability also at very high temperature. Therefore, this technology represents a valid alternative to typical Si MOSFET and IGBT power devices for many applications [1]. The prospects for strong growth in SiC devices are high and are even stimulated by the increasing sales of plug-in hybrid and electric vehicles. Compared to the well mature Si technology, the field reliability of SiC devices must be demonstrated for various applications, and a voltage-derating design guideline needs to be established. This is especially important for applications in which reliability is extremely critical, such as the automotive and aerospace applications. Hence, it becomes imperative to apply in-depth studies on the SiC devices performance and operating limits, especially when they are used in very critical applications, for instance when they are integrated into three-phase traction inverters powering electric motors. In fact, because power conversion systems equipped with a SiC device potentially provide higher current density than one with Si devices, which leads to a larger thermal ripple, they need more stringent requirements for the package materials.

Studies on the state-of-the-art SiC MOSFET's reliability evaluation and failure mode analysis were carried out in [2–4]; these studies pointed out the evolution and improvements as well as the future challenges of this promising device technology. The electro-thermal co-simulation approach based on a PSpice-based model, including temperature dependency and a Simulink-based thermal network, was even proposed in [5] for SiC MOSFETs. Some model quantities have been obtained by FEM simulation for more accurate results, and a MATLAB script has been used to manage and interface the data from the different simulation tools. An accelerated power cycling test platform using a current source converter for SiC-MOSFET power modules was also presented in [6], where the junction temperature variations of the devices were monitored without the removal of silicone gel. Moreover, the analysis was used to examine some failure precursors and then to estimate the useful lifetime of SiC MOSFET modules. A comparison in the area of device reliability accounting for condition monitoring and active thermal control as well as the lifetime was recently carried out [7]. A method to obtain the thermal impedance of a SiC module by combining optical measurement and multi-physics simulations was proposed in [8], where the measurement of the junction temperature was performed by using fiber optic instead of temperature-sensitive electrical parameters. Several major achievements and novel architectures in SiC modules packaging were analyzed in [9], where the authors reported an accurate survey of the materials by considering their coefficient of thermal expansion and their proper combination to reduce the thermal stress in the material interfaces. The impact of different pulse width modulation control techniques on the power losses and thermal stress on SiC power modules used in a three-phase inverter was investigated in [10]. The advantages and problems due to the use of SiC MOSFET in a traction inverter were discussed in [11] to provide the guidelines for a viable solution. Electrical and thermal issues, safety and reliability problems, and challenges due to device paralleling and layout were analyzed by exploiting on-field experience in the industry sector, i.e., experimental tests, finite element analysis, and circuit simulations.

To reduce the total design and maintenance cost and to guarantee the service continuity as well as the human safety, it is also required to have an accurate prediction of the remaining lifetime prediction of power converters, which can help to prevent unwanted failures and generate better maintenance plans. While the reliability and lifetime prediction of silicon (Si) semiconductor device based power converters have been widely investigated in the literature [12–14], SiC MOSFETs are facing new reliability challenges. Hence, the design of more reliable SiC power converters requires an accurate lifetime prediction as well as online monitoring strategies for real-time lifetime prediction [15]. Different approaches can be applied to estimate the lifetime of the SiC power devices and power converters [16].

In the context discussed so far, this work presents a lifetime prediction method of SiC power MOSFETs integrated into three-phase inverters supplying a three-phase induction machine. The analysis exploits a suitable developed simulation tool realized in MATLAB and Simulink to keep the computational burden low, and it also provides a modular structure of the proposed procedure. The electric drive, the Coffin–Manson relation, the rainflow counting method, and Miner's rule are suitable for the analysis and are combined to calculate the lifetime prediction [17–28]. As it is simple and reliable, Miner's rule is the most widely used fatigue life prediction technique in this field.

In the proposed approach, the off-line prediction of the lifetime is performed, starting from the data obtained, by exploiting suitable modeling of the electric drive, operating at the conditions provided by the mission profile (MP) of the application. In the investigated case, several tests are carried out using 650 V, 45 A SiC power MOSFETs by customizing the MPs with suitable weights based on the operating conditions. Hence, the proposed procedure allows for the prediction of the behavior of the thermal stresses affecting the power switches in a wide operating range. Moreover, the option of creating several customized MPs allows us to quickly carry out many analyses and then making a comparison among them. In the following, a step-by-step description of the developed activity is provided, along with the underlying advantages and limits of the approach.

2. Framework of the MATLAB-Simulink Tool for SiC MOSFET Lifetime Estimation

The tool provides a lifetime estimation by performing the main steps summarized in Figure 1 and listed below:

1. Simulation of the motor drive to carry out the necessary electrical quantities required to compute the power losses calculation in semiconductor power devices is described as follows: I_{on}, ma_n, $cos(\varphi)_n$, and fe_n, where the parameters I_{on}, ma_n, $cos(\varphi)_n$, and fe_n represent the maximum output current of one phase of the inverter (I_{on}) (considering a balanced three-phase system); the amplitude modulation index (ma_n); the power factor ($cos(\varphi)_n$); and the output electrical frequency (fe_n). All quantities are evaluated at steady-state conditions.
2. The temporal weight assignment is decided here, as well as the grouping of the motor drive operating conditions.
3. Power losses calculation is performed by considering the inverter driven by a space vector pulse width modulation (SVPWM) strategy. The average value of the dissipated power is computed at each switching period of the carrier signal.
4. Junction temperature estimation is performed by using a thermal Foster network, where the power losses represent the inputs of the thermal network.
5. The rainflow counting theory is applied to the junction temperature profiles obtained from the previous step to determine the number of cycles N_c associated with the n-th operating condition of the drive.
6. The number of cycles to failure N_f is calculated by applying the Coffin–Manson law.
7. Lifetime estimation is performed by exploiting Miner's rule.

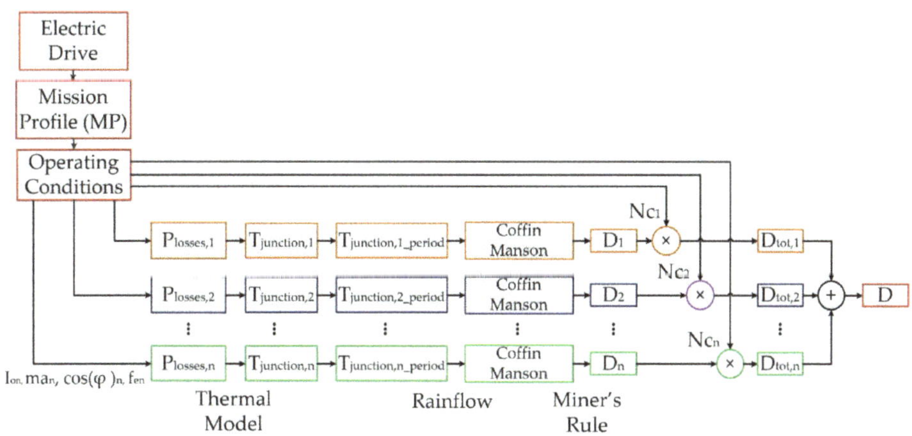

Figure 1. Workflow of the lifetime prediction starting from the electric drive system mission profile.

2.1. Generation of the Mission Profile

2.1.1. Motor Drive System

The electric drive was modelled on Simulink, which is a MATLAB-based graphical programming environment and was exploited to model the entire electric drive. The model includes a three-phase asynchronous machine whose characteristics are shown in Table 1. The motor is supplied by an ideal three-phase inverter and controlled according to an indirect field oriented control (IFOC) [16]. The mission profile was carried out through the execution of a series of simulations by setting a matrix of reference speeds and reference torques values. The results of the simulations represent the dataset establishing the operating conditions of the inverter and thus power devices, according to the considered mission profile. The Simulink model of the electric drive is shown in Figure 2. It consists of different grouped blocks: vector control IFOC, modulation block SVPWM (space vector pulse width modulation), a three-phase inverter realized with ideal power switches, and the asynchronous motor. The block "Power Factor" performs the calculation of the quantity $cos(\varphi(t))$, while the block "To Workspace" allows the average values of the modulation index $ma(t)$, the power factor $cos(\varphi)(t)$, electric frequency $f_e(t)$ and the maximum current of the phase a, $i_{as}(t)$ to be available, thus allowing us to extract the following quantities:

- $I_0 \rightarrow$ maximum value of the generic motor phase current;
- $ma_{avg} \rightarrow$ average value of the amplitude modulation index at the steady-state;
- $cos(\varphi)_{avg} \rightarrow$ average value of the power factor at steady-state;
- $f_{e\_avg} \rightarrow$ average value of the fundamental frequency at steady-state.

Table 1. Technical specification of the induction motor.

Parameters	Symbols	Value	u.m.
Pole pairs	P	2	/
Nominal power	A_n	15	kVA
Nominal voltage	V_n	400	V
Nominal frequency	$f_{e,n}$	50	Hz
Nominal speed	n_n	1460	rpm
Stator resistance	R_s	214.7	mΩ
Stator leakage inductance	L_{ls}	991	µH
Equivalent rotor resistance	R'_r	220.5	mΩ
Equivalent rotor leakage inductance	L'_{lr}	991	µH
Mutual inductance	L_m	64.19	mH
Inertia coefficient	J	0.102	kg·m^2
Friction coefficient	F	0.009541	N·m·s

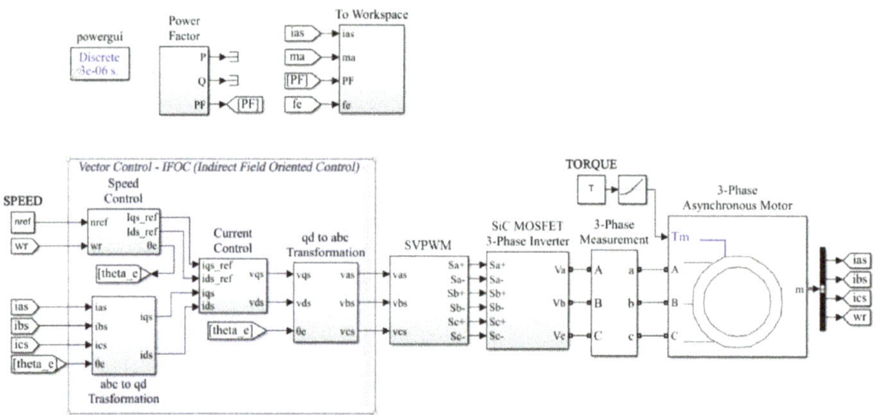

Figure 2. Simulink model of the electric drive system.

The simulations were carried out by using a sampling time $T_s = 3\times10^{-6}$ s, an input dc voltage of the inverter $V_{dc} = 560$ V, and a switching frequency of the power switches equal to $f_{sw} = 10$ kHz.

Figure 3 shows some of the results carried out from these simulations. In particular, it displays the waveforms of the above-listed electrical quantities when the reference speed and reference torque of the drive are 1400 rpm and 80 Nm, respectively. The blue traces represent the instantaneous values, while the orange traces represent the steady-state values. A campaign of simulations in which the speed is varied from 0 to 1400 rpm and the torque from 0 to 80 Nm was executed, allowing us to identify the variation ranges of the electrical quantities, which are summarized in the surfaces shown in Figure 4. These surfaces represent the input dataset required in the next step.

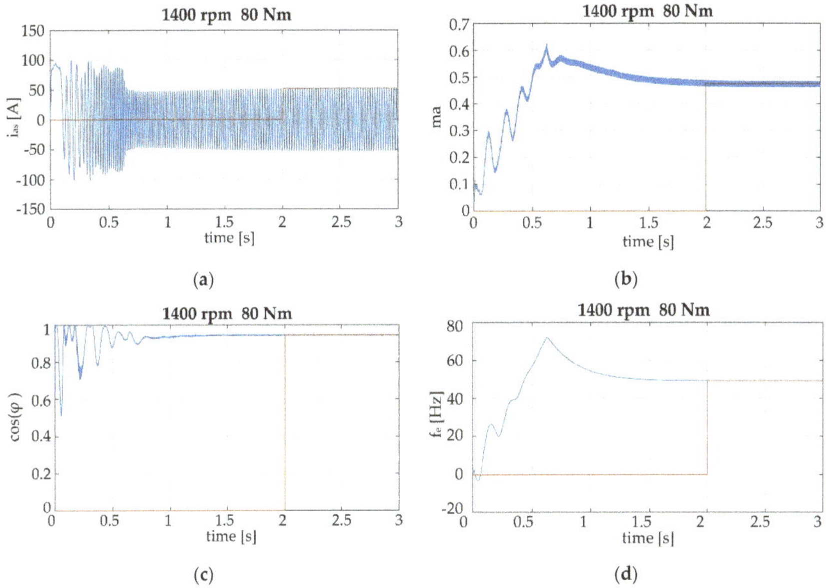

Figure 3. Results of the electric drive model operated at 1400 rpm and 80 Nm: (**a**) motor phase current a; (**b**) amplitude modulation index; (**c**) power factor; (**d**) output frequency.

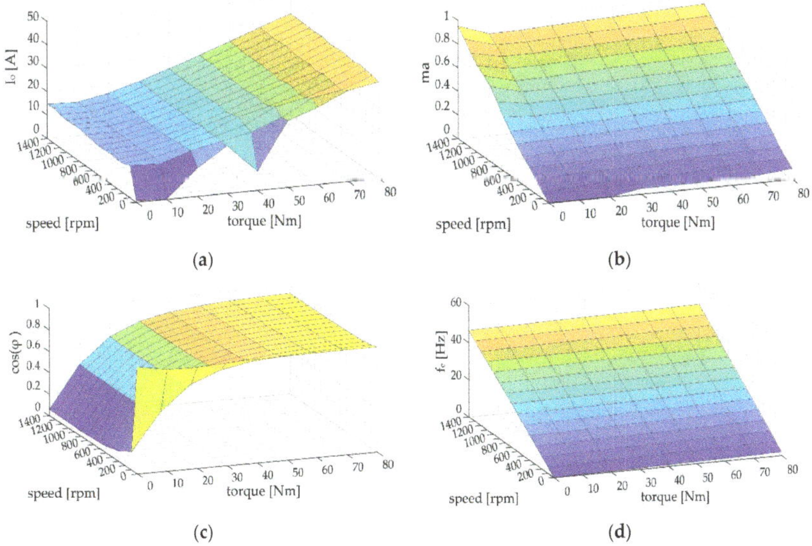

Figure 4. Results of the electric drive model as a function of the reference speed and electromagnetic torque: (**a**) maximum current; (**b**) amplitude modulation index; (**c**) power factor; (**d**) frequency.

2.1.2. Weights Assignment

For each operating condition, expressed in terms of speed and torque, a unique set of current, modulation index, power factor, and frequency values were obtained. A customizable MP, i.e., related to a specific use of the electric drive, can be obtained by assigning a time interval to each operating condition, according to a preset probability of occurrence. In this study, the sum of gaussian distributions, given by Equation (1), was used to achieve a customized time interval weights distribution, and was normalized so that their sum is equal to 100%.

$$f(x,y) = \sum_{i=1}^{n} A_i \cdot exp\left[-\left(\frac{(x - x_{0,i})^2}{2\cdot\sigma_{x,i}^2} + \frac{(y - y_{0,i})^2}{2\cdot\sigma_{x,i}^2}\right)\right] \quad (1)$$

where:

- $A_i \rightarrow$ peak amplitude of the i-th Gaussian distribution;
- $x_{0,i}$ and $y_{0,i} \rightarrow$ central values of x-axis and y-axis of the i-th Gaussian distribution;
- $\sigma_{x,i}$ and $\sigma_{y,i} \rightarrow$ standard deviation of x-axis and y-axis of the i-th distribution;
- x and $y \rightarrow$ torque and speed input vectors, respectively.

The above expression was applied to the dataset carried out in the previous step, by substituting the values of speed and torque to the quantities x and y respectively, while $x_{0,i}$ and $y_{0,i}$ are the central values of each Gaussian distribution. In practice, Equation (1) provides the sum of Gaussian surfaces, which establishes the weight that has to be assigned to each operating condition of the drive; in the proposed procedure, these surfaces were generated by using a MATLAB algorithm. Figure 5 shows two examples of distribution when the index $i = 1$ (single Gaussian distribution) and $i > 1$ (multiple Gaussian distributions).

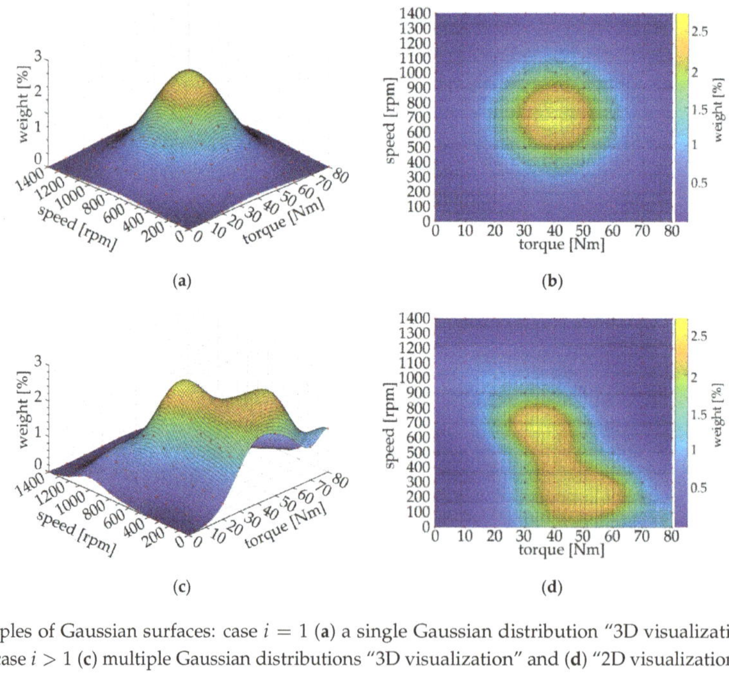

Figure 5. Examples of Gaussian surfaces: case $i = 1$ (**a**) a single Gaussian distribution "3D visualization" and (**b**) "2D visualization"; case $i > 1$ (**c**) multiple Gaussian distributions "3D visualization" and (**d**) "2D visualization".

2.1.3. Grouping of the Motor Drive Operating Conditions

After having assigned the weights to each operating condition of the drive according to the MP, a grouping process is required to group similar operating conditions, thus reducing the computational burden required to the entire lifetime estimation procedure. The grouping is performed by initially splitting in a certain number of intervals the values of I_0, ma_{avg}, $cos(\varphi)_{avg}$, fe_avg, defining the minimum, maximum, step, and central value of each interval, according to Table 2. Then, the values of the electrical quantities extracted from the simulations of the electric drive are inserted in the corresponding intervals. A suitable number of intervals have to be chosen to guarantee a good compromise between the representative conditions of the drive and the computational efforts needed to compute the lifetime of the SiC power device [22,23].

Table 2. Grouping of the electric drive operating conditions.

	I_o [A]	ma	cos(φ)	f_e [Hz]
Min	0	0	0.05	0.0
Max	43.1	0.95	0.99	48.7
Step	10	0.2	0.2	10
Central value of each interval	5-15-25-35-45	0.1-0.3-0.5-0.7-0.9	0.1-0.3-0.5-0.7-0.9	5-15-25-35-45

A MATLAB function was implemented to perform the grouping procedure by assigning each working condition of the drive to a specific set data, as seen in Figure 6. All empty intervals were discarded, and they are not included in the calculation of power losses. Obviously, such an approach can be adopted even when there are data from a real MP. Even in case of handling a huge number of operating condition datasets, this mechanism enables a significant reduction of operating conditions for which the power losses have to be calculated, thus reducing the lifetime prediction time.

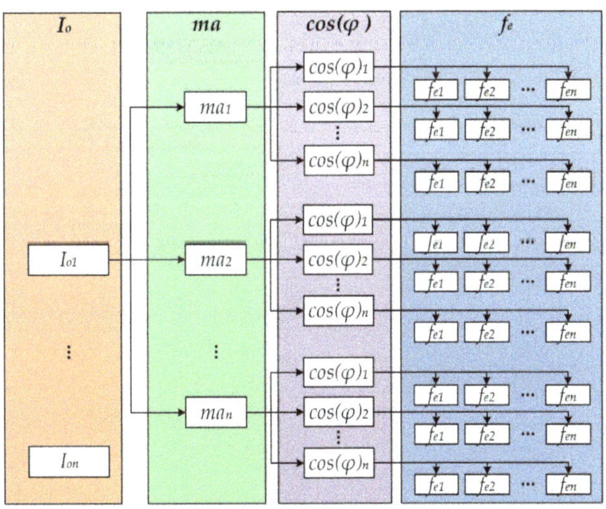

Figure 6. Grouping procedure where each motor drive operation is included in a specific subset.

2.2. *SiC MOSFET Power Losses Computation*

The power losses of each SiC device composing the inverter are carried out from the dataset obtained in the grouping procedure. In this analysis, the lifetime estimation is based on solder degradation, strictly correlated to the ΔT_j variations of junction temperatures in the fundamental period of the stator voltages; therefore, the calculation of the power losses was implemented by considering the average value of the power, over each switching period T_{sw}. The following equations show the method used for the calculation [22–25]:

$$DC^n = \left\{ \begin{array}{l} DC^n_{MOS} = 0.5 \cdot [1 + m_a \cdot sin(2\pi f_e \cdot nT_{sw})] \;\; se \; 0 \leq nT_{sw} < \frac{T_e}{2} \\ DC^n_{SBD} = 0.5 \cdot [1 + m_a \cdot sin(2\pi f_e \cdot nT_{sw})] \;\; se \; \frac{T_e}{2} \leq nT_{sw} < T_e \end{array} \right\} \tag{2}$$

$$DC^n = \left\{ \begin{array}{l} DC^n_{MOS} = 0.5 \cdot [1 + m_a \cdot sin(2\pi f_e \cdot nT_{sw})] \;\; se \; 0 \leq nT_{sw} < \frac{T_e}{2} \\ DC^n_{SBD} = 0.5 \cdot [1 + m_a \cdot sin(2\pi f_e \cdot nT_{sw})] \;\; se \; \frac{T_e}{2} \leq nT_{sw} < T_e \end{array} \right\} \tag{3}$$

$$I^n_o = I_o \cdot \sin(2\pi f_e \cdot nT_{sw} - \varphi) \tag{4}$$

where:

- DC^n_{MOS} , DC^n_{SBD} → duty cycles of MOSFET and diode in the n-th switching period;
- $V^n_{on,MOS}(I^n_o)$, $V^n_{on,SBD}(I^n_o)$ → voltage drops on MOSFET and diode in the n-th switching period;
- $E_{sw,MOS}(I^n_o)$, $E_{sw,SBD}(I^n_o)$ → switching energy losses on MOSFET and diode in the n-th;
- f_{sw} and T_{sw} → switching frequency and switching period;
- f_e and T_e → fundamental frequency and period of the electrical quantities;
- I_o and φ → peak amplitude of the current and phase shift angle.

The voltage drops were carried out from the output characteristics of the SiC devices contained in the datasheets, as well as the switching energy loss and the reverse conduction characteristic. The characteristics of the SiC device under test are shown in Figure 7.

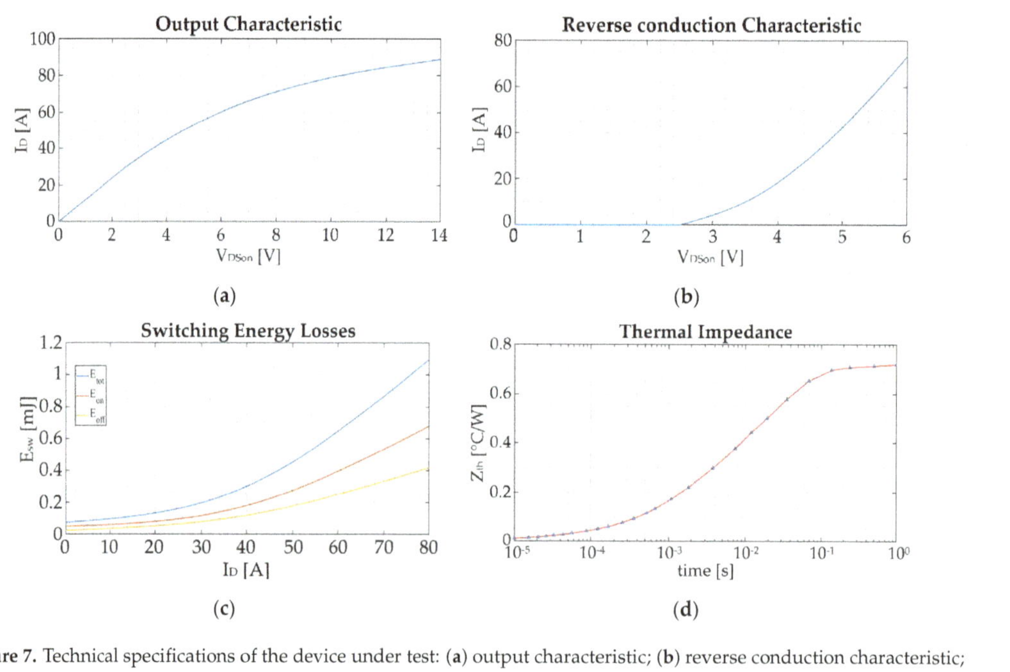

Figure 7. Technical specifications of the device under test: (**a**) output characteristic; (**b**) reverse conduction characteristic; (**c**) switching energy losses; (**d**) thermal impedance.

Figure 8 shows the power losses curve of a SiC device in a fundamental period, given for different operating conditions.

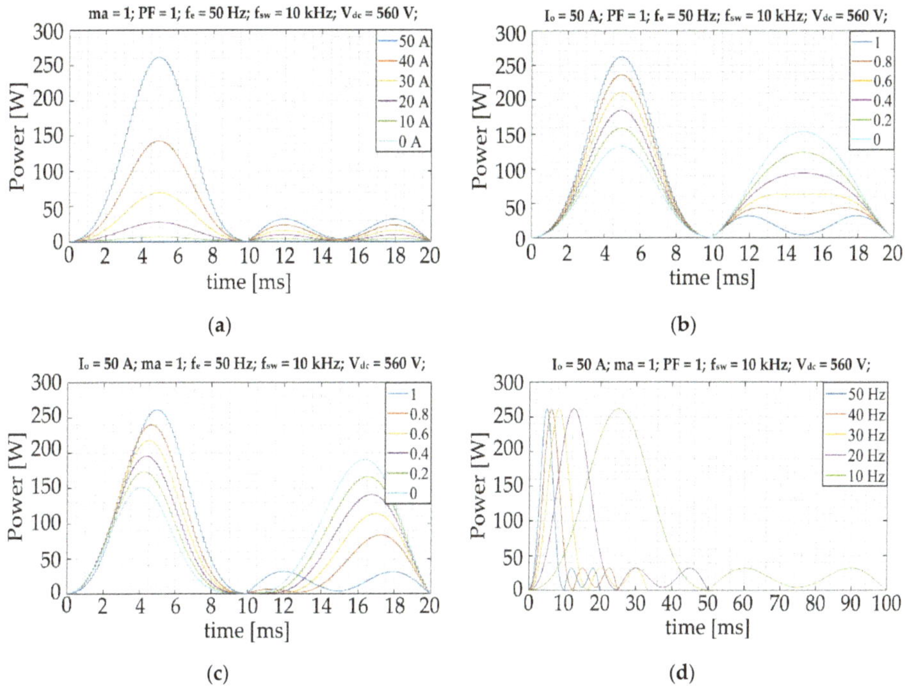

Figure 8. Power losses estimated for different operating conditions: (**a**) different currents; (**b**) different modulation indexes; (**c**) different power factors; (**d**) different fundamental frequencies.

2.3. Junction Temperature Estimation

The junction temperatures of the SiC Mosfets under test were estimated by exploiting a linear thermal model, whose generalized mathematical representation is given by [26]:

$$\begin{bmatrix} T_j^1(t) \\ T_j^2(t) \\ T_j^3(t) \\ \vdots \\ T_j^n(t) \end{bmatrix} = \begin{bmatrix} Z_{th}^{11}(t) & Z_{th}^{21}(t) & \cdots & Z_{th}^{n1}(t) \\ Z_{th}^{12}(t) & Z_{th}^{22}(t) & \cdots & Z_{th}^{n2}(t) \\ Z_{th}^{13}(t) & Z_{th}^{23}(t) & \cdots & Z_{th}^{n3}(t) \\ \cdots & \cdots & \cdots & \cdots \\ Z_{th}^{1n}(t) & Z_{th}^{2n}(t) & \cdots & Z_{th}^{nn}(t) \end{bmatrix} \begin{bmatrix} P_1(t) \\ P_2(t) \\ P_3(t) \\ \vdots \\ P_n(t) \end{bmatrix} + [T_a(t)] \tag{5}$$

where:

- $T_j(t) \rightarrow$ junction temperature;
- $Z_{th}(t) \rightarrow$ thermal impedance (see Figure 7d);
- $P(t) \rightarrow$ dissipated power;
- $T_a \rightarrow$ ambient temperature.

Given that in this analysis we are considering discrete SiC power devices, the thermal coupling with other devices can be assumed to be negligible, and thus the terms of mutual coupling can be nullified. Hence, Equation (5) can be rewritten as follows:

$$T_j\,(t) = [Z_{th}(t) \cdot P(t)] + T_a(t) \tag{6}$$

According to Equation (6), the computation of the junction temperature $T_j(t)$ requires the knowledge of the thermal impedances $Z_{th}(t)$. The last is graphically provided in the datasheet of the power switch, but a circuital representation of $Z_{th}(t)$ is required to easily simulate variable power losses profiles. This goal is reached by combining a curve-fitting procedure followed by the implementation of an equivalent Foster thermal network, as intently described in the following step.

2.3.1. Curve Fitting

Starting from the thermal impedance profile $Z_{th}(t)$, it is possible to implement a curve-fitting procedure on it by using Equation (7), where it is assumed to achieve a profile of $Z_{th}(t)$ with a Foster thermal network, whose parameters $R_{th,n}$ and $C_{th,n}$ constitute the n-th pole of the network [26,27].

$$\begin{cases} Z_{th}(t) = \sum_{n=1}^{N_p} R_{th,n} \cdot \left(1 - e^{-\frac{t}{\tau_n}}\right) \\ \tau_n = R_{th,n} \cdot C_{th,n} \end{cases} \quad (7)$$

where:

- $R_{th,n} \rightarrow$ resistance value of the n-th pole;
- $C_{th} \rightarrow$ capacitance value of the n-th pole;
- $\tau_n \rightarrow$ dissipated power;
- $T_a \rightarrow$ ambient temperature;
- $N_p \rightarrow$ number of poles.

It is important to underline the choice of the number of poles. A low number of poles does not guarantee the necessary accuracy, while a high number of poles will certainly be more accurate but results in an increase of the computational efforts. By applying the curve fitting algorithm to the SiC devices under tests, the best compromise has been obtained with seven poles, as shown in Figure 9.

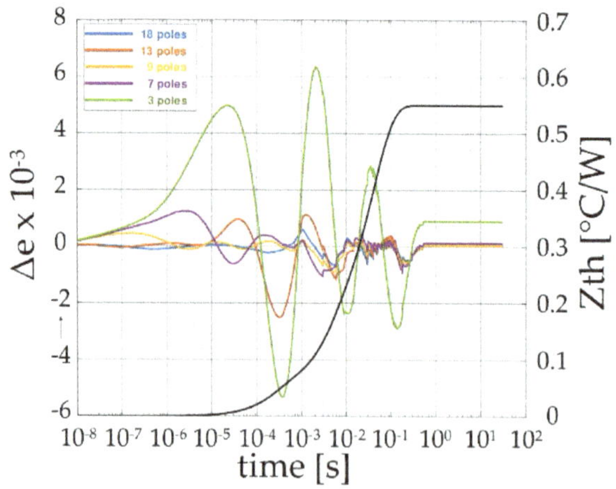

Figure 9. Comparison of curve fitting using a different number of poles.

2.3.2. Foster Network

The Foster network can be realized starting from the $R_{th,n}$ and $C_{th,n}$ parameters obtained from the previous step, reproducing the thermal behavior of the device under test. Figure 10 displays the Foster network realized to emulate the $Z_{th}(t)$ of Figure 7d.

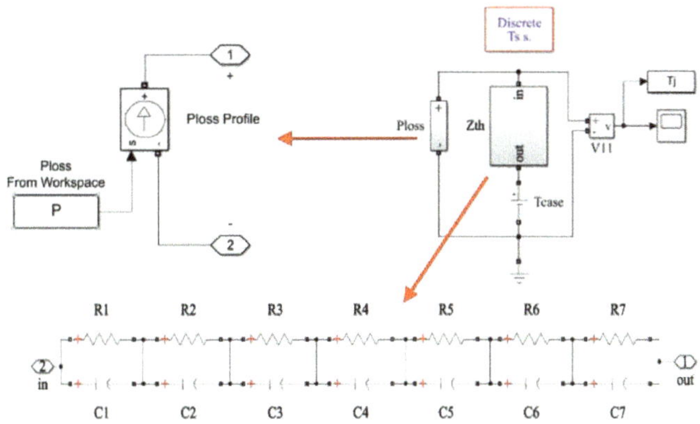

Figure 10. Simulink model of the Foster network.

2.3.3. Examples of Junction Temperature Profiles

Figure 11 displays the temperature trends of the die junction temperature for different motor drive operating conditions. It should be noted that a significant temperature variation can be experienced during the fundamental period of the phase motor current, leading to high thermal stresses. Moreover, different power losses distributions can be appreciated at varying of the load conditions, which is consistent with the theoretical analysis.

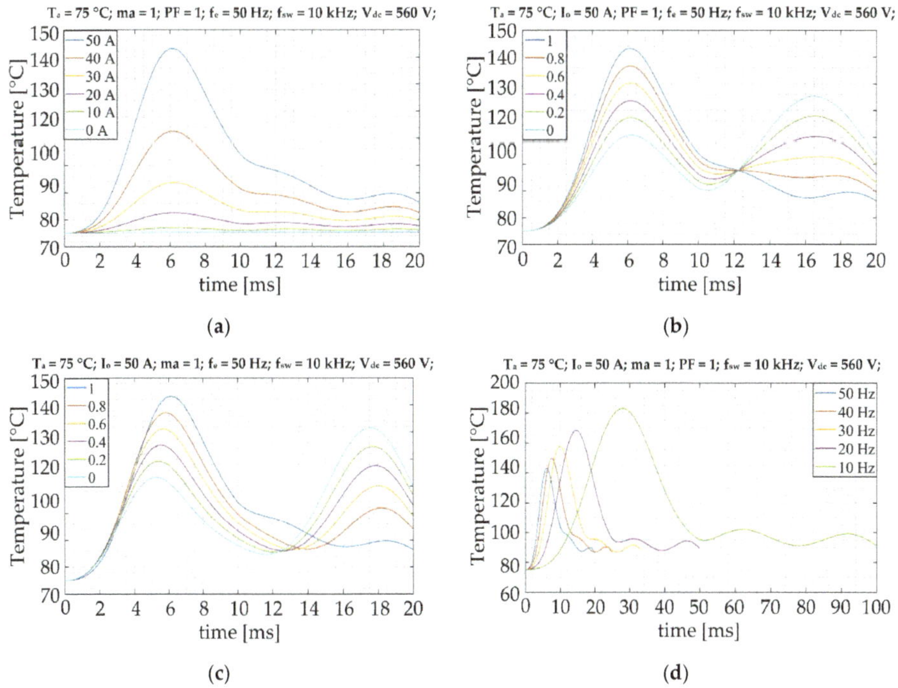

Figure 11. Temperature trends carried out for: (**a**) different currents; (**b**) different modulation indexes; (**c**) different power factors; (**d**) different fundamental frequencies.

2.4. Rainflow Counting

The rainflow-counting (or cycle counting) algorithm is used to count the number of cycles present in a generic waveform. It is a statistical method for the analysis of the random load process, and its counting principle is carried out based on the stress-deformation behavior of the material, as shown in Figure 12. It combines load reversals by defining hysteresis cycles. Each of them has a range of deformation and average stress that can be compared with the constant amplitude.

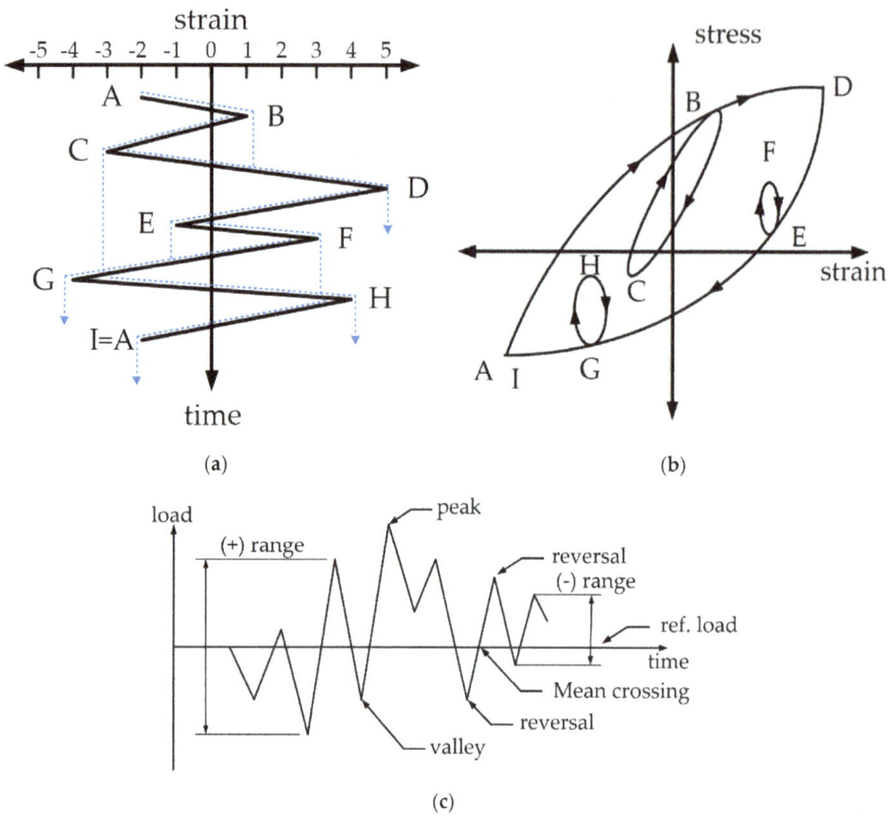

Figure 12. Graphical representation of the rainflow method: (**a**) rainflow on a load profile; (**b**) stress–strain chart; (**c**) main parameters of the rainflow.

The rules for the rainflow counting method are the following: Let X be the range taken into consideration; let Y be the previous range, adjacent to X; let S be the starting point in the history of the load profile [17,18,28]:

1. Read the next peak or valley. If the data are not available, go to step 6.
2. If there are less than three peaks and/or valleys (i.e., A, B, C of Figure 12), go to step 1. Form the X and Y ranges using the three most recent peaks and valleys that have not been discarded.
3. Compare the absolute values of the X and Y ranges:
    a. If $X < Y$, go to step 1.
    b. If $X \geq Y$, go to step 4.
4. If range Y contains starting point S, go to step 5. Otherwise, count range Y as a cycle, discard the peak and valley of Y and go to step 2.

5. Count the Y interval as a half cycle, discard the first point (peak or valley) of the Y range, move the starting point to the second point in the Y interval, and go to step 2.
6. Count any range that was not previously counted as a half cycle.

Figure 12 summarizes the rainflow counting procedure.

The load history of Figure 12a is used to illustrate the process. Details of the cycle counting are as follows [28]:

1. S = A; Y = | A–B | ; X = | B–C | ; X > Y. Y contains S, that is, point A. Count | A–B | as one-half cycle and discard point A; S = B.
2. Y = | B–C | ; X = | C–D | ; X > Y. Y contains S, that is, point B. Count | B–C | as one-half cycle and discard point B; S = C.
3. Y = | C–D | ; X = | D–E | ; X < Y.
4. Y = | D–E | ; X = | E–F | ; X < Y.
5. Y = | E–F | ; X = | F–G | ; X > Y. Count | E-F | as one cycle and discard points E and F (note that a cycle is formed by pairing range E-F and a portion of range F-G).
6. Y = | C–D | ; X = | D–G | ; X > Y; Y contains S, that is, point C. Count | C-D | as one-half cycle and discard point C. S = D.
7. Y = | D–G | ; X = | G–H | ; X < Y.
8. Y = | G–H | ; X = | H–I | ; X < Y. End of data.
9. Count | D–G | as one-half cycle, | G–H | as one-half cycle, and | H–I | as one-half cycle.
10. End of counting (see Table 3 for a summary of the cycles counted in this example).

Table 3. Example of rainflow counting referred to Figure 12a [28].

Range	Cycle Counts	Events
1	0	/
2	0	/
3	0.5	A–B
4	1.5	B–C, G–H
5	0	/
6	0.5	H–I
7	0	/
8	1.0	C–D, G–H
9	0.5	D–G
10	0	/

After having satisfied the criterion for counting cycles N_c, the amplitude and the corresponding mean values of stress or strain for each cycle were calculated. A matrix with information on the cycle, mean value, and amplitude of stress or strain forming a cycle is thus created.

Example of Rainflow Counting

Figure 13 show an example of implementation of rainflow procedure on a temperature profile carried out in a specific drive operating condition, by applying the previous steps to the SiC MOSFETs under test; the temperature profile is first linearized to form peaks and valleys. The cycle count N_c is carried out by considering the jumps of the junction temperature ΔT_j and the average temperature T_m in each cycle.

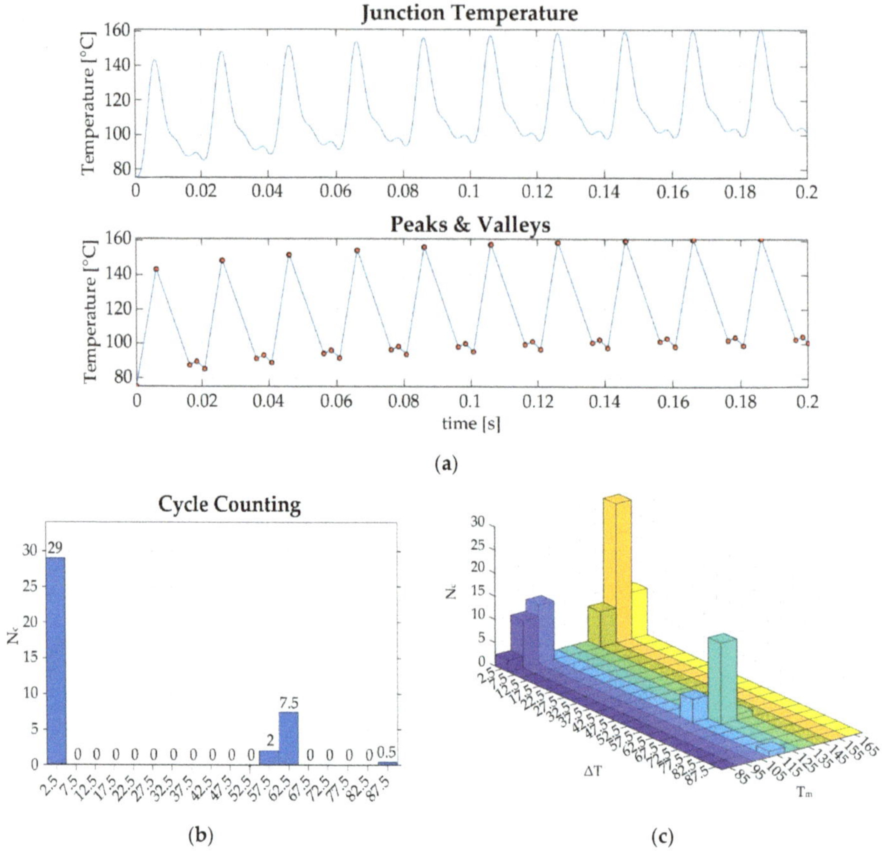

Figure 13. Example of rainflow: (**a**) linearized temperature profile in peaks and valleys; (**b**) cycle counting neglecting the average temperature; (**c**) cycle counting considering the average temperature.

2.5. *Coffin–Manson's Equation*

The main factors influencing the life of the power devices under test are the jumps in the junction temperature ΔT_j and the average junction temperature T_m; in fact, the estimate of the number of thermal cycles to failure N_f is based on these two parameters.

Equation (8) shows the correlation between the number of cycles to failure with thermal jumps and the average temperature of the cycles [17,19,20]

$$N_f = a \cdot \left(\Delta T_j\right)^{-n} \cdot e^{\frac{E_a}{k_B \cdot T_m}} \tag{8}$$

where:

- $N_f \rightarrow$ number of cycles to failure;
- $a \rightarrow$ experimental coefficient;
- $\Delta T_j \rightarrow$ jumps of junction temperature;
- $n \rightarrow$ experimental coefficient;
- $k_B \rightarrow$ Boltzmann constant;
- $E_a \rightarrow$ activation energy.

The parameters a and n of Equation (8) are determined according to the power cycling tests. Figure 14a shows a type of power cycling (PC_{sec}) in which an input stress is applied for a period $t_{on} < 15\ s$. In this case, the tests exert a stress in the interconnection areas close to the chips (bond wire, die-attach). On the contrary, Figure 14b shows a type of power cycling PC_{min} in which a stress is applied for a period $t_{on} > 15\ s$, stressing the solder layer. The resulting data of these tests give us information on the reliability of the devices under examination under specific operating conditions, and it is normally provided by the manufacturer [17,20,21].

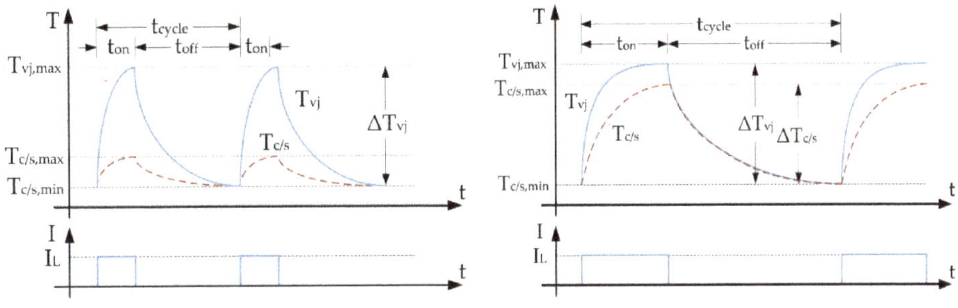

Figure 14. Types of power cycling: (**a**) fast power cycling PC_{sec}; (**b**) slow power cycling PC_{min}.

An example of Coffin–Manson curves that were calculated by considering the following parameters [20] are displayed in Figure 15.

- $a = 3.71 \times 10^{13}$
- $n = 10.122$
- $k_B = 8.617 \times 10^{-5}\ \frac{eV}{K}$
- $E_a = 0.814$ eV

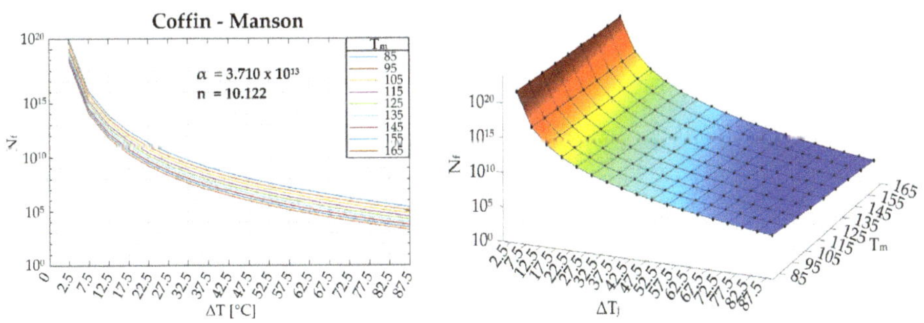

Figure 15. Coffin–Manson equation representation: (**a**) 2D curves (each one at a fixed T_m); (**b**) 3D Scheme.

2.6. Miner's Rule

The lifetime estimation is determined according to the Miner's rule, or the theory of the damage accumulation. In practice, the damage is associated with each stress condition by making the ratio between the number of thermal cycles N_c and the relative N_f associated with the same pair of ΔT_j and T_m. Extending this reasoning to all operating conditions, it is possible to assign damage to each of them. Palmgreen–Miner's theory of linear cumulative damage, called Miner's rule, is given by [17,19,20]:

$$D = \sum_{i=1}^{k} \frac{N_{c,i}}{N_{f,i}} \tag{9}$$

$$T = \frac{1}{D} \tag{10}$$

where:

- $D \rightarrow$ total damage;
- $N_{c,i} \rightarrow$ number of cycles of the considered thermal profile associated with the i-th operating condition of the drive;
- $N_{f,i} \rightarrow$ number of cycles to failure associated with the specific operating condition i-th of the drive;
- $k \rightarrow$ total number of stresses.

By computing the $N_{c,i}/N_{f,i}$ ratios for all the i-th working points of the electric drive, the overall damage of SiC power devices can be predicted. If the damage is less than 1, then the power device is capable of handling the thermal stresses to which it was subjected in the entire mission profile; otherwise (if $N_{c,i}/N_{f,i}$ is greater than or equal to 1), the device is considered faulty. Parameter T, which would be the inverse of D, provides an important indication of how many cycles the device still has before failure. For "cycle", we must consider the stress input to the rainflow; in our case, one cycle corresponds to one second of junction temperature at steady-state. To obtain the lifetime, Equation (11) was used:

$$Lifetime = \sum_{i=1}^{k} \frac{1}{3600 \cdot h \cdot 365 \cdot w_i \cdot D_i} \tag{11}$$

where:

- $h \rightarrow$ hours of work per day;
- $w_i \rightarrow$ weight corresponding to the i-th operating condition;
- $D_i \rightarrow$ damage corresponding to the i-th operating condition, considering one second of junction temperature in steady-state condition;
- $k \rightarrow$ total number of stresses.

Therefore, the lifetime is estimated considering a predetermined number of hours of work per day.

3. Case Studies

In this section, the aforementioned life prediction procedure was applied to different operating scenarios for the drive so far considered: low speed–high torque, medium/high speed–low torque, and mixed operation. The goal of the following activity is to analyze the differences in lifetime estimation when different operating scenarios of the drive are considered. Figures 17, 19, and 21 show the lifetimes of each scenario as a function of daily working hours. For each scenario, daily use of the drive of 2 h is considered, although the working hours of the drive can be highly variable depending on the application.

3.1. Case Study 1: Low Speed–High Torque

This scenario is characterized by low speeds and a wide range of torques. Figure 16 shows the weight distribution considered for this mission profile, where the frequency of rotor speed is higher for values included in the range of 0–700 rpm. The conditions for which the speed is 0 rpm have a considerable weight; this could represent the numerous standing starts that normally occur in electric traction in the case of an urban cycle.

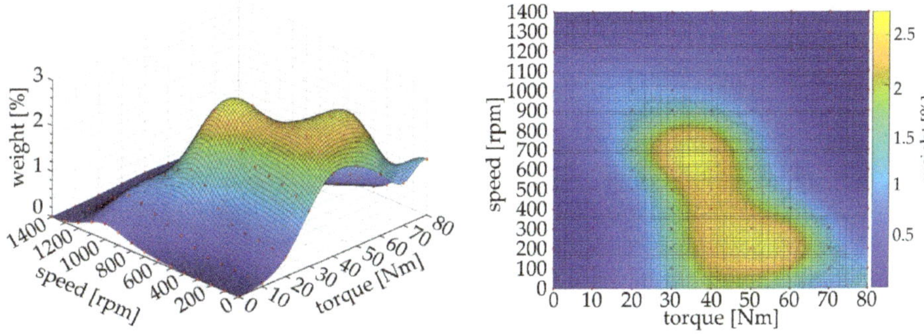

(a) (b)

Figure 16. Weight distribution of case study 1: (**a**) 3D surface; (**b**) 2D surface.

Figure 17 shows the lifetime for this scenario, obtained by applying the step-by-step procedure defined in the previous section. Assuming an average use of the motor drive equal to 2 h per day, the lifetime of the SiC MOSFET composing the inverter of the electric drive is estimated to be about 12 years.

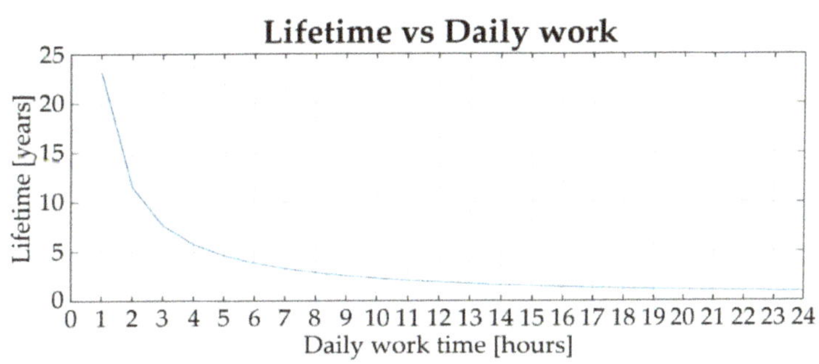

Figure 17. Lifetime estimation of case study 1.

3.2. *Case Study 2: Medium/High Speed–Low Torque*

In this scenario, the drive is mainly operated at medium-high speeds and low torques. A higher lifetime is expected because of the low torques, and thus low currents, which should lead to a lower degradation of the SiC power device under examination. Figure 18 shows the weight distribution associated to this scenario, which is projected towards medium-high speeds; in particular, a higher percentage of electric drive operation is centered around 700–800 rpm, even though the motor operation is observed up to 1400 rpm. The electromagnetic torque is normally kept quite low in this working profile, i.e., mainly around 20–30 Nm.

Figure 19 shows the lifetime estimation of case study 2. As expected, for the same hours of use, the lifetime of the SiC power MOSFET installed in the three-phase inverter is higher than in scenario 1. For instance, when an average daily use of the drive equal to 2 h is considered, the lifetime of the SiC devices is about 28 years.

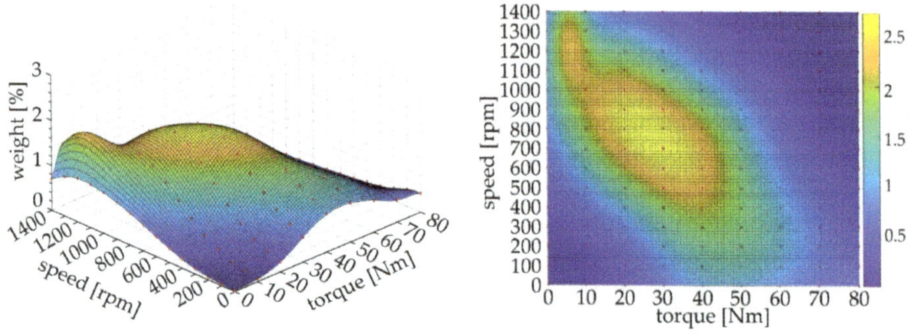

(**a**) (**b**)

Figure 18. Weight distribution of case study 2: (**a**) 3D surface; (**b**) 2D surface.

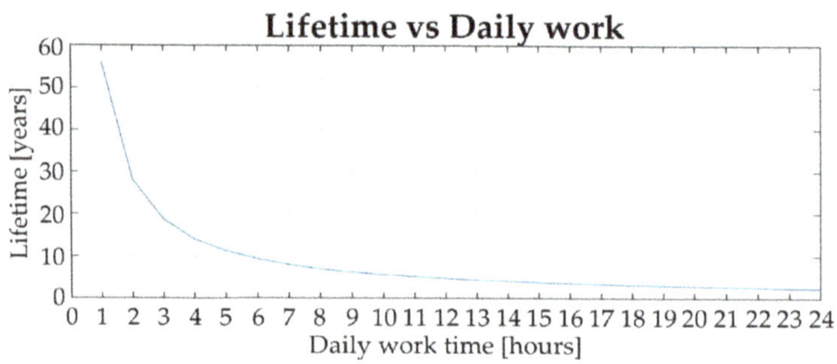

Figure 19. Lifetime estimation of case study 2.

3.3. *Case Study 3: Mixed Operation*

In the case of mixed-use, an intermediate situation between the two previous scenarios is considered here. The speed and torque ranges are much wider than the previous cases. Figure 20 shows the weight distribution, and it can be noted how wide the speed range is, i.e., from 200 to 1000 rpm, while the torque is mostly between 15 and 45.

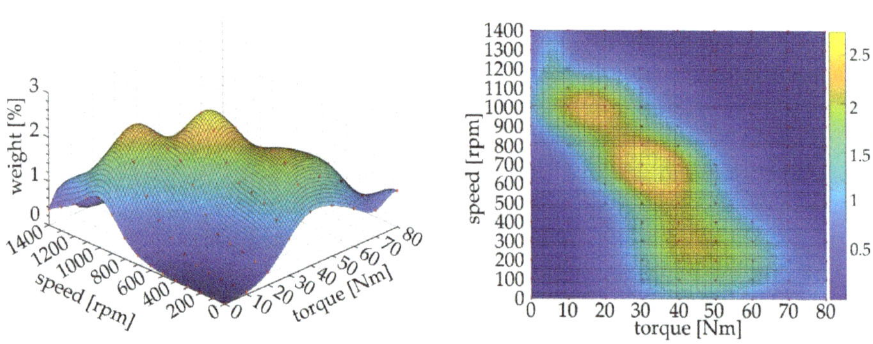

(a) (b)

Figure 20. Weight distribution of the mixed-use: (**a**) 3D surface; (**b**) 2D surface.

Figure 21 shows the lifetime estimation of case study 3. In this case, an intermediate lifetime of previous use was found, as expected.

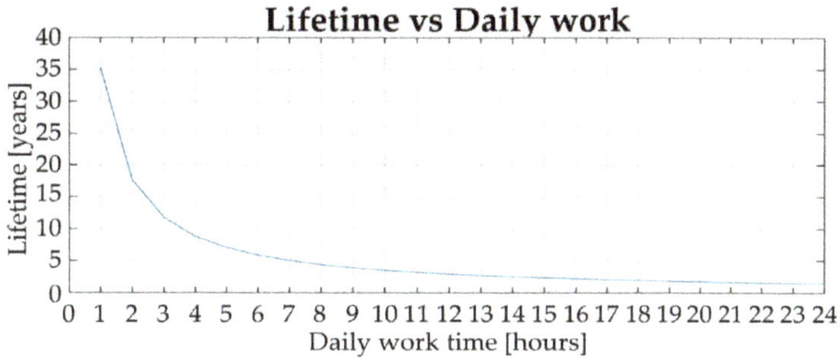

Figure 21. Lifetime estimation of mixed operation.

3.4. Results Assessment

According to the previous analysis, it is possible to note a variable lifetime that is consistent with the considered working scenarios. In fact, by assuming a daily work of the SiC-based inverter equal to 2 h, the proposed procedure provided the following lifetimes:

- Case study 1: lifetime 12 years
- Case Study 2: lifetime 28 years
- Case Study 3: lifetime 18 years

On the other hand, the same procedure allows for evaluating the number of working hours per day associated with each considered scenarios, when a lifetime equal to 15 years is imposed as a constraint for the SiC devices. The results are as follows:

- Case study 1: 1.6 h/day
- Case study 2: 2.5 h/day
- Case study 3: 3.8 h/day

Figure 22 shows the comparison between the lifetime results.

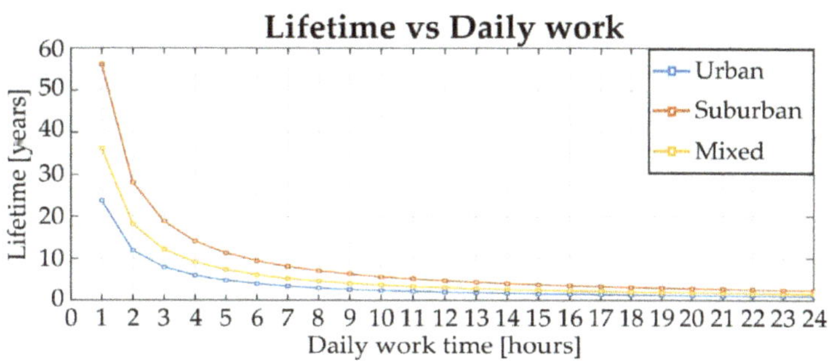

Figure 22. Comparison of the lifetime in case study 1, case study 2, and case study 3.

4. Conclusions

A detailed analysis of a lifetime estimation procedure devoted to discrete MOSFETs was presented in this paper. Such a technique was set with the aim of keeping the computational efforts and simulation times limited and low while providing an effective evaluation of thermal stresses and their effects on the power devices composing the electronic converter. The method is of a general application, and it can be used for any mission profile generated by the electric drive model. This allowed us to compare several operational states of the drive and to predict the worse operating conditions for the power devices. The proposed approach can be extended to different operating scenarios and power devices technologies, even though the main focus of this study is the SiC technology. In the case of multi-chip modules, it is possible to easily extend the analysis also, while considering the effects of mutual thermal coupling.

Author Contributions: Conceptualization and methodology, C.B., G.S. (Giacomo Scelba) and G.S. (Giuseppe Scarcella); validation and data curation, C.B.; software and visualization C.B.; writing—original draft preparation, C.B., G.S. (Giacomo Scelba) and S.A.R.; writing—review and editing, C.B., G.S. (Giacomo Scelba), S.A.R. and M.C.; supervision, G.S. (Giuseppe Scarcella) and M.C.; project administration, G.S. (Giacomo Scelba) All authors have read and agreed to the published version of the manuscript.

Funding: This work has been partially supported by the Italian Ministry for Economic Development (MISE), under the project "M9"—C32F18000100008 and by the University of Catania under the interdepartmental project PIA.CE.RI. 2020-2022 Line 2-TMESPEMES.

Conflicts of Interest: The authors declare no conflict of interest.

References

1. Pulvirenti, M.; Montoro, G.; Nania, M.; Scollo, R.; Scelba, G.; Cacciato, M.; Scarcella, G.; Salvo, L. *2018 IEEE Energy Conversion Congress and Exposition (ECCE)*; IEEE: Portland, OR, USA, 2018; pp. 1895–1902.
2. Gonzalez, J.O.; Wu, R.; Jahdi, S.; Alatise, O. Performance and Reliability Review of 650 V and 900 V Silicon and SiC Devices: MOSFETs, Cascode JFETs and IGBTs. *IEEE Trans. Ind. Electron.* **2020**, *67*, 7375–7385. [CrossRef]
3. Zhang, L.; Yuan, X.; Wu, X.; Shi, C.; Zhang, J.; Zhang, Y. Performance Evaluation of High-Power SiC MOSFET Modules in Comparison to Si IGBT Modules. *IEEE Trans. Power Electron.* **2019**, *34*, 1181–1196. [CrossRef]
4. Ceccarelli, L.; Reigosa, P.D.; Iannuzzo, F.; Blaabjerg, F. A survey of SiC power MOSFETs short-circuit robustness and failure mode analysis. *Microelectron. Reliab.* **2017**, *76–77*, 272–276. [CrossRef]
5. Ceccarelli, L.; Bahman, A.S.; Iannuzzo, F.; Blaabjerg, F. *A Fast Electro-Thermal Co-Simulation Modeling Approach for SiC Power MOSFETs*; IEEE APEC: Portland, OR, USA, 2017.
6. Luo, H.; Iannuzzo, F.; Blaabjerg, F.; Turnaturi, M.; Mattiuzzo, E. Aging Precursors and Degradation Effects of SiCMOSFET Modules Under Highly Accelerated Power Cycling Conditions. In Proceedings of the 2017 IEEE Energy Conversion Congress and Exposition (ECCE), Cincinnati, OH, USA, 1–5 October 2017; pp. 2506–2511.
7. Nguyen, M.H.; Kwak, S. Enhance Reliability of Semiconductor Devices in Power Converters. *Electronics* **2020**, *9*, 2068. [CrossRef]
8. Kim, M.-K.; Yoon, S.W. Thermal Impedance Characterization Using Optical Measurement Assisted by Multi-Physics Simulation for Multi-Chip SiC MOSFET Module. *Micromachines* **2020**, *11*, 1060. [CrossRef] [PubMed]
9. Seal, S.; Mantooth, H.A. High Performance Silicon Carbide Power Packaging—Past Trends, Present Practices, and Future Directions. *Energies* **2017**, *10*, 341. [CrossRef]
10. Cougo, B.; Morais, L.; Segond, G.; Riva, R.; Duc, H.T. Influence of PWM Methods on Semiconductor Losses and Thermal Cycling of 15-kVA Three-Phase SiC Inverter for Aircraft Applications. *Electronics* **2020**, *9*, 620. [CrossRef]
11. Mauromicale, G.; Raciti, A.; Rizzo, S.A.; Susinni, G.; Abbatelli, L.; Buonomo, S.; Cavallaro, D.; Giuffrida, V. SiC Power Modules for Traction inverter in Automotive Application. In Proceedings of the Annual Conference of the IEEE Industrial Electronics Society, Lisbon, Portugal, 14–17 October 2019.
12. Zhou, D.; Wang, H.; Blaabjerg, F. Mission Profile Based System-Level Reliability Analysis of DC/DC Converters for a Backup Power Application. *IEEE Trans. Power Electron.* **2018**, *33*, 8030–8039. [CrossRef]
13. Khosroshahi, A.; Abapour, M.; Sabahi, M. Reliability evaluation of conventional and interleaved DC–DC boost converters. *IEEE Trans. Power Electron.* **2015**, *30*, 5821–5828. [CrossRef]
14. Ni, Z.; Lyu, X.; Yadav, O.P.; Singh, B.N.; Zheng, S.; Cao, D. Overview of Real-Time Lifetime Prediction and Extension for SiC Power Converters. *IEEE Trans. Power Electron.* **2020**, *35*, 7765–7794. [CrossRef]
15. Hirschmann, D.; Tissen, S.; Schroder, S.; De Doncker, R.W. Reliability prediction for inverters in hybrid electrical vehicles. *IEEE Trans. Power Electron.* **2007**, *22*, 2511–2517. [CrossRef]
16. Novotny, D.W.; Lipo, T.A. *Vector Control and Dynamics of AC Drives*; Oxford University Press Inc.: New York, NY, USA, 1996.
17. Wang, B.; Cai, J.; Du, X.; Zhou, E.L. Review of Power Semiconductor Device Reliability for Power Converters. *CPSS Trans. Power Electron. Appl.* **2017**, *2*, 101–117. [CrossRef]
18. Leddy, L.; Reddy, G.; Tolbert, L.M.; Ozpineci, B.; Pinto, O.P. Rainflow Algorithm-Based Lifetime Estimation of Power Semiconductors in Utility Applications. *IEEE Trans. Ind. Appl.* **2015**, *51*, 3368–3375.
19. Letor, R.; Russo, S.; Crisafulli, R. *Life Time Prediction and Design for Reliability of Smart Power Devices for Automotive Exterior Lighting*; STMicroelectronics: Catania, Italy, 2008.
20. Lai, W.; Chen, M.; Alatise, O.; Xu, S.; Mawby, P. Low ΔTj Stress Cycle Effect in IGBT Power Module Die-Attach Lifetime Modeling. *IEEE Trans. Power Electron.* **2016**, 6575–6585. [CrossRef]

21. ABB Switzerland Ltd Semiconductors. *Load-Cycling Capability of HiPak IGBT Modules*; Application Note 5SYA 2043-04; ABB Switzerland Ltd.: Lenzburg, Switzerland, 2004; pp. 1–8.
22. Ma, K.; Blaabjerg, F. Multi-Timescale Modelling for the Loading Behaviours of Power Electronics Converter. In Proceedings of the 2015 IEEE Energy Conversion Congress and Exposition (ECCE), Montreal, QC, Canada, 20–24 September 2015; pp. 5749–5756.
23. Ceccarelli, L.; Kotecha, R.M.; Bahman, A.S.; Iannuzzo, F.; Mantooth, H.A. Mission-Profile-Based Lifetime Prediction for a SiC mosfet Power Module Using a Multi-Step Condition-Mapping Simulation Strategy. *IEEE Trans. Power Electron.* **2019**, *34*, 9698–9708. [CrossRef]
24. Ahmed, M.H.; Wang, M.; Hassan, M.A.S.; Ullah, I. Power Loss Model and Efficiency Analysis of Three-Phase Inverter Based on SiC MOSFETs for PV Applications. *IEEE Access* **2019**, *7*, 75768–75781. [CrossRef]
25. Schonberger, J. Averaging Methods for Electrical-Thermal Converter Models. In Proceedings of the 2011 14th European Conference on Power Electronics and Applications, Birmingham, UK, 30 August–1 September 2011; pp. 1–8.
26. Allegra, A.; Scelba, G.; Cacciato, M.; Foti, S.; Scarcella, G.; Salerno, N.; Cavallaro, D.; Bazzano, A.; Aiello, G. *Thermal Equivalent Circuit Model of Multi-Die SiC Power Modules*; 2020 ELEKTRO: Taormina, Italy, 2020; pp. 1–6.
27. Chen, Z.; Gao, F.; Yang, C.; Peng, T.; Zhou, L.; Yang, C. *Converter Lifetime Modeling Based on Online Rainflow Counting Algorithm*; IEEE: Vancouver, BC, Canada, 2019; pp. 1743–1748.
28. ASTM American Society for Testing and Materials, Standard Practices for Cycle Counting in Fatigue Analysis. Designation: E 1049–85 (Reapproved 1997). In *Annual Book of ASTM Standards*; ASTM: West Conshohocken, PA, USA, 2014; pp. 1–10.

electronics

MDPI

Article

Energy-Saving Research on New Type of LED Sensor Lamp with Low-Light Mode

Chun-Te Lee * **and Ping-Tsan Ho**

Department of Leisure and Sport Management, Cheng-Shiu University, Kaohsiung City 833, Taiwan; K0630@gcloud.csu.edu.tw

* Correspondence: charter@gcloud.csu.edu.tw; Tel.: +886-918-013-340

Received: 21 August 2020; Accepted: 5 October 2020; Published: 10 October 2020

check for updates

Abstract: In general, the sensor lamps in the corridors, stairwells, or toilets of buildings will change from completely dark to full brightness when someone passes by. It will make the human eyes feel very uncomfortable, and when the sensor lamp is completely dark, the whole corridor and stairwell will be dark, making women and children feel insecure at night. If the lighting is changed to be sensor-less, there is a serious problem of wasted energy. To solve this dilemma, we developed a new type of "LED sensor lamp with low-light mode" that changes the original "full dark mode" to "low-light mode". As such, when someone approaches the sensor lamp, their eyes will not be uncomfortable with the momentary illumination. Furthermore, when no one passes by, the sensor lamp will stay in low-light mode, so that people returning home at night no longer have to go through dark corridors, thereby achieving safety, aesthetics, and energy-saving purposes. This new sensor lamp's power consumption in low-light mode is only 1/10 of the high-light mode, but its brightness can be up to half of the high-light mode, making it very suitable for parking lots, corridors, stairways, or toilets of buildings. It only requires the replacement of the lamp but not the original lamp socket, yet the basic brightness can be maintained. Take the general 15W T8 LED lamp (sensor-less) as an example: if it is replaced by this new type of sensor lamp, and the place where it is installed is rarely passed by people, the power saving rate will be as high as 90%. Assuming that there are 12 passers by per hour, the saving rate is still 81%.

Keywords: sensor lamp; low-light mode; high-light mode

1. Introduction

Global warming keeps worsening and climate anomalies are occurring increasingly frequently around the world. After the Kyoto Protocol, which aims to curb the global greenhouse gas emissions, came into force on 16 February 2005, the innovation of energy technologies and the change of social behavior for energy conservation will be an effective strategy for carbon dioxide reduction [1]. Due to the growing influence of global warming and the rapid changes in the ecological environment, many unusual ecological crises have started to occur all over the world. Therefore, the search for "renewable energy" is a matter of great importance to all governmental and non-governmental organizations. For the U.S., the UK, and China, the goal is for renewable energy to account for at least 20% of total electricity generation by 2020, and it is expected that renewable energy will contribute more than 50% of total electricity generation in some countries by 2050 [2–4].

Energy-saving, smart building, and green building have recently become compelling topics, and lighting control design is tailored to the needs of each location in buildings to reduce power consumption [5,6]. Artificial lighting accounts for a significant portion of the world's electrical energy consumption. In office buildings in particular, artificial lighting may account for up to 40% of the total energy consumed [7]. In fact, the use of LED technology in lighting systems can save energy [8].

The development and application of LED technology are changing rapidly, the use thereof can reduce the required power value of lighting, enhance their durability, as well as achieving environmental protection and low power consumption [9–12]. LED lighting will eventually replace traditional lighting due to environmental protection, energy-saving and economic cost considerations. In addition to its long life, environmental protection, and safety advantages, LED lighting is more energy efficient than traditional lighting, consuming only half the power of a fluorescent lamp and one eighth the power of a light bulb [13].

According to the latest report "2018 Commercial Lighting, Smart Lighting and Panel Light Market Report" from LEDinside, a green-energy division of the market research firm TrendForce, the global LED commercial lighting market has continued to expand in recent years due to the energy efficiency and cost effectiveness of LED luminaires, which are gradually recognized by consumers. The global LED indoor commercial lighting market would reach USD 15.87 Billion in 2018, accounting for 42% of the global LED lighting production value [14].

Although the luminous efficacy of LED lighting is excellent, twice as good as fluorescent lamp, take the underground parking lot of the building for example: it must be lit 24 h a day and its area is very large, so even if all the lights are replaced with LED lighting, the electricity bill is still very high. Furthermore, due to safety or aesthetic reasons, many buildings must maintain a basic level of brightness in the corridors and stairwells at night, resulting in untold amounts of wasted electricity.

To implement "energy-saving and carbon reduction", many schools and government agencies have replaced all the lights in the corridors, stairwells, or toilets with sensor lamps. However, most lights must be controlled by an external circuit for the sensors to function. When sensing someone passing by, it will change from full darkness to full brightness, so the human eyes would feel very uncomfortable. Moreover, when the sensor lamp is full dark, the whole corridor and stairway will be dark, which will make people feel insecure when returning home at night. However, if we change the lighting to be sensor-less, we will have to face the problem of wasting energy again.

To solve this dilemma, we developed a new type of "LED sensor lamp with low-light mode" that changes the original "full dark mode" to "low-light mode". As such, when someone approaches the sensor lamp, their eyes will not be uncomfortable with the momentary illumination. When no one passes by, the sensor lamp stays in low-light mode, so that people returning home at night no longer must go through dark corridors, thereby achieving safety, aesthetics, and energy-saving purposes.

In addition to indoor lighting applications, this technology is also suitable for outdoor lighting or solar-powered LED lighting systems. Stand-alone solar power systems are used in remote areas not accessible by grid power [15]. The system must be designed with a good matching between the installed capacity of solar photovoltaic module and battery capacity to obtain a proper loss of load probability (LLP) [16]. Therefore, the power consumption of the solar-powered LED lighting system must be as low as possible. This product's power consumption is only one-tenth of that of general lighting, so it is very suitable for stand-alone solar power LED lighting systems.

The concept of energy-saving in that article [17] is the same with ours; however, the structure of their LED light is suitable for large-scale area, and ours is for private and smaller one. They must set up a server and complicated wirings to detecting illumination and user movement to control the intensity of LED light. It costs a lot money and waste of time to save 58% energy consumptions. Nevertheless, our design could save more energy consumptions without setting up any server and wirings but only change T8 LED tube to our new type of sensor lamp.

Conventional LED sensor lamps have a common problem: they only have two modes—full dark mode and full light mode, which makes the human eyes very uncomfortable. Imagine what it would be like if the whole corridor is sensor-lit. The corridors of the five-star hotels in Europe are all equipped with sensor lamps to save energy. When customers enter the corridor, they will see darkness in front of them, and they must go forward and then the lamps in the corridor will be lit up one by one. Although this has the effect of saving electricity, human eyes may feel uncomfortable when the lamps are first turned on.

The following are the problems of ordinary sensor lamps:

1. They have only two modes: full dark mode and full light mode, which is very uncomfortable for the human eyes.
2. In the full dark mode, the entire corridor is dark, making women and children feel insecure when returning home at night.
3. The trigger would not be activated again when the sensor lamp is on.

2. New Type of LED Sensor Energy-Saving Lamp with Low-Light Mode

General sensor lamps are fully illuminated when people pass by and completely dimmed when there is no one passing by. Although it can achieve the purpose of energy-saving, it also produces the problems mentioned earlier. This work aims to improve this technical problem to develop a new generation of more energy-efficient "New type of LED sensor energy-saving lamp with low-light mode".

2.1. Introduction of Works

We have developed a new type of "LED sensor energy-saving lamp with low-light mode", which changes the original "full dark mode" to "low-light mode". As such, when someone approaches the sensor lamp, his/her eyes will not have adjustment issue. When no one passes by the corridor and stairwell, the sensor lamp will be half-lit, so when someone passes by, he/she does not have to face a dark corridor in front of them, thereby achieving the purpose of safety, aesthetics and energy-saving.

The new LED sensor energy-saving lamp we developed enters low-light mode when no one is around, the brightness is about half of the high light, and the power consumption is only about one-tenth thereof and adjustable. Meanwhile, when someone is passing by, it will enter the full brightness high-light mode. In addition, we also changed our sensor lamp from single triggering to continuous triggering, which will continue to be triggered as long as the user is still around. Therefore, this new type of LED sensor energy-saving lamp is very suitable for parking lots, corridors, stairwells, or toilets in buildings. It can maintain basic brightness, and the power saving rate is up to 90% (assuming that there are seldom people passing by the installation site, and ignoring the power consumption of the circuit, with only the power consumption of the lamp tubes is calculated. It will be described in detail in the next section). It completely eliminates the shortcomings of conventional LED sensor lamps.

The LED particles have a higher energy conversion efficiency (the efficiency of converting electrical energy to light energy) at a lower light level, making LED ideal for use in low-light requirements [18]. As a result, we only need to supply 10% of power for the LED particles to reach nearly half the brightness. Furthermore, in most cases, the light decay of LED is caused by high temperature, but this new LED sensor energy-saving lamp mostly operates in low-light mode, and the current is only 10%, so the temperature is very low. Therefore, it can reduce the light decay of LED and its life will be about twice to three times that of the general LED. Presently, there are three types of products we created: recessed light, T5 mountain ceiling lamp and T8 lamp tube. The first two are with external circuits on the bulb or lamp tube (Figure 1 shows a T5 mountain ceiling lamp).

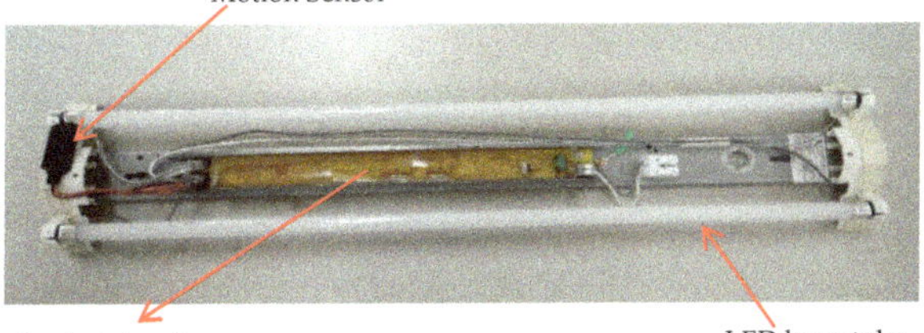

Figure 1. LED sensor energy-saving lamp (T5 mountain ceiling lamp).

Another latest product of ours puts the control circuit and motion sensor into a T8 LED tube with no external circuit required (Figure 2). As such, users can remove the original T8 lamp tubes (fluorescent or LED lamps) and replace them with our new T8 LED sensor energy-saving lamp:

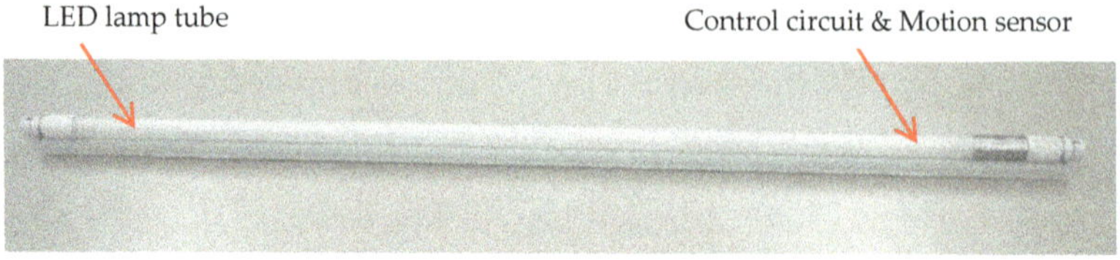

Figure 2. LED sensor energy-saving lamp (T5 mountain ceiling lamp).

2.2. System Block Diagram

Below is the system block diagram of the "New type of LED sensor lamp with low-light mode" to illustrate how the circuit design achieves the energy-saving effect:

The bridge rectifier converts the main supply voltage of 110 V/220 V AC power into pulsating DC power, and then the voltage detection chip determines if it is 110 V or 220 V. If the voltage is 110 V, SW_1 and SW_2 will be ON, SW_3 will be OFF, and LED_1 and LED_2 are connected in parallel. If the voltage is 220 V, SW_1 and SW_2 will be OFF, SW_3 will be ON, and the LED_1 and LED_2 are connected in series, so that this sensor lamp can receive the full voltage of 110 V/220 V.

The constant current circuit makes judgments according to the detection results of the microwave sensor. If someone is passing by (in high-light mode), a high signal will be sent to CC_1 and CC_2, then CC_1 and CC_2 will provide a current of 100%. If no one is detected (low-light mode), a low signal will be sent to CC_1 and CC_2, and CC_1 and CC_2 will only supply a current of 10% (this circuit is designed to be flexible in setting the current with the jumper, and the setting range is 10% to 20%).

In addition, the output of the bridge rectifier in Figure 3 is pulsating DC instead of stabilized DC voltage because the bridge rectifier has a better power factor (PF) when pulsating DC is adopted. In the CE (Conformite Europeenne) safety regulations, it is stipulated that if a LED light consumes more than 25 W of power, the PF value must be greater than 0.9. If the power consumption is between 5 W and 25 W, the PF value must be greater than 0.7. If the power consumption is less than 5 W, the PF value only needs to be greater than 0.5.

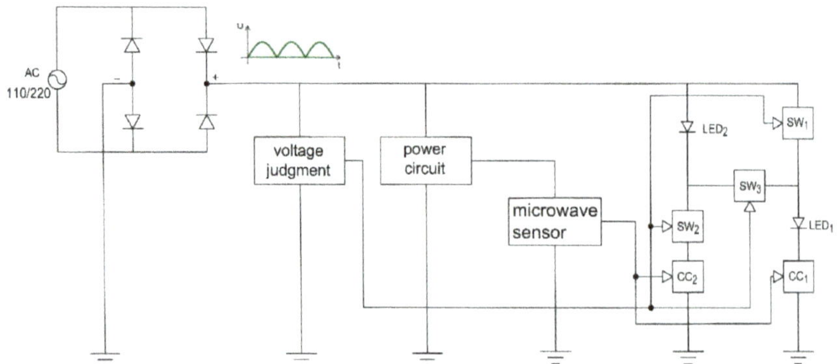

Figure 3. System block diagram of LED Sensor lamp.

3. Energy-Saving Analysis

3.1. Calculation of Power Consumption

Since this new type of sensor lamp has high-light mode and low-light mode, the calculation of power consumption is more complicated unlike the simple calculation of general sensor-less LED lamp. This new type of sensor lamp enters low-light mode when no one is passing by, and its power consumption is only one-tenth of that of high-light mode. When someone is passing by, the sensor lamp will be fully lit for 30 s, so the factors affecting the power consumption are: the duration of the activation, the frequency of people passing by, whether a continuous trigger is activated, and so on. However, to derive the equation, we assume that people do not stay long when passing by and it takes 30 s for the next person to pass by, i.e., there will be no continuous trigger. Furthermore, the durations that people pass by are equal (if they are not equal, the total power is still the same, but the formula is more difficult to derive) and dispersed.

Let us use the below figure (Figure 4) "Power variation diagram for high-light mode and low-light mode" to illustrate how to calculate the hourly power consumption of this new type of sensor lamp.

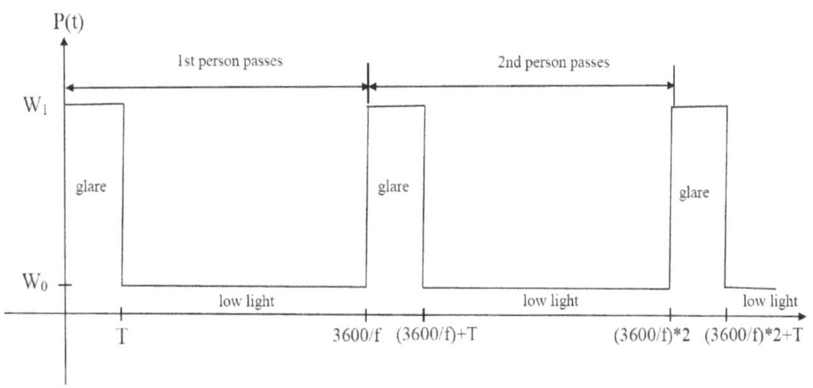

Figure 4. Power variation diagram for high-light mode and low-light mode.

P(t) illustrates the power change between the high-light mode and the low-light mode, and the meanings of other parameters are as follows:

T: The number of seconds in which the high-light mode is maintained when someone passes by. The current setting is 30 s.

f: Number of passer-by per hour, i.e., the number of times that high-light mode is trigger per hour.

W_1: Power for high-light mode. The current power for the high-light mode of this product is 15 W (the power of the lamp tube).

W_0: Power for low-light mode. The current power for the low-light mode of this product is 1.5 W (the power of the lamp tube).

Since it was previously assumed that "the durations that people pass by are equal and dispersed", the duration in which a person passes by is 3600/f (for f > 0). For the first person, duration between 0 and T seconds is in the high-light mode, the duration between T and 3600/f seconds is the low-light mode, so we can calculate the consumed power per lamp per hour (E_{ph}) as follows:

$$\text{if } f = 0: \quad E_{ph} = (3600 \times W_0)\,J = (3600 \times W_0)/3.6 \times 10^6 \text{ kw h} \tag{1}$$

$$\text{if } f > 0:$$
$$\begin{aligned} E_{ph} &= (f \times T/3600) \times W_1 + (3600 - (f \times T)/3600) \times W_0 \, J \\ &= ((f \times T/3600) \times W_1 + (3600 - (f \times T))/3600) \times W_0)/3.6 \times 10^6 \text{ kw h} \end{aligned} \quad (2)$$

In addition, we assume that each lamp will be on for 8 h a day, meaning that the lamp will be on for a total of 2920 h (8 × 365) per year, so the annual power consumption (E_{py}) of this new sensor lamp is:

$$E_{py} = \sum E_{py} = E_{ph} \times 2920 \text{ kw h} \quad (3)$$

For an ordinary sensor-less 15W T8 LED lamp (operates only in 15W full light mode), the power consumption per lamp per year is (15W/1000) × 2920 = 43.8 kw h. If the "ordinary" sensor-less 15W T8 LED lamp is replaced with our "new type of sensor lamp", we can calculate the Energy-Saving Rate (ESR), kilowatt-hour saved per lamp per year (D_{py}), cost saved per year (M_{py}) assuming it costs NT$4 per kilowatt-hour), and the carbon emission reduced per year (C_{py}) as follows (1 KW·h can generate 0.623 kg CO_2e of carbon emission):

$$ESR = (15 - E_{ph})/15 \times 100\% \quad (4)$$

$$D_{py} = 43.8 - E_{ph} \times 2920 \text{ kw h} \quad (5)$$

$$M_{py} = (43.8 - E_{ph} \times 2920) \text{ kw h} \times 4 \text{ dollar/kw h} \quad (6)$$

$$C_{py} = (43.8 - E_{ph} \times 2920) \text{ kw h} \times 0.623 \text{ kg } CO_2\text{e/kw h} \quad (7)$$

3.2. Calculation of Energy-Saving Rate

Since the ESR of this new type of sensor lamp is closely related to the number of times of high-light mode is triggered per hour (f), this section investigates the relationship between "ESR" and "number of times of high-light mode is triggered per hour (f)" and calculates the maximum and minimum values of the ESR.

First of all, we know from Equation (4) that "energy-savings rate" is inversely proportional to "the amount of electrical energy consumed per lamp per hour (E_{ph})", and we also know from Equations (1) and (2) that E_{ph} is directly proportional to the number of times of high-light mode is triggered per hour (f). As such, the ESR is inversely proportional to the number of times of high-light mode is triggered per hour (f), meaning that the more times the high light is triggered, the lower the ESR.

Now, let us look at the range of the number of times of high-light mode is triggered per hour (f). The minimum value is 0, which means no one passes by in an hour, and the maximum value is 3600/T, which means that T seconds after someone passes by each time, another person passes by immediately without interruption (that is, the high-light mode is on the whole hour). Therefore, the range of f is $0 \leq f \leq 3600/T$. The T-value of this product is 30 s, so the range of f here is $0 \leq f \leq 120$. We bring different f-values into the E_{ph} formula to calculate the E_{py}, ESR, D_{py}, M_{py}, and C_{py} values of each lamp with different f-values (compared with the general sensor-less 15W T8 LED lamp), and the below table (Table 1) only lists the important portions (f-values with an interval of 6):

Table 1. E_{py}, ESR, D_{py}, M_{py}, and C_{py} values for each lamp with different f-values.

f	E_{ph} (kw h)	E_{py} (kw h)	ESR (%)	D_{py} (kw h)	M_{py} (NT$)	C_{py} (kgCO_2e)
0	0.0015	4.3800	90.00%	39.4200	157.680	24.559
1	0.0016	4.7085	89.25%	39.0915	156.366	24.354
6	0.0022	6.3510	85.50%	37.4490	149.796	23.331
12	0.0029	8.3220	81.00%	35.4780	141.912	22.103
18	0.0035	10.2930	76.50%	33.5070	134.028	20.875
24	0.0042	12.2640	72.00%	31.5360	126.144	19.647
30	0.0049	14.2350	67.50%	29.5650	118.260	18.419
36	0.0056	16.2060	63.00%	27.5940	110.376	17.191

Table 1. *Cont.*

f	E_{ph} (kw h)	E_{py} (kw h)	ESR (%)	D_{py} (kw h)	M_{py} (NT$)	C_{py} (kgCO_2e)
42	0.0062	18.1770	58.50%	25.6230	102.492	15.963
48	0.0069	20.1480	54.00%	23.6520	94.608	14.735
54	0.0076	22.1190	49.50%	21.6810	86.724	13.507
60	0.0083	24.0900	45.00%	19.7100	78.840	12.279
66	0.0089	26.0610	40.50%	17.7390	70.956	11.051
72	0.0096	28.0320	36.00%	15.7680	63.072	9.823
78	0.0103	30.0030	31.50%	13.7970	55.188	8.596
84	0.0110	31.9740	27.00%	11.8260	47.304	7.368
90	0.0116	33.9450	22.50%	9.8550	39.420	6.140
96	0.0123	35.9160	18.00%	7.8840	31.536	4.912
102	0.0130	37.8870	13.50%	5.9130	23.652	3.684
108	0.0137	39.8580	9.00%	3.9420	15.768	2.456
114	0.0143	41.8290	4.50%	1.9710	7.884	1.228
120	0.0150	43.8000	0.00%	0.0000	0.000	0.000

From the above table, it can be seen that 90% of the maximum value of the ESR occurs when f = 0 (meaning no one has passed by for 1 h), and the same is true for other values of E_{py}, D_{py}, M_{py}, and C_{py}. The ESR decreases linearly with the increase of f. When f = 120 (meaning that people pass by continuously for 1 h), the ESR decreases to 0%. However, this is unlikely to happen, as sensor lamps are usually installed in areas where people are less likely to pass by. The diagram below (Figure 5) shows the relationship between the ESR and the f value.

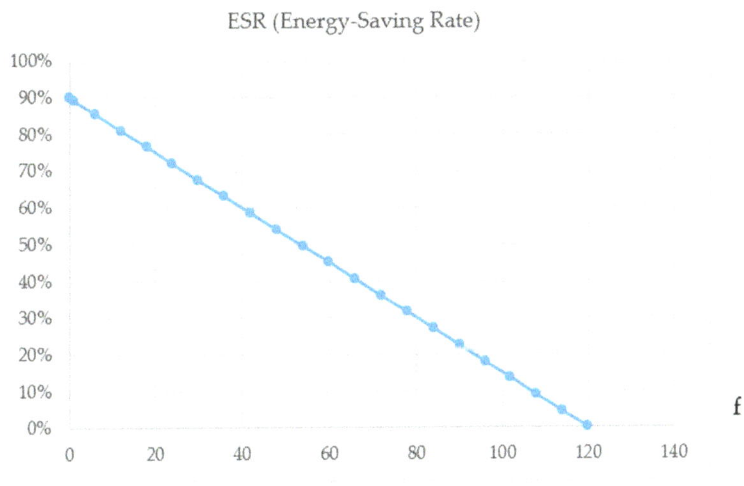

Figure 5. Relationship between energy-saving rate (ESR) and f.

3.3. Energy-Saving Examples

We would like to use the example of Cheng Shiu University for energy-saving analysis. Most of the lights in the corridors, parking lots, stairwells, and toilets in its campus buildings are 15W T8 LED lamps (sensor-less), and it is estimated that there are about 3000 of them. Assuming that all of them are replaced with this new type of LED sensor energy-saving lamps, we will show the energy-saving effect and the expected effect by the "reduction of power consumption in the same period of one year".

Since the brightness of the 15W T8 LED sensor lamp (in high-light mode) is the same as that of the general 15W T8 LED sensor-less lamp, the school can replace the original 15W T8 LED lamp (sensor-less lamp) with this product. Although they can simply replace the lamp tubes without replacing the lamp sockets, the basic brightness can also be maintained, so the energy-saving efficiency is expected to be good.

We will use 3000 lamps as the basis for estimating the "reduction of power consumption in the same period of one year", but in order to calculate the power consumption in a more accurate way, we assume that there will be 12 people passing by each lamp per hour (about 1 person passing by every 5 min, which is a more average number because the use rate of lamps in 1st floor is very high but that for the other floors will be very low).

We can calculate it by directly referencing to Table 1 (taking the E_{py}, D_{py}, M_{py}, and C_{py} values when f = 12):

1. The power consumption of 3000 lamps per year (when f = 12, E_{py} value = 8.322): 8.322 × 3000 = 24,966 kW h
2. 15W T8 LED tube (sensor-less, 15W in full brightness mode only): 15 × 8 × 365 × 3000 = 131,400,000 W h = 131,400 kW h
3. The annual reduction in power consumption with the 3000 lamps (when f = 12, D_{py} = 35.478): 35.478 × 3000 = 106,434 kW h
4. The reduced cost per year for using the 3000 lamps (when f = 12, M_{py} = 141.912): 141.912 × 3000 = NT$425,736
5. Carbon emissions reduced per year by using the 3000 lamps (when f = 12, C_{py} = 22.103): 22.103 × 3000 = 66,308.4 kg CO_2e

Therefore, all we have to do is to replace all 3000 lamps in the corridors, parking lots, stairwells, and toilets of all the campus buildings with the new type of LED sensor energy-saving lamps, and we will get the estimated results as follows Table 2 (assuming the lamps are lit for 8 h a day, with 12 people passing by per hour and the electricity costs at NT$ 4 per kW h. Furthermore, the above calculation is for the lamps only, with the electricity for the control circuit being excluded):

Table 2. Reduced power consumption, cost, and carbon emissions by Cheng Shiu University compared to the same period of the previous year.

Power Consumption Before Improvement (kW·h/yr)	Power Consumption After Improvement (kW·h/yr)	Saved Energy (kW·h/yr)	Energy-Saving Rate (%)	Savings in Electricity Charges (NT$/yr)	CO_2 Emissions Reduced (kg CO_2e)
131,400	24,966	106,434	81%	NT$ 425,736	66,308.4

From the above table, we can save 106,434 kW h of electricity per year, equivalent to NT$425,736 in electricity bill, by replacing all the 3000 lamps in all the buildings with the new type of LED sensor energy-saving lamps. The ESR of 81% (assuming 12 people passing by per hour) is a significant benefit to the school.

On the same conditions, compared to our new type of sensor lamp, the conventional LED sensor lamp could save NT$472,800 electricity charge. It only saves NT$47,064 than ours; however, the full dark mode of the conventional LED sensor lamp is insecure and uncomfortable to our eye, which is the problem we consider.

4. Circuit Testing

The new type of T8 LED sensor energy-saving lamp has been sent to a fair-minded third-party laboratory for testing to prove that the power consumption of the lamp in low-light mode is only one-tenth of that of the high-light mode, but its brightness can be up to half of the brightness of the high-light mode. Its product test report is as shown below (Figures 6 and 7):

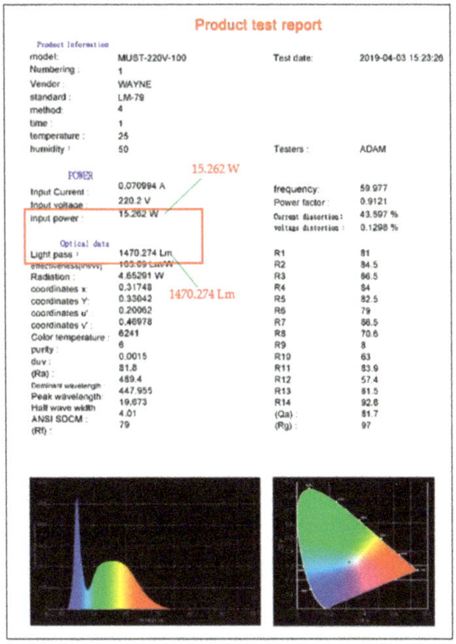

Figure 6. Test data in high-light mode.

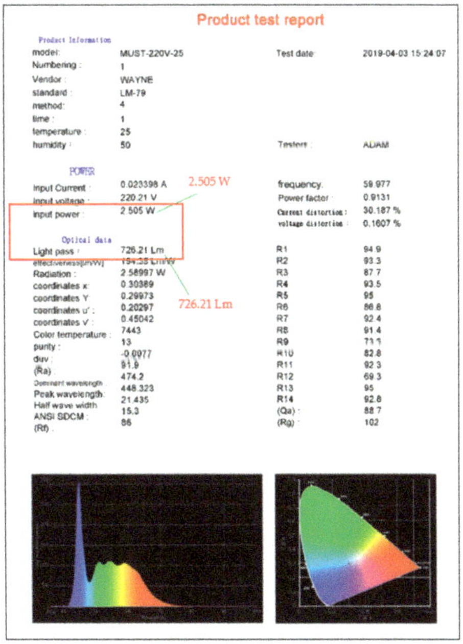

Figure 7. Test data in low-light mode.

From the test data in Figures 6 and 7, it can be seen that the luminous flux in high-light mode is 1470.274 Lm, and the luminous flux in low-light mode is 726.21 Lm, meaning that the brightness in low-light mode is about 49.39% of that in high-light mode. The input power in high-light mode is 15.262 W, and the input power in low-light mode is 2.505 W. The brightness in low-light mode is about 16.41% of that in high-light mode, which is significantly different from the 10% we designed. This is because the input power is measured together with those of the control circuit and the motion sensor, and the power of these two is about 1 W. If those are deducted, it is close to 10% ((2.505 − 1)/(15.262 − 1) = 10.55%).

In addition, we also purchased an illuminometer and power meter to conduct our own measurement experiments. The brand of the illuminometer is TASI, the model number is TA8121, and its maximum resolution is 0.1 LUX. The brand of the power meter is WANF, the model number is WF-D02A, and its maximum resolution is 0.1 W. Below are the measurement results taken with the TASI TA8121 illuminometer (5 cm away from the lamp tube) (Figure 8):

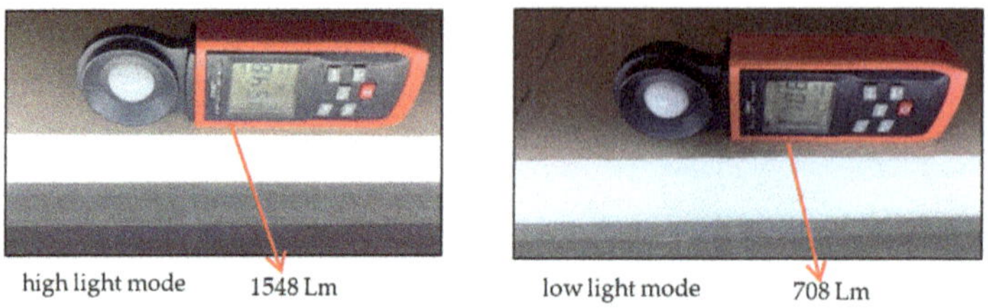

Figure 8. Luminous flux in high-light mode and low-light mode.

The above experiment results show that the luminous flux in high-light mode is 1548 Lm, and that in the low-light mode is 708 Lm. The luminous flux in low-light mode is about 45.74% of that in high-light mode, which is not much different from the 49.39% in the results obtained by the laboratory. Below are the results we measured with our WANF WF-D02A power meter (Figure 9):

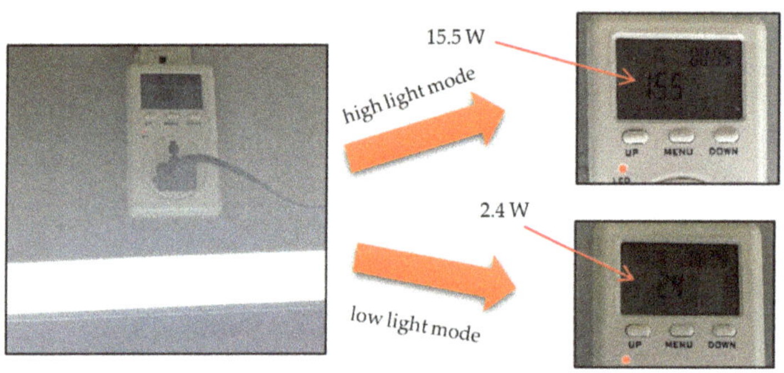

Figure 9. Power consumption in high-light mode and low-light mode.

The power measured in the above-mentioned experiments is 15.5 W in the high-light mode and 2.4 W in the low-light mode, and the brightness of the low-light mode is about 15.48% of that of the high-light mode, which is significantly different from the 10% we designed. Similarly, it is because the input power is measured together with those of the control circuit and the motion sensor, and the power of the two is about 1 W. If those are deducted, it is about 10% ((2.4 − 1)/(15.5 − 1) = 9.66%).

5. User Satisfaction Analysis

In this study, a total of 249 people completed the questionnaire, of which 46% were males and 54% were females. User satisfaction is divided into four aspects: "overall", "parking lots", "corridors, stairwells" and "toilets", and the analysis results are shown in Table 3. The average ranking of the overall satisfaction shows that the top three satisfaction ratings of the sensor lamp are "I think this sensor lamp can help maintain a certain level of brightness in the corridors and stairwells without wasting electricity" (m = 4.8), "I think the function of low-light mode (half brightness) is very good" (m = 4.77), "I think this sensor lamp is very energy efficient" (m = 4.76), and "I would recommend this sensor lamp to others" (m = 4.76).

Table 3. Satisfaction Analysis Table.

Question	Mean	Standard Deviation	Number of Lamps	Mean Ranking
1. I think this sensor lamp can help maintain a certain level of brightness in the corridors and stairwells without wasting electricity	4.8	0.489	249	1
2. I think the function of low-light mode (half brightness) is very good	4.77	0.516	249	2
3. I think this sensor lamp is very energy efficient	4.76	0.587	249	3
4. I would recommend this sensor lamp to others	4.76	0.528	249	4
5. I feel that when this sensor lamp is fully lit, it will not cause discomfort to human eyes for momentary illumination.	4.74	0.568	249	5
6. When this sensor lamp is installed in a "parking lot", I think the lamp stays fully lit for just the right length of time (currently 30 s) when someone passes by.	4.37	0.708	225	6
7. When this sensor lamp is installed in "corridors and stairwells", I think its sensitivity and response distance is appropriate.	4.36	0.685	228	7
8. When this sensor lamp is installed in "corridors and stairwells", I think the lamp stays fully lit for just the right length of time (currently 30 s) when someone passes by	4.34	0.704	224	8
9. When this sensor lamp is installed in "parking lots", I think its sensitivity and response distance is appropriate.	4.34	0.702	222	9
10. When this sensor lamp is installed in "corridors and stairwells", I think the brightness of this lamp in low-light mode is just right.	4.32	0.677	226	10
11. When this sensor lamp is installed in "parking lots", I think the brightness of this lamp in low-light mode is just right.	4.32	0.732	222	11
12. When this sensor lamp is installed in the "toilets", I think the brightness of this lamp in low-light mode is just right.	4.27	0.701	224	12
13. When the light is installed in "toilets", I think its sensitivity and response distance is appropriate.	4.25	0.663	225	13
14. When this sensor lamp is installed in the "toilet", I think the lamp stays fully lit for just the right length of time (currently 30 s) when someone passes by.	4.22	0.739	227	14

In terms of the satisfaction for the "parking lots", "When this sensor lamp is installed in a "parking lot", I think the lamp stays on for just the right length of time (currently 30 s) when someone passes by" (m = 4.37)" is the most popular among consumers. In terms of satisfaction for "corridors and stairwells", "When this sensor lamp is installed in "corridors and stairwells", I think its sensitivity and response distance are appropriate" (m = 4.36) is the most popular among users. In terms of satisfaction for the "toilets", "When this sensor lamp is installed in the "toilets", I think the brightness of this lamp in low-light mode is just right" (m = 4.27) is the most popular among consumers.

According to the above results, the satisfaction rate of this sensor lamps installed in different areas is in the order of parking lot > corridors, stairwells > toilets, and "Full brightness when someone passes by and maintain for 30 s" is the most popular among consumers. To conclude, this sensor lamp can maintain brightness and provide proper response while saving energy, and it is satisfactory to the consumers.

We design 4 modes of intensity as 65%, 50%, 35%, and 20% of high-light mode to experiment the most appropriate intensity in our low-light mode. The experimental result of 20 subjects as below: 12 subjects responded positively to 50% intensity of high-light mode, 2 subjects responded positively to 65% intensity of high-light mode, 2 subjects responded positively to 35% intensity of high-light mode, and 4 subjects responded positively to 20% intensity of high-light mode. It will be seen from this result that the 50% intensity of high-light mode is the most appropriate intensity for alleviating the discomfort of brightness variation.

6. Discussion

From the above discussion, it is known that the power consumption of this new type sensor lamp in low-light mode is only one-tenth of that in high-light mode, but its brightness can be up to half of the high-light mode, which is very suitable for the sensor lighting in the parking lots, corridors, stairwells, and toilets. It only requires the replacement of the lamp but not the original lamp socket, yet the basic brightness can be maintained. Take the general 15W T8 LED lamp (sensor-less) as an example, if it is replaced by this new type of sensor lamp, and the place where it is installed is rarely passed by people, the power saving rate will be as high as 90%. Assuming that there are 12 passers-by per hour, the saving rate is still 81%.

This work has already been adopted in many schools, including Cheng-Shu University, I-Shou University, National Chiayi Economics Vocational High School, Chiayi County Jhuci Senior High School, Budai Junior High School, Guang Rong Elementary School and so on, and it is expected to be extended to various schools nationwide.

In this paper, we use Cheng Shiu University as an example of the energy-saving analysis with the assumption that there are 12 people passing by per hour. If the frequency of passer-by is higher, the ESR will become less than 81%. Conversely, the energy-saving will be higher than 81% if the frequency of people passing by is lower. In addition, when we carried out the energy-saving analysis, we only calculated the power consumed by the lamp tube but not the circuit (about 1 W). If we include the power consumption of the circuit in the calculation (the power measured in the experiment is 15.5 W in high-light mode and 2.4 W in low-light mode, and it is assumed that the installation site is seldom passed by), the ESR is still at (15.5 − 2.4)/15.5 = 84.5%.

Author Contributions: C.-T.L. focus on the writing and integrating articles for research conceptual architecture, literature review, and product analysis. P.-T.H. writes data acquisition, statistical analysis, and data analysis. Besides, he adjusts the overall writing of the article. All authors have read and agreed to the published version of the manuscript.

Funding: This research received no external funding.

Conflicts of Interest: The authors declare no conflict of interest.

References

1. Sin-Rong, C. Effect of Light Emitting Diode Lumileds and Photovoltaic on Electric system in Taiwan. Master's Thesis, National Cheng Kung University, Tainan, Taiwan, 2007.
2. U.K. Government. *Meeting the Energy Challenge-A White Paper on Energy, Department of Trade and Industry*; U.K. Government: London, UK, 2007.
3. China Daily. *China Eyes 20% Renewable Energy by 2020*; China Daily: Beijing, China, 2009.

4. Lee, K.C.; Li, S.; Hui, S.Y. A Design Methodology for Smart LED Lighting Systems Powered By Weakly Regulated Renewable Power Grids. *IEEE Trans. Smart Grid* **2011**, *2*, 548. [CrossRef]
5. Li, D.H.W.; Cheung, K.L.; Wong, S.L.; Lam, T.N.T. An analysis of energy-efficient light fittings and lighting controls. *Appl. Energy* **2010**, *87*, 558. [CrossRef]
6. Tan, Y.K.; Huynh, P.T.; Wang, Z. Smart personal sensor network control for energy saving in DC grid powered LED lighting system. *IEEE Trans. Smart Grid* **2013**, *4*, 669. [CrossRef]
7. US Energy Information Administration. *Commercial Building Energy Consumption Survey*; U.S. Energy Information Administration: Washington, DC, USA, 2003.
8. Ozcelik, M.A. Light sensor control for energy saving in DC grid smart LED lighting system based on PV system. *J. Optoelectron. Adv. Mater.* **2016**, *18*, 468–474.
9. Farsakoglu, O.F.; Atik, I. Analysis of the factors affecting operation and efficiency of power LED drivers and circuit design. *Optoelectron. Adv. Mat.* **2015**, *9*, 1356.
10. Chiu, J.H.; Lo, Y.K.; Chen, J.T.; Cheng, S.J.; Chung, Y.L.; Mou, S.C. A high-efficiency dimmable LED driver for low-power lighting applications. *IEEE Trans. Ind. Electron.* **2010**, *57*, 735. [CrossRef]
11. Manuel, A.; Lamar, D.G.; Sebastian, J.; Balacco, D.; Aguissa, D.; Manuel, A.; Lamar, D.G.; Sebastian, J.; Balacco, D.; Aguissa, D. High-Efficiency LED Driver without Eletrolityc Capacitor for Street Lighting. *IEEE Trans Ind. Appl.* **2013**, *49*, 127.
12. Li, S.; Chen, H.; Tan, S.C.; Hui, S.Y.; Waffenschmidt, E. Power flow analysis and critical design issues of retrofit light-emitting diode (LED) light bulb. *IEEE Trans. Power Electron.* **2015**, *30*, 3830. [CrossRef]
13. Bureau of Energy, MOEA. New Century Power-Saving Environmental Lighting. Available online: https://energymagazine.itri.org.tw/Cont.aspx?CatID=30&ContID=542 (accessed on 20 May 2019).
14. LEDinside. The Global LED Indoor Commercial Lighting Market Would Reach USD15.87 Billion in 2018, Accounting for 42% of the Global LED Lighting Production Value. Available online: https://www.ledinside.com.tw/research/20180427-35233.html (accessed on 24 March 2018).
15. Huang, B.J.; Wu, M.S.; Hsu, P.C.; Chen, J.W.; Chen, K.Y. Development of high-performance solar LED lighting system. *Energy Convers. Manag.* **2010**, *51*, 1669–1675. [CrossRef]
16. Hadj, A.A.; Chenlo, F.; Benghanem, M. Loss-of-load probability of photovoltaic water pumping systems. *Sol. Energy* **2004**, *76*, 713–723. [CrossRef]
17. Yoonsik, U.; Hong, I.; Kim, G.; Lee, B.; Park, S. Design and implementation of power-aware LED light enabler with location-aware adaptive middleware and context-aware user pattern. *IEEE Trans. Consum. Electron.* **2010**, *56*, 231–239.
18. DT-LUX LED. Latest News. Query on LED Energy-Saving Lighting Subsidy Application in Taiwan: Community and Service Industry Can Apply for a Maximum Subsidy of NT$750 per Lamp. Available online: http://www.dt-lux.com.tw/LED%20FAQ/LED%20FAQ.html (accessed on 28 May 2019).

electronics

MDPI

Article

Multi-Chip IGBT Module Failure Monitoring Based on Module Transconductance with Temperature Calibration

Chenyuan Wang, Yigang He *, Chuankun Wang, Lie Li and Xiaoxin Wu

School of Electrical Engineering and Automation, Wuhan University, Wuhan 430072, China; 2014302540237@whu.edu.cn (C.W.); 2019202070020@whu.edu.cn (C.W.); 2017282070166@whu.edu.cn (L.L.); 2015302540160@whu.edu.cn (X.W.)

* Correspondence: yghe1221@whu.edu.cn

Received: 5 August 2020; Accepted: 21 September 2020; Published: 23 September 2020

Abstract: The Insulated Gate Bipolar Transistor (IGBT) is the component with the highest failure rate in power converters, and its reliability is a critical issue in power electronics. IGBT module failure is largely caused by solder layer fatigue or bond wires fall-off. This paper proposes a multi-chip IGBT module failure monitoring method based on the module transconductance, which can accurately monitor IGBT module chip failures and bond wire failures. The paper first introduces the failure mechanism and module structure of the multi-chip IGBT module; then, it proposes a reliability model based on the module transconductance and analyzes the relationship between chip failure, bond wire failure, and the transmission characteristic curve of the IGBT module. Finally, the module transconductance under chip failure and bond wire failure is measured and calculated through simulation, and the temperature is calibrated, which can eliminate the influence of temperature on health monitoring. The results show that the method has a high sensitivity to chip failures and bond wire failures, can realize the failure monitoring of multi-chip IGBT modules, and is of great significance for improving the reliability of power converters.

Keywords: multi-chip IGBT module; bond wire; module transconductance; temperature calibration; failure monitoring; reliability

1. Introduction

The Insulated Gate Bipolar Transistor (IGBT) module is a power electronic integrated module composed of multiple IGBT chips, diode chips, solder layers, bond wires, ceramic copper-clad substrates, heat dissipation base plates, and power terminals. IGBT modules are mainly used in inverters, frequency converters, uninterruptible power supplies, wind power, and solar power generation. They are the core components of power converters and have one of highest failure rates out of all the power converters [1]. To reduce the failure rate and improve the reliability of the power converter, it is of great significance to monitor the failure status of the IGBT module [2].

At present, many scholars at home and abroad are committed to the reliability research of IGBT modules, most of which focus on the online measurement of IGBT health-sensitive parameters (HSPs) or thermosensitive electrical parameters (TSEPs). Researchers measure the collector, gate, emitter, and appropriate Kelvin terminals of the IGBT module to obtain the external electrical parameters of the IGBT module, and realize the health monitoring of the IGBT module through the fault characteristic parameters [3–5]. Reference [6,7] proposed measuring IGBT saturation voltage drop $V_{CE(sat)}$ during the operation of the power converter. This method must ensure that the measured current is constant during operation, but it is difficult to achieve in practice. Reference [8–10] proposes using the on-state voltage drop $V_{CE(ON)}{}^{a}$ under load current to monitor the shedding of IGBT bond wires in real time. Still, this parameter has a weak anti-interference ability. In reference [11], the on-state voltage drop $V_{CE(on)}{}^{B}$ at the inflection point is used to monitor the aging and shedding of bond wires, which has strong resistance to junction temperature interference. However, the measurement of the on-state voltage drop $V_{CE(ON)}$ needs to consider the high-voltage isolation problem, and the measurement is from thousands of volts in the off phase to a few volts in the on phase. The measurement range is broad, and the measurement circuit must consider a large V_{CE} voltage swing. There are certain difficulties in practical applications. Reference [12] uses on-state resistance $R_{CE(ON)}$ to monitor the aging degree of the IGBT module package online. Although it has high sensitivity, its resistance to junction temperature interference is low, and the first bond wire lift-off in the IGBT module cannot be detected. Reference [13–15] uses the peak gate current I_{GPeak} to monitor the shedding of the bond wire in the parallel IGBT module. Although this parameter has a strong resistance to changes in junction temperature, detecting the peak gate current value is difficult in practical applications. In addition, the gate current is relatively small, and the measurement is performed on the side close to the IGBT module. The detection circuit also needs to deal with the challenge of electromagnetic interference. In reference [16], the grid emitter voltage variation dV_{CE}/dt was used to detect the drop of the bond wire during the process of turning the circuit on and off. This method could not identify the first bond wire lift-off. Reference [17] proposes using the gate-emitter pre-threshold voltage $V_{GE(pre\text{-}th)}$ to monitor IGBT chip failures in multi-chip IGBT modules. This method can only monitor chip failures and cannot monitor a small number of bond wire fall-off failures. Most of the above research methods can only monitor one of the failures of the IGBT bond wire lift-off or the IGBT chip failure, and it cannot be monitored at the same time. Temperature-sensitive electrical parameters need to be considered when used as fault characteristic parameters. However, in the above studies, temperature-sensitive electrical parameters have not been normalized to eliminate the impact of temperature on failure monitoring.

Aiming at the shortcomings of the existing research methods, this paper proposes using the IGBT module transconductance as the characteristic quantity to identify both the chip failure and the bond wire failure in the multi-chip IGBT module and realize the failure monitoring of the IGBT module. The paper first analyzes the failure mechanism and module structure of the multi-chip IGBT module; then, it proposes a reliability model based on the module transconductance and analyzes the relationship between chip failure, bond wire failure, and the IGBT module transmission characteristic curve (i_C-u_{GE}) Finally, the quantitative relationship between chip failure and bond wire failure and module transconductance is measured and calculated through simulation. Furthermore, the temperature is normalized, which can eliminate the influence of temperature on failure monitoring. The results show that the method has good sensitivity to chip failures and bond wire failures, can realize the failure monitoring of multi-chip IGBT modules, and is of great significance for improving the reliability of power converters.

The sections of this article are arranged as follows. In Section 2, the failure mechanism and module structure of the multi-chip IGBT module are introduced in detail. In Section 3, a reliability model based on the module transconductance is constructed. Section 4 is based on the reliability model to monitor the chip failure and bond wire failure status of the DIM800NSM33-F IGBT module. The results are discussed and analyzed in Section 5.

2. Multi-Chip IGBT Module Failure Mechanism and Structure

The failure of the solder layer and the falling off of the bond wires are the main aging failure mechanisms of IGBT modules. The bond wire is mainly used to connect the chip (IGBT chip or diode chip) and the terminal (gate or emitter). The main reason for the bond wire to fall off is the inconsistent coefficient of thermal expansion (CTE) of each layer of the semiconductor device. Figure 1 shows the IGBT module packaging structure and the coefficient of thermal expansion of each layer of material. The IGBT chip and the diode chip are electrically connected by bond wires, as are the chip and the copper substrate. During the normal operation of the IGBT, the heat generated by the power loss of the semiconductor chip will pass through the multi-layer structure to the heat sink to produce a junction temperature fluctuation ΔT, which causes the bond wire to generate shear stress at its welding point, and the changing stress produces cracks, which leads to the bond wires fall-off [18,19]. Initially, cracks are generated at both ends of the bond wire; then, they are gradually extended to the middle until the bond wire completely falls off. Once a bond wire in the module falls off, the current passing through the remaining bond wires will increase immediately, thereby accelerating the fall-off process of the remaining bond wires [20]. The deformation ε_t of the IGBT due to the fluctuation of the junction temperature can be expressed by Equation (1):

$$\varepsilon_t = L(\alpha_{Al} - \alpha_{Si}) \cdot \Delta T \quad (1)$$

where α_{Al} and α_{Si} are the CTEs of aluminum and silicon, L is the length of the aluminum bond wire; ΔT is the junction temperature fluctuation.

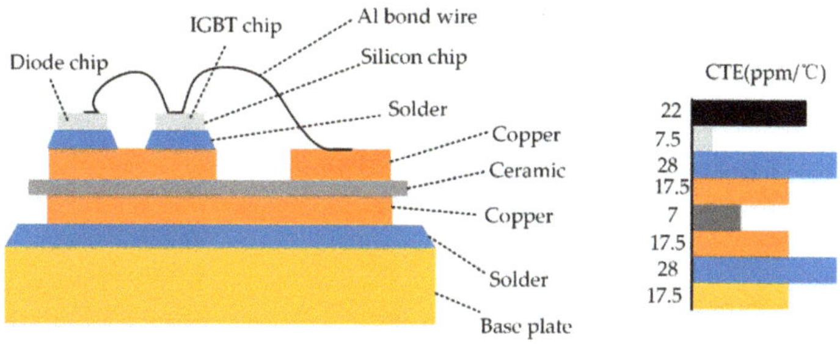

Figure 1. Insulated Gate Bipolar Transistor (IGBT) module structure and material coefficient of thermal expansion (CTEs).

For multi-chip IGBT modules, in addition to the parasitic parameters of the cell chip, stray parameters such as the substrate, board lining, solder layer, and bond wire in the module packaging process need to be considered [21]. Among the stray parameter changes caused by IGBT failure, the bond wire fall-off is the most typical. Therefore, the chip fault and bond wire fault studied in this paper are caused by the bond wire fall-off. If the bond wires of a certain unit IGBT chip are all falling off, the number of effective IGBT chips connected in parallel in the IGBT module will be reduced, and the fault type is chip failure at this time. If the bond wire of a certain unit IGBT chip falls off, the number of effective bond wires on the chip will be reduced, and the fault type is bond wire failure at this time. The reliability of multi-chip IGBT modules depends on the health of all parallel IGBT chips and bond wires. Therefore, it is very important to evaluate the overall health of the multi-chip IGBT module by monitoring IGBT chip failure and bond wire failure [22].

Figure 2 shows the equivalent diagram of the internal structure of the multi-chip IGBT module. There are n IGBT chips in parallel. This article takes the DIM800NSM33-F IGBT module as the research object. This module is a commercial module. The physical map and the internal equivalent circuit of a single chip in the module are shown in Figure 3. The module is a 16-unit IGBT module. Each chip has eight emitter bond wires, and there are 128 emitter bond wires in total. Refer to the data sheet for the basic parameters of the module, as shown in Table 1. The acceptable number of chip losses for IGBT modules depends on the working environment and working requirements. Generally, 10% chip losses in IGBT modules and 70% bond wire lift-off on the chip are considered acceptable [23]. Therefore, for the DIM800NSM33-F IGBT module, the detection of two chip failures in 16 chips is set as the safety margin. The failure simulation is the situation of the first two chips that failed; the first chip that failed is called "Chip 1", the second chip that failed is called "Chip 2".

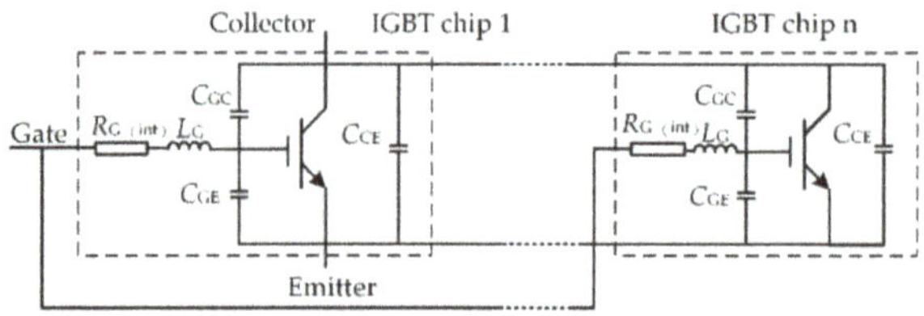

Figure 2. Internal structure diagram of multi-chip IGBT module.

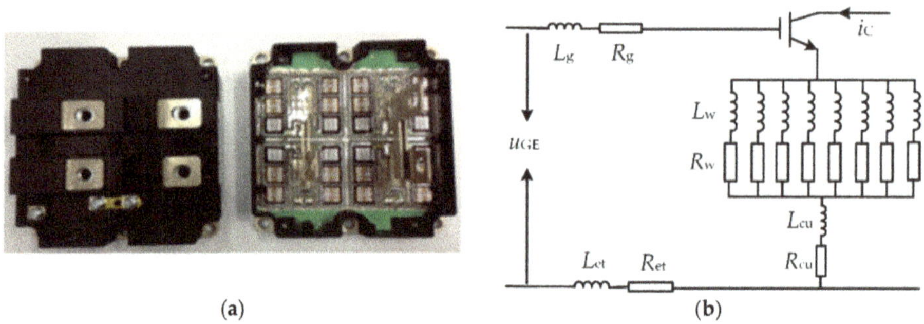

Figure 3. The physical diagram of the IGBT module and the internal equivalent circuit diagram of a single chip: (**a**) physical diagram; (**b**) equivalent circuit diagram.

Table 1. Basic parameters of IGBT module.

Gate Voltage Source V_{GG}	Internal Gate Resistance $R_{G(int)}$	Gate-Emitter Capacitance C_{GE}	Gate-Collector Capacitance C_{GC}	Gate Inductance L_G
±15 V	2.16 mΩ	8.8625 nF	0.1375 nF	12 nH

3. Reliability Model Based on Module Transconductance

The transmission characteristic curve of the IGBT module usually adopts Auto Curve Tracer to test, the horizontal axis is the gate-emitter voltage u_{GE}, and the vertical axis is the collector current i_C, which is affected by temperature. The general definition of transconductance refers to the slope of the transmission characteristic curve of the IGBT chip, which is the characteristic parameter of the IGBT chip itself. For the entire IGBT module, its module transconductance is affected by IGBT chip fatigue and bond wire failure. The transmission characteristic of IGBT refers to the response function of i_C to u_{GE} changes at different temperatures. The gradient of the transmission characteristic at a given temperature is called the transconductance of the device at that temperature. IGBT is a device that uses input voltage to control output current. The ratio of output current to input voltage is represented by transconductance, which also reflects the gain of IGBT. The chip transconductance g_{mc} reflects the sensitivity of i_c to u_{GE}, and the module transconductance g_m reflects the sensitivity of i_c to u_i. This article mainly focuses on the failure research of the IGBT module chip and bond wire, which belongs to module package level failure: a slow and gradual process.

The relationship between IGBT chip fatigue, bond wire lift-off, and module transconductance is analyzed in detail below. According to the working principle of the IGBT chip, when the IGBT chip operates in the active area, the governing equation of the collector current i_c is shown in Equation (2) [24]:

$$i_c = \frac{\mu_{ni} C_{OX} Z}{2L(1-\alpha_{pnp})}(u_G - U_{GE(th)})^2 = A(u_{GE} - U_{GE(th)})^2 \tag{2}$$

where μ_{ni} is the electron migration speed, C_{OX} is the oxide layer capacitance, Z and L are the length and width between the gate-emitter of the MOSFET, α_{pnp} is the current gain of the PNP transistor,

$U_{GE(th)}$ is the threshold voltage of the IGBT chip; the above parameters are all determined by the chip structure and material, but they are all affected by the junction temperature T_j of the chip.

For simplification, the parameter A is used to characterize the parameters related to the physical size and structure of the chip in Equation (2), and the chip transconductance g_{mc} is the sensitivity of i_c to u_{GE} as shown in Equation (3):

$$g_{mc} = \frac{di_c}{du_{GE}} = 2A(u_{GE} - U_{GE(th)}). \tag{3}$$

From Figure 3b, we can see that for a single IGBT branch, the circuit equation during the switching transient device is Equation (4):

$$u_G = u_i - i_C R_w - L_G \frac{di_G}{dt} - i_G R_G - (L_{cu} + L_{et}) \frac{d(i_G + i_C)}{dt} - (i_G + i_C)(R_{cu} + R_{et}) \tag{4}$$

where R_w is the equivalent parasitic resistance of bond wires; R_{cu} and L_{cu} are the equivalent parasitic resistance and inductance of the copper layer; u_i is the gate drive voltage of the IGBT module; and i_G is the gate current of the IGBT module. L_{et} is the equivalent stray inductance of the IGBT emitter lead, and R_{et} is the equivalent stray resistance of the IGBT emitter lead.

The parasitic resistance of the IGBT is about 40 μΩ, the parasitic inductance is about 30 nH, and the bonding wire resistance is about 50 mΩ. The parasitic resistance and all parasitic inductances are relatively small compared to the bonding wire resistance. In practical applications, di_c/dt is about 3000 A/μs, and di_g/dt is about 0.2 A/μs. When the IGBT works in the safe active region, the collector current i_c and the gate current i_g remain almost constant. Considering that the measurement of the transconductance g_m of the detection parameter module in this paper is when the IGBT is working in the safe active area, Equation (4) can be simplified to Equation (5):

$$u_G = u_i - i_c R_w. \tag{5}$$

When u_i is equal to $U_{GE(th)}$, the current flowing through the IGBT is zero. In order to study the influence of IGBT module transconductance and bond wire resistance R_W on the module transconductance, we substitute Equation (5) into Equation (1) and perform appropriate transformations to obtain the relationship between i_c and u_i as Equation (6):

$$i_c = \frac{2AR_w\left(u_i - U_{GE(th)}\right) + 1 - \sqrt{4AR_w\left(u_i - U_{GE(th)}\right) + 1}}{2AR_w{}^2}. \tag{6}$$

The module transconductance g_m of the IGBT module reflects the sensitivity of i_c to u_i, which can be derived as Equation (7):

$$g_m = \frac{di_c}{du_i} = \frac{1}{R_w}\left(1 - \frac{1}{\sqrt{4AR_w\left(u_i - U_{GE(th)}\right) + 1}}\right). \tag{7}$$

Comparing Equations (2) and (6), we can see that the relationship between chip transconductance g_{mc} and module transconductance g_m is as shown in Equation (8):

$$g_m = \frac{1}{R_w + 1/g_{mc}}. \tag{8}$$

The transconductance of the IGBT module is not only related to the transconductance of the IGBT chip itself but also to the health of the parallel chip and the bond wires. The module transconductance will decrease as R_w increases during the aging failure process. From Equation (2), we can get Equation (9):

$$u_i - U_{GE(th)} = \frac{Ri_c\sqrt{A} + \sqrt{i_c}}{\sqrt{A}}. \tag{9}$$

Then, we put Equation (9) into Equation (7) to get the module transconductance, as shown in Equation (10). The transconductance expression of this module has nothing to do with the threshold voltage affected by the gate oxide layer.

$$g_m = \frac{1}{R_w}\left(1 - \frac{1}{\sqrt{4Ai_cR_w{}^2 + 4\sqrt{Ai_c}R_w + 1}}\right) \tag{10}$$

Among the failure modes of power devices, package-related failures are the most common and are generally considered to be the main factors affecting the life of IGBTs. It can be seen from Equation (10) that the module transconductance is a function of parameter R_w, parameter A, and collector current i_c. Parameter A characterizes the physical size and structural parameters of the chip, and it is generally believed that these parameters will not change with aging. In this case, the bond wire lift-off is the only aging factor that affects the module transconductance change. For multi-chip IGBT modules, the chip characteristics and bond wire resistance on each branch may not be consistent, and the gate oxide degradation of each chip may also be different. For simplicity, it is assumed that the chip branches are consistent, and the aging process is also consistent, so as to eliminate the influence of the threshold voltage $U_{GE(th)}$.

The junction temperature will affect the electrical characteristics of the IGBT. The dependence of the physical material parameters of the silicon-based semiconductor on temperature determines the temperature dependence of its operating characteristics. The lifetime of carriers increases with the decrease of temperature and mobility, and the charge storage in the drift region decreases with the increase of temperature, so the IGBT switching process is affected by temperature. In the IGBT module, both the parameter R_w and the parameter A are affected by the junction temperature, and the transconductance of the module under the same collector current can be expressed as Equation (11):

$$g_m = f(R_w, T). \tag{11}$$

In actual engineering, with the aging of the IGBT module, the equivalent resistance of the module will change accordingly. At the same time, the junction temperature T_j will also fluctuate during the normal operation of the power converter [25,26]. The influence of fluctuation on the transconductance of the module can be expressed by Equation (12):

$$\begin{cases} \Delta g_{mi\_T} = f(T_i, R_{w0}) - f(T_0, R_{w0}) \\ \Delta g_{mi\_Rw} = f(T_0, R_{wi}) - f(T_0, R_{w0}) \end{cases} \tag{12}$$

where T_0 is the selected reference temperature; R_{w0} is the equivalent resistance of bonding wire in the healthy state of the IGBT module; T_i is the temperature after change; and R_{wi} is the bond wire equivalent resistance after module aging failure.

Figure 4 shows the transmission characteristic curve before and after the IGBT module bonding wire lift-off. As the bond wires fall off, the collector current gradually decreases in the active area. At the same collector current i_C measurement point, the gate-emitter voltage after the bond wire falling off increases from u_{GE1} to u_{GE2}, and the power loss of the IGBT module will also increase, which will lead to an increase in junction temperature and failure of the bond wire. The resulting junction temperature difference is $T_2 - T_1$. Therefore, when the module transconductance is used to monitor the health of a multi-chip parallel IGBT module, it is necessary to eliminate the effect of junction temperature. The extraction of module transconductance requires the chip to work in the active area, and it is also necessary to consider that the IGBT module is located in the safe operating area (SOA).

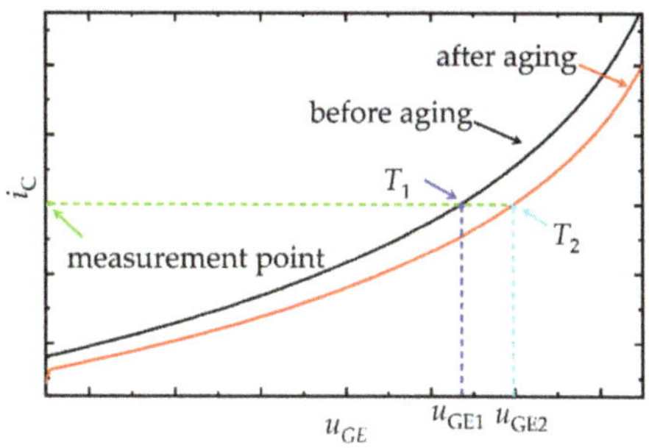

Figure 4. Transmission characteristic curve of the IGBT module before and after aging.

4. Model and Test

Figure 5 shows the schematic diagram of multi-chip IGBT module failure monitoring based on module transconductance. According to the test circuit, the collector current and gate emitter voltage are measured to obtain the transmission characteristic curve, and then the transconductance of the module is calculated according to the transmission characteristic curve. The DC bus power supply voltage of the test circuit is 1800 V, and the load is an inductive load of 400 μH. The tested IGBT module is driven by the gate driver with a voltage pulse of approximately −15 to +15 V. Between the ideal pulse voltage power supply and the IGBT gate terminal, 3.9 Ω turn-on gate resistance $R_{G(ext),on}$ and 6.2 Ω turn-off gate resistance $R_{G(ext),off}$ are used. In this paper, the test of the transmission characteristic curve under the condition of chip failure and bonding wire failure is completed in Matlab/Simulink.

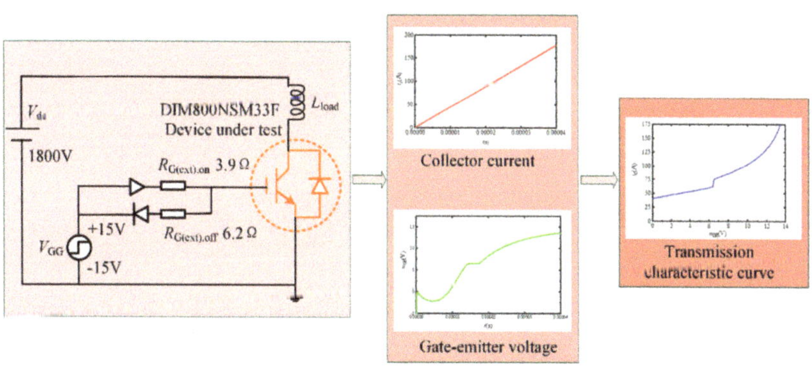

(a) (b) (c)

Figure 5. Schematic diagram of failure monitoring: (**a**) Test circuit; (**b**) Measurement i_C and u_{GE}; (**c**) Acquisition of transmission characteristic curve.

4.1. IGBT Chip Failure

The type of chip failure studied in this paper is that all bond wires on a chip fall off or the IGBT chip is completely fatigued due to the optimus effect caused by overcurrent, overvoltage, overheating, or beyond the shutdown safe working area. The research object DIM800NSM33-F IGBT module of this paper is a multi-chip parallel IGBT module. All parallel chip branches are regarded as the same, and chip failure is simulated by changing the number of parallel chip branches. Considering the safety margin of the module under study—two chip failures, the chip position has a relatively small influence, so the influence of the chip failure position is not considered for the time being. Select the reference temperature T_0 = 2 °C, and consider the safety margin of the selected IGBT module. In the above model, the IGBT module is tested under three different health conditions: "health", "one chip failure", and "two chip failure". The transmission characteristic curve of the IGBT module under different chip fault conditions is shown in Figure 6. With the increase of the number of chip failures, under the same collector current i_C, the gate-emitter voltage u_{GE} gradually increases; under the same gate-emitter voltage u_{GE}, the collector current i_C gradually decreases. The slope of the transmission characteristic curve of the IGBT module also decreases gradually. That is, the transconductance of the IGBT module decreases gradually. When the gate voltage u_{GE} is less than the threshold voltage $V_{GE(th)}$, the IGBT is in the off state. In most of the collector current range after the IGBT is turned on, i_C and u_{GE} have a good positive correlation. Considering that the IGBT works in the active area and is located in the safe working area, and the transmission characteristic curves have obvious differences under different fault conditions, the u_{GE} value is selected as 12 V to calculate the transconductance value of the module. According to the transmission characteristic curve, the transconductance value of the IGBT module under different chip fault states is calculated, as shown in Table 2. It can be seen that the transconductance of the module gradually decreases with the increase of IGBT failure chips, and the transconductance of the module decreases by about 6.2% when an IGBT chip fails. When two IGBT chips in the module fail, the transconductance of the module is reduced by 12.472%. At this time, the health status of the module has exceeded the safety margin, and the module is close to failure. It should be repaired or replaced in time to improve the reliability of the power converter.

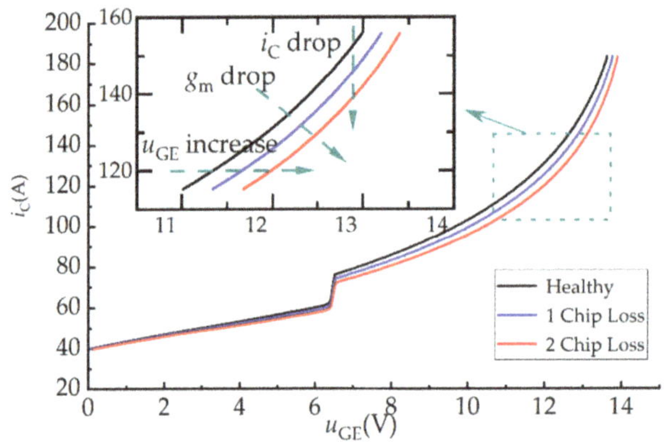

Figure 6. Transmission characteristic curves of IGBT modules under chip failure states.

Table 2. Transconductance of IGBT modules under chip failure states.

Number of IGBT Chip Failures	IGBT Module Transconductance g_m (A/V)	Percentage Change (%)
0	19.1843	0
1	17.9982	−6.243
2	16.8176	−12.472

When the IGBT is aging and failing, it will cause changes in parameters such as the thermal resistance and heat capacity, which will change the junction temperature of the IGBT, which will affect the transconductance parameters of the IGBT module. This article attributes the effect of aging failure to the effect of temperature on the transconductance, so it is necessary to exclude the effect of temperature on the module transconductance. When the IGBT module was in a healthy state, that is, when there was no chip failure and no bond wire fall-off, we measured and calculated the transconductance of the IGBT module at 25, 50, 75, 100, and 125 °C, as shown in Table 3. When the module is in a healthy state, as the temperature increases, the transconductance of the IGBT module gradually decreases. Figure 7 shows the relationship between the IGBT module transconductance and temperature. It can be seen from the figure that the module transconductance and temperature show a very good linear relationship, so the linear fitting can be used to obtain the module transconductance and temperature. The relationship, as shown in Equation (13), has a linear fit of up to 0.998. In order to eliminate the influence of temperature, the module transconductance at different temperatures can be calibrated to 25 °C according to Equation (13). At this time, the module transconductance is only related to the module's health status.

$$g_{\mathrm{mT,j\_25\,°C}} = g_{\mathrm{mT,j}} + 0.00664\,(T - 25) \tag{13}$$

where T is the temperature, j is the module's different health states, $g_{\mathrm{mT,j}}$ is the transconductance at different temperatures and different health states, and $g_{\mathrm{mT,j\_25\,°C}}$ is the module's transconductance calibrated to 25 °C.

Table 3. Transconductance of the IGBT module in a healthy state at different temperatures.

Temperature (°C)	IGBT Module Transconductance g_m (A/V)
25	19.1843
50	19.0316
75	18.8554
100	18.6828
125	18.5292

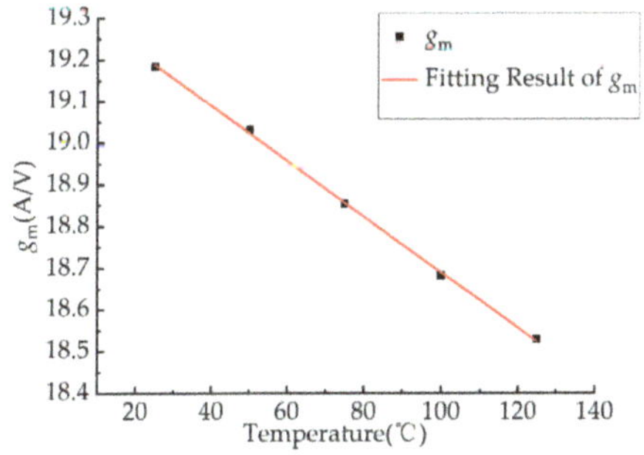

Figure 7. The relationship between IGBT module transconductance and temperature.

We tested different IGBT chip fault types at 25, 50, 75, 100, and 125 °C and calculated the IGBT module transconductance, as shown in Table 4. It can be seen that the module transconductance decreases as the number of IGBT chips increases, and it decreases as the temperature increases. Figure 8a shows the module transconductance under different temperatures and health conditions. Within a certain temperature range, the module transconductance can be used to determine the number of IGBT module chip failures and realize the health status monitoring of the IGBT module.

Table 4. Transconductance of IGBT modules under different temperatures and healthy states.

Number of IGBT Chip Failures	IGBT Module Transconductance g_m (A/V)				
	T = 25 °C	T = 50 °C	T = 75 °C	T = 100 °C	T = 125 °C
0	19.1843	19.0212	18.8554	18.6937	18.5268
1	17.9982	17.8328	17.6823	17.5135	17.3416
2	16.8176	16.6553	16.4876	16.3231	16.1587

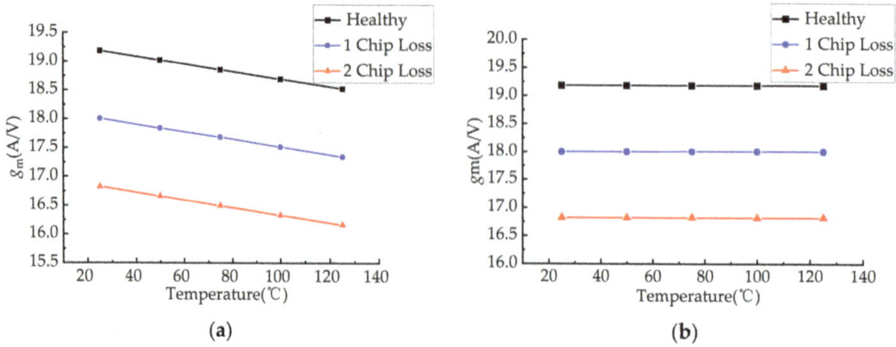

Figure 8. Transconductance of the IGBT module with chip failure: (**a**) before temperature calibration (**b**) after temperature calibration.

In order to monitor the health status of the IGBT module more accurately and eliminate the influence of temperature, the module transconductance at different temperatures is calibrated to the reference temperature T = 25 °C according to Equation (13). Figure 8b shows the module transconductance after temperature calibration. Compared with the module transconductance that has not undergone temperature correction, g_m after temperature calibration is only related to the module's health status. The latter can more intuitively and clearly judge the type of chip failure and more accurately assess the health of the IGBT module status.

4.2. Bond Wire Failure

The bond wire failure studied in this paper is caused by the bond wire lift-off. The DIM800NSM33-F IGBT module has eight bond wires on each IGBT chip. Considering the maximum safety margin of the IGBT module, the following will test and analyze different types of bond wire faults on Chip 1 and Chip 2.

For Chip 1, the state of the bond wire is i (i = 0, 1, 2, ..., 8) bond wire lift-off. When the reference temperature T_C = 25 °C, the transmission characteristic curve of the measurement of Chip 1 under different bond wire failure states is shown in Figure 9. After the eighth bond wire lift-off, the transmission characteristic curve is completely separated from other curves, and Chip 1 is completely invalid at this time. This is because all the bond wires fall off so that the chip is completely faulty and invalid, and the electrical characteristics of the IGBT module have changed significantly, making the transmission characteristic curve clearly separated. According to the transmission characteristic curve, the module transconductance corresponding to the number of bond wire shedding is calculated, as shown in Table 5. It can be seen from Table 5 that the module transconductance gradually decreases with the increase in the number of bond wires falling off. Each time a bond wire lifts off, the module transconductance decreases about 0.753% on average. When the last bond wire falls off, the transconductance of the module changes significantly.

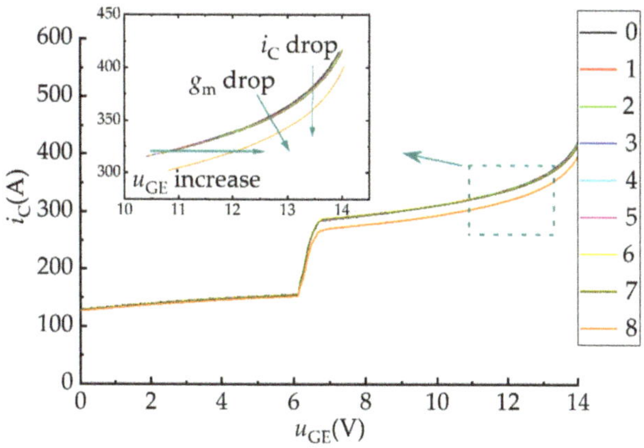

Figure 9. Transmission characteristic curves of Chip 1 under bond wire failure states.

Table 5. Module transconductance under bond wire failure states of Chip 1.

Number of Chip 1 Bond Wires Lift-Off	IGBT Module Transconductance g_m (A/V)	Percentage Change (%)
0	19.1843	0
1	19.0469	−0.723
2	18.8914	−1.542
3	18.8065	−1.988
4	18.7641	−2.211
5	18.6086	−3.030
6	18.4774	−3.721
7	18.3015	−4.646
8	18.0772	−5.827

The same measurements and calculations are done for different bond wire fault states of Chip 1 at different temperatures, and the module transconductance is obtained, as shown in Table 6. It can be seen from Figure 10a that under the same bond wire state, the temperature is different, the module transconductance value is also different, and the temperature has a great influence on the type of bond wire failure of the monitoring of Chip 1, and it is impossible to accurately determine the number of Chip 1 bond wires falling off. For example, when Chip 1 is in a healthy state, the decrease of the module transconductance caused by the temperature increase may be misjudged as a bond wire failure state.

Table 6. Module transconductance of Chip 1 in bond wire fault states at different temperatures.

Number of Bond Wires Lift-Off	IGBT Module Transconductance g_m(A/V)				
	T = 25 °C	T = 50 °C	T = 75 °C	T = 100 °C	T = 125 °C
i = 0	19.1843	19.0187	18.8754	18.6753	18.5412
i = 1	19.0469	18.8852	18.7231	18.5498	18.3864
i = 2	18.8914	18.7316	18.5576	18.4135	18.2418
i = 3	18.8065	18.6451	18.4832	18.3218	18.1517
i = 4	18.7641	18.5935	18.4426	18.2687	18.1102
i = 5	18.6086	18.4506	18.2775	18.1174	17.9483
i = 6	18.4774	18.3079	18.1621	17.9789	17.8235
i = 7	18.3015	18.1427	17.9812	17.8056	17.6518
i = 8	18.0772	17.9185	17.7509	17.5903	17.4186

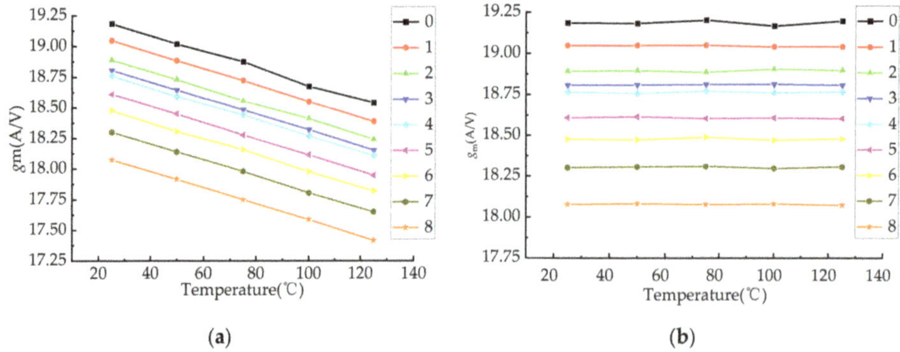

Figure 10. Transconductance of Chip 1 bonding wire fault module: (**a**) before temperature calibration and (**b**) after temperature calibration.

In order to reduce the error rate of bond wire fault monitoring and improve the accuracy of IGBT module health monitoring, it is necessary to eliminate the influence of temperature. Therefore, Equation (13) can be used to calibrate the module transconductance at different temperatures to the reference temperature T_C = 25 °C, as shown in Table 7; then, the temperature calibrated module transconductance can be used to achieve IGBT module health condition monitoring. The module transconductance interval is different under different bond wire failure states; that is, the module transconductance difference is different, and the percentage of change is different. When the last bond wire falls off, the module transconductance interval is the largest, and the module transconductance change is the largest. The module transconductance and junction temperature show a good linear relationship under the bonding wire failure state. Figure 10b shows the transconductance of the IGBT module after temperature correction. It can be seen from the figure that the calibrated module transconductance can accurately distinguish the chip bonding wire off at any temperature, and it is basically not affected by temperature changes. It can accurately monitor the health status of the IGBT module.

Table 7. Module transconductance of Chip 1 bond wire fault state after temperature calibration.

Number of Bond Wires Lift-Off	IGBT Module Transconductance g_m (A/V)				
	T = 25 °C	T = 50 °C	T = 75 °C	T = 100 °C	T = 125 °C
i = 0	19.1843	19.1824	19.2028	19.1664	19.1960
i = 1	19.0469	19.0489	19.0505	19.0409	19.0412
i = 2	18.8914	18.8953	18.885	18.9046	18.8966
i = 3	18.8065	18.8088	18.8106	18.8129	18.8065
i = 4	18.7641	18.7572	18.7700	18.7598	18.7650
i = 5	18.6086	18.6143	18.6049	18.6085	18.6031
i = 6	18.4774	18.4716	18.4895	18.4700	18.4783
i = 7	18.3015	18.3064	18.3086	18.2967	18.3066
i = 8	18.0772	18.0822	18.0783	18.0814	18.0734

Similarly, in order to monitor the health status of the bond wires of Chip 2, when the temperature is 25 °C, the transmission characteristic curves of different bond wires of Chip 2 under fault conditions are measured, as shown in Figure 11. The change rule of the transmission characteristic curve is the same. According to the calculation of the transmission characteristic curve, the module transconductance corresponding to the number of bond wires lift-off is shown in Table 8. It can be seen from Table 8 that the module transconductance gradually decreases with the increase in the number of bond wires lift-off. For each bond wire lift-off, the module transconductance decreases by about 0.812% on average. The change rule is basically the same as that of the Chip 1 bond wire failure state.

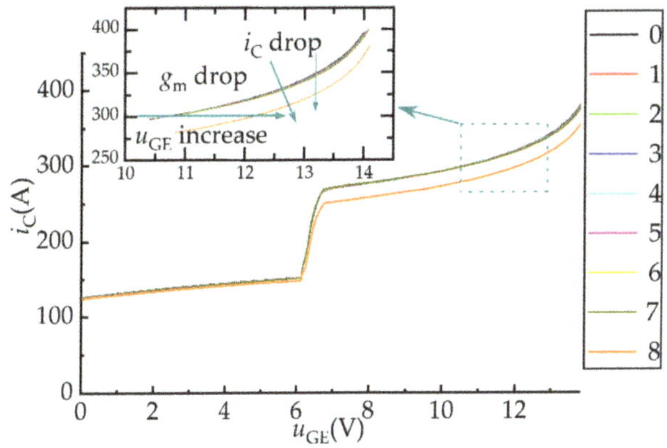

Figure 11. The transmission characteristic curve of Chip 2 under bond wire failure states.

Table 8. Module transconductance of Chip 2 under bond wire failure states.

Number of Bond Wires Lift-Off	IGBT Module Transconductance g_m (A/V)	Percentage Change (%)
0	18.0147	−6.156
1	17.9369	−6.565
2	17.8167	−7.198
3	17.7036	−7.793
4	17.6258	−8.203
5	17.4279	−9.244
6	17.2299	−10.286
7	17.0743	−11.105
8	16.9188	−11.924

In order to study the influence of temperature on the monitoring of the fault state of the bond wire of Chip 2, the transmission characteristic curves of different bond wire fault states are measured at temperatures of 25, 50, 75, 100, and 125 °C, and the IGBT is obtained by calculation. The module transconductance is shown in Table 9. It can be seen from Figure 12a that with the increase of temperature, the module transconductance decreases gradually, which is consistent with the change of the number of bond wires lift-off. The increase of IGBT module transconductance caused by temperature rise is easy to be misjudged as the falling off of bond wire, so the influence of temperature change will interfere with the monitoring of the failure state of bond wires.

Table 9. Module transconductance of Chip 2 with bond wire failure at different temperatures.

Number of Bond Wire Lift-Off	IGBT Module Transconductance g_m (A/V)				
	T = 25 °C	T = 50 °C	T = 75 °C	T = 100 °C	T = 125 °C
i = 0	18.0147	17.8479	17.6754	17.5316	17.3618
i = 1	17.9369	17.7682	17.5897	17.4429	17.2865
i = 2	17.8167	17.6581	17.4962	17.3315	17.1589
i = 3	17.7036	17.5454	17.3705	17.2183	17.0535
i = 4	17.6258	17.4651	17.3326	17.1287	16.9842
i = 5	17.4279	17.2645	17.0898	16.9465	16.7706
i = 6	17.2299	17.0731	16.8956	16.7498	16.5802
i = 7	17.0743	16.9305	16.7510	16.5345	16.4832
i = 8	16.9188	16.7554	16.6031	16.4010	16.2972

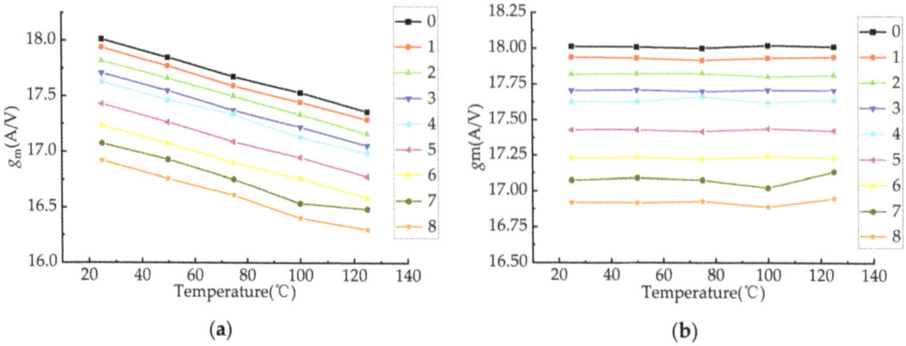

Figure 12. Module transconductance of Chip 2 with bond wire failure: (**a**) before temperature calibration and (**b**) after temperature calibration.

In order to eliminate the interference of temperature factors on health status monitoring, Equation (13) is used to normalize the module transconductance at different temperatures to the reference temperature T_c = 25 °C, as shown in Table 10. Then, the module transconductance after temperature calibration is used to monitor the health status of the IGBT module. Figure 12b shows the transconductance of the IGBT module after temperature correction. It can be seen from the figure that the calibrated module transconductance can accurately distinguish the chip bond wire lift-off condition at any temperature and monitor the health status of the IGBT module.

Table 10. Module transconductance of Chip 2 with bond wire failure after temperature normalization.

Number of Bond Wires Lift-Off	IGBT Module Transconductance g_m (A/V)				
	T = 25 °C	T = 50 °C	T = 75 °C	T = 100 °C	T = 125 °C
i = 0	18.0147	18.0116	18.0028	18.0227	18.0166
i = 1	17.9369	17.9319	17.9171	17.9340	17.9413
i = 2	17.8167	17.8218	17.8236	17.80226	17.8137
i = 3	17.7036	17.7091	17.6979	17.7094	17.7083
i = 4	17.6258	17.6288	17.6600	17.6198	17.6390
i = 5	17.4279	17.4282	17.4172	17.4376	17.4254
i = 6	17.2299	17.2368	17.2230	17.2409	17.2350
i = 7	17.0743	17.0942	17.0784	17.0256	17.1380
i = 8	16.9188	16.9191	16.9305	16.8921	16.9520

4.3. Result Verification

In the reference [27] "Research on Accelerated Aging and Aging Characteristic Parameters of IGBT Modules", the author chooses multi-chip parallel IGBT modules as the research object and uses the method of artificial wire cutting to simulate the falling off of the IGBT module bonding wires. Among them, the experimental test results of Chip 1 falling off zero, one, two, three, and four bonding wires are shown in Table 11. The simulation results of this article are also recorded in Table 11 for comparison.

It can be seen from Table 11 that the transconductance g_m of the two modules are similar, and with the increase in the number of bond wires falling off, the change trend and rate of change of the transconductance g_m of the module are also very close; that is, the parallel IGBT module experiment result is similar to the parallel IGBT module simulation. The results are basically the same, with only slight differences, and the number of bond wires falling off can be judged by the percentage of transconductance change. Therefore, based on the experimental data, this paper can preliminarily show that the theoretical method and simulation results are correct and effective.

Table 11. Verification of simulation results.

Number of Bond Wires Lift-Off	Transconductance g_{m1} Experiment Results (A/V)	g_{m1} Percentage Change (%)	Transconductance g_{m2} Simulation Results (A/V)	g_{m1} Percentage Change (%)
0	22.528	0	19.184	0
1	22.326	−0.90	19.047	−0.723
2	22.259	−1.20	18.891	−1.542
3	21.998	−2.35	18.807	−1.988
4	21.730	−3.54	18.764	−2.211

5. Result Analysis

According to the above theoretical analysis and test results, it can be known that the chip failure and bond wire failure in the multi-chip IGBT module will cause the module transmission characteristic curve u_{GE}-i_C to change, and the module transconductance value calculated from the transmission characteristic curve can be used as characteristic parameters for health monitoring of the IGBT module. Since the temperature dependence of the module transconductance can conceal its fault characteristics or reduce its ability to monitor bond wire or chip faults, temperature calibration processing is required to eliminate temperature effects. The temperature-corrected module transconductance can effectively monitor chip failures and bond wire failures in IGBT modules.

For the IGBT module selected in this article, when one IGBT chip fails, the value of the module transconductance decreases by 6.243%. When two IGBT chips fail, the value of the module transconductance decreases by 12.472%. At this time, it can be basically determined that the IGBT module is in an aging failure state. It means that system failure will occur at any time, and the IGBT module should be repaired or replaced in time. If 70% of the bond wires on one IGBT chip fall off, it can be determined that the chip is close to aging failure [28]. When six bond wires in Chip 1 lift off, the value of the module transconductance is reduced by 3.721%, which means that Chip 1 is about to fail; when six bond wires lift off in Chip 2, the value of the module transconductance decreases by 10.286%, which means that the system is at risk of failure, Chip 2 is about to age and fail, and the module should be degraded or repaired.

Table 12 compares different state monitoring methods. Under the same test object and condition, the pre-threshold voltage $V_{GE(pre\text{-}th)}$ during the gate-emitter turn-on transient, collector-emitter voltage change rate dV_{CE}/dt, and gate peak current $I_{G(peak)}$ have high sensitivity, but the three fault characteristic parameters are affected by temperature, and $V_{GE(pre\text{-}th)}$ and dV_{CE}/dt cannot be used to monitor bond wire faults. dV_{CE}/dt has a weak anti-interference ability. The $I_{G(peak)}$ measurement circuit needs to consider electromagnetic interference. The advantages of the method proposed in this article are as follows. (1) It can monitor chip faults and bonding wire faults at the same time, and it has good sensitivity. (2) This method has low sampling signal frequency and short measurement time, and it can ignore external heat sources and other electrical signals. It has a strong anti-interference ability such as the influence of external noise. (3) The parameter module mentioned in this article has a good linear relationship between the transconductance and temperature. This method is added to the temperature calibration process, which can be more accurately achieved under different junction temperatures IGBT module fault status monitoring. The method proposed in this paper can be used for in situ monitoring of the fault status of multi-chip parallel IGBT modules. It can be used to perform in situ monitoring when the converter is shut down (such as when the fan cuts off the wind speed, when the electric vehicle is stopped, etc.). Limitations and challenges include the stability of in situ monitoring devices, the lack of a standardized analysis of fault diagnosis, and condition assessment technology. However, this method cannot locate the specific location of chip faults and bond wire faults in the IGBT module and will continue to study the real-time online location and monitoring of IGBT module faults in the future.

Table 12. Comparison of IGBT module chip fault monitoring methods.

Method	1 Chip Failure	70% of the Bonding Wires Fall Off	Affected by Temperature	Anti-Interference Ability
$V_{GE(pre\text{-}th)}$	485 mV (8.182%)	-	yes	strong
dV_{CE}/dt	113.175 V/μs (7.545%)	-	yes	weak
$I_{G(peak)}$	0.1734 A (−9.372%)	0.04135 A (−2.235%)	yes	strong
This article	1.1861 A/V (−6.243%)	0.7069 A/V (−3.721%)	no	strong

6. Conclusions

This paper presents a method for monitoring the health of a multi-chip IGBT module based on module transconductance. According to the reliability model based on the module transconductance, the transmission characteristic curve of the selected IGBT module is measured in the test circuit, and its module transconductance value is calculated. The results show that the method can accurately identify the type of chip failure and the type of bond wire failure, and after temperature calibration, it can identify the aging failure process of the IGBT module without the influence of temperature. The method proposed in this article can simultaneously monitor the chip failure and the bond wire failure in the multi-chip IGBT module, and it is not affected by temperature. Compared with the existing monitoring method, it has better comprehensive characteristics, can realize the failure monitoring of the IGBT module, and improve the power inverter reliability.

Author Contributions: Conceptualization, methodology, software, validation, formal analysis, investigation, resources, C.W. (Chenyuan Wang) and Y.H.; data curation, writing—original draft preparation, C.W. (Chuankun Wang); writing—review and editing, X.W.; visualization, L.L. All authors have read and agreed to the published version of the manuscript.

Funding: This research was supported by the National Natural Science Foundation of China (Grant No. 51977153, 51977161, and 51577046), the State Key Program of National Natural Science Foundation of China (Grant No. 51637004), the national key research and development plan "important scientific instruments and equipment development" of China (Grant No.2016YFF0102200), the Equipment research project in advance of China (Grant No.41402040301).

Conflicts of Interest: The authors declare no conflict of interest.

References

1. Schulze, H.; Niedernostheide, F.; Pfirsch, F.; Baburske, R. Limiting Factors of the Safe Operating Area for Power Devices. *IEEE Trans. Electron Devices* **2013**, *60*, 551–562. [CrossRef]
2. Ghimire, P.; de Vega, A.R.; Beczkowski, S.; Rannestad, B.; Munk-Nielsen, S.; Thogersen, P. Improving power converter reliability: Online monitoring of high-power IGBT modules. *IEEE Ind. Electron. Mag.* **2014**, *8*, 40–50. [CrossRef]
3. Wang, Z.; Qiao, W. An Online Frequency-domain Junction Temperature Estimation Method for IGBT Modules. *IEEE Trans. Power Electron.* **2015**, *30*, 4633–4637. [CrossRef]
4. Bruckner, T.; Bernet, S. Estimation and Measurement of Junction Temperatures in a Three-level Voltage Source Converter. *IEEE Trans. Power Electron.* **2007**, *22*, 3–12.
5. Eleffendi, M.A.; Johnson, C.M. Application of Kalman Filter to Estimate Junction Temperature in IGBT Power Modules. *IEEE Trans. Power Electron.* **2016**, *31*, 1576–1587.
6. Choi, U.; Blaabjerg, F.; Jørgensen, S.; Munk-Nielsen, S.; Rannestad, B. Reliability Improvement of Power Converters by Means of Condition Monitoring of IGBT Modules. *IEEE Trans. Power Electron.* **2017**, *32*, 7990–7997.
7. Beczkowski, S.; Ghimre, P.; de Vega, A.R.; Munk-Nielsen, S.; Rannestad, B.; Thogersen, P. Online Vce Measurement Method for Wear-out Monitoring of High Power IGBT Modules. In Proceedings of the 15th European Conference on Power Electronics and Applications (EPE), Lille, France, 2–6 September 2013; pp. 1–7.

8. Smet, V.; Forest, F.; Huselstein, J.; Rashed, A.; Richardeau, F. Evaluation of Vce Monitoring as a Real-time Method to Estimate Aging of Bond wire-IGBT Modules Stressed by Power Cycling. *IEEE Trans. Ind. Electron.* **2013**, *60*, 2760–2770.
9. Ji, B.; Pickert, V.; Cao, W.; Zahawi, B. In Situ Diagnostics and Prognostics of Wire Bonding Faults in IGBT Modules for Electric Vehicle Drives. *IEEE Trans. Power Electron.* **2013**, *28*, 5568–5577.
10. Oh, H.; Han, B.; Mccluskey, P.; Han, C.; Youn, B.D. Physics-of-failure, Condition Monitoring, and Prognostics of Insulated Gate Bipolar Transistor Modules: A review. *IEEE Trans. Power Electron.* **2015**, *30*, 2413–2426.
11. Singh, A.; Anurag, A.; Anand, S. Evaluation of Vce at Inflection Point for Monitoring Bond Wire Degradation in Discrete Packaged IGBTs. *IEEE Trans. Power Electron.* **2017**, *32*, 2481–2484.
12. Eleffendi, M.A.; Johnson, C.M. In-service Diagnostics for Wire-bond Lift-off and Solder Fatigue of Power Semi-conductor Packages. *IEEE Trans. Power Electron.* **2017**, *32*, 7187–7198.
13. Baker, N.; Dupont, L.; Munk-Nielsen, S.; Iannuzzo, F.; Liserre, M. IR Camera Calidation of IGBT Junction Temperature Measurement via Peak Gate Current. *IEEE Trans. Power Electron.* **2017**, *32*, 3099–3111.
14. Baker, N.; Munk-Nielsen, S.; Iannuzzo, F.; Liserre, M. IGBT Junction Temperature Measurement via Peak Gate Current. *IEEE Trans. Power Electron.* **2016**, *31*, 3784–3793. [CrossRef]
15. Mandeya, R.; Chen, C.; Pickert, V.; Naayagi, R.T. Prethreshold Voltage as A Low-component Count Temperature Sensitive Electrical Parameter without Self-heating. *IEEE Trans. Power Electron.* **2018**, *33*, 2787–2791. [CrossRef]
16. Kexin, W.; Mingxing, D.; Linlin, X.; Jian, L. Study of bonding wire failure effects on external measurable signals of IGBT module. *IEEE Trans. Device Mater. Reliab.* **2014**, *14*, 83–89. [CrossRef]
17. Mandeya, R.; Chen, C.; Pickert, V.; Naayagi, R.T.; Ji, B. Gate-Emitter Pre-threshold Voltage as a Health-Sensitive Parameter for IGBT Chip Failure Monitoring in High-Voltage Multichip IGBT Power Modules. *IEEE Trans. Power Electron.* **2019**, *35*, 9158–9169. [CrossRef]
18. Du, M.; Kong, Q.; Ouyang, Z.; Wei, K.; Hurley, W.G. Strategy for Diagnosing the Aging of an IGBT Module by ON-State Voltage Separation. *IEEE Trans. Electron. Devices* **2019**, *66*, 4858–4864. [CrossRef]
19. Musallam, M.; Johnson, C.M. Real-time Compact Thermal Models for Health Management of Power Electronics. *IEEE Trans. Power Electron.* **2010**, *25*, 1416–1425. [CrossRef]
20. Hui, H.; Mawby, P.A. A Lifetime Estimation Technique for Voltage Source Inverters. *IEEE Trans. Power Electron.* **2013**, *28*, 4113–4119.
21. Luo, H.; Chen, Y.; Sun, P.; Li, W.; He, X. Junction Temperature Extraction Approach with Turn-Off Delay Time for High-Voltage High-Power IGBT Modules. *IEEE Trans. Power Electron.* **2016**, *31*, 5122–5132. [CrossRef]
22. Sun, P.; Gong, C.; Du, X.; Peng, Y.; Wang, B.; Zhou, L. Condition Monitoring IGBT Module Bond Wires Fatigue Using Short-Circuit Current Identification. *IEEE Trans. Power Electron.* **2017**, *32*, 3777–3786. [CrossRef]
23. Chen, C.; Pickert, V.; Al-Greer, M.; Jia, C.; Ng, C. Localization and Detection of Bond Wire Faults in Multichip IGBT Power Modules. *IEEE Trans. Power Electron.* **2020**, *35*, 7804–7815. [CrossRef]
24. Wang, K.; Zhou, L.; Sun, P.; Du, X. Monitoring Bond Wires Defects of IGBT Module Using Module Transconductance. *IEEE J. Emerg. Sel. Top. Power Electron.* **2020**, in press. [CrossRef]
25. Gao, B.; Yang, F.; Chen, M., Ran, L.; Ullah, I.; Xu, S.; Mawby, P.A. Temperature Gradient-Based Potential Defects Identification Method for IGBT Module. *IEEE Trans. Power Electron.* **2017**, *32*, 2227–2242. [CrossRef]
26. Hu, K.; Liu, Z.; Du, H.; Ceccarelli, L.; Iannuzzo, F.; Blaabjerg, F.; Tasiu, I.A. Cost-Effective Prognostics of IGBT Bond Wires with Consideration of Temperature Swing. *IEEE Trans. Power Electron.* **2020**, *35*, 6773–6784. [CrossRef]
27. Li, Y. Research on Accelerated Aging and Aging Characteristic Parameters of IGBT Modules. Ph.D. Thesis, Chongqing University, Chongqing, China, 2018.
28. Reigosa, P.D.; Wang, H.; Yang, Y.; Blaabjerg, F. Prediction of Bond Wire Fatigue of IGBTs in a PV Inverter Under a Long-Term Operation. *IEEE Trans. Power Electron.* **2016**, *31*, 7171–7182.

electronics

MDPI

Article

Investigation of Reverse Recovery Current of High-Power Thyristor in Pulsed Power Supply

Jiufu Wei, Zhenxiao Li and Baoming Li *

National Key Laboratory of Transient Physics, Nanjing University of Science and Technology, Nanjing 210094, China; wjfnjust@163.com (J.W.); lizhxnjust@126.com (Z.L.)

* Correspondence: baomingli@njust.edu.cn; Tel.: +86-139-5181-9662

Received: 1 July 2020; Accepted: 10 August 2020; Published: 12 August 2020

Abstract: The instantaneous overvoltages from the load side can cause damages of high-power thyristors in conventional pulsed power supply topologies, especially in cases of numerous pulse-forming units that operate together with discharge time intervals. The instantaneous overvoltages from the load side, which leads to high reverse recovery currents in high-power thyristors, can be induced by load mutations in the electromagnetic launching field. This paper establishes circuit models of PPS topologies, and investigates effects of the initial voltage of the energy-storage capacitor, the discharge time intervals, and the load resistance on the reverse recovery currents in high-power thyristors. To overcome the shortcomings of conventional PPS topologies, an improved PPS topology is developed. The improved PPS topology applies coupling inductor and resistance-capacitance snubber techniques, which can absorb the surge energy from the load side and reduce the reverse recovery currents in high-power thyristors. The simulation technique has been applied to validate theoretical analysis and the proposed model.

Keywords: PPS; high-power thyristors; reverse recovery currents; electromagnetic launching field

1. Introduction

Pulsed power supply (PPS) is widely applied in the production of nano-powder [1], drilling of hard rocks [2], electrothermal-chemical gun [3], electromagnetic railgun [4–10], etc. There are various energy storage ways for PPS, such as capacitor bank, homopolar generator-inductor, explosive magnetic flux compression [11]. The PPS investigated in this paper is based on the capacitor bank, which is mainly composed of high-power energy-storage capacitors (hereinafter referred to as energy-storage capacitors) [4], high-power semiconductor devices [12–20], and high-power pulse-shaping inductors (hereinafter referred to as pulse-shaping inductors). High-power semiconductor devices include high-power thyristors (hereinafter referred to as thyristors) [12–17] and high-power fast recovery diodes (hereinafter referred to as fast recovery diodes) [18–20], as shown in Figure 1.

The investigated PPS is widely applied in the electromagnetic launching (EML) field [3–10]. There are two types of conventional PPS topologies, which are defined as type I and II PPS topologies in this paper, respectively. Each PPS topology contains multiple pulse-forming units (PFUs) in parallel in practical applications. Compared with the devices such as energy-storage capacitors and pulse-shaping inductors, the semiconductor devices in conventional PPS topologies are easier to suffer from overvoltage and high current change rate *di*/*dt* in the reverse recovery phases, which inevitably leads to damages of them [6]. The overvoltage and high current change rate *di*/*dt* can be induced by load mutation in the EML experiment [5]. The load mutation can lead to an open circuit or sharp increase in resistance of the load side, which results in instantaneous overvoltage and high current change rate *di*/*dt* in parts of pulse-shaping inductors. Due to the randomness of load mutation, a typical phenomenon in a multiple uninterruptible discharge process is that the thyristors in conventional PPS topologies are not always but randomly damaged, especially in the case of numerous PFUs operating together with discharge time intervals, although all the initial conditions and devices' parameters keep the same.

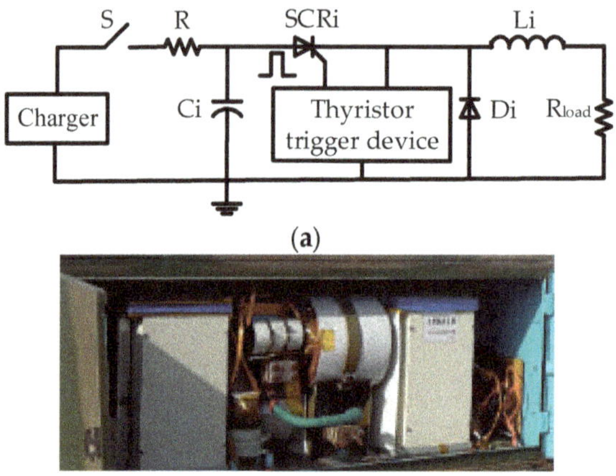

(b)

Figure 1. Type I pulse-forming unit (PFU) topology and its photograph. (**a**) Type I PFU topology; (**b**) photograph of type I PFU topology.

This paper establishes circuit models of PPS topologies. Then, the reverse recovery current characteristics of the thyristors are explored. Effects of the initial voltage of the energy-storage capacitor, the discharge time intervals, and the load resistance on the reverse recovery currents are systematically investigated. It is found that the reverse recovery currents in the thyristors in the type I PPS topology are larger than those in the type II PPS topology under the same working conditions, indicating that the type II PPS topology is more helpful to reduce the breakdown risks of the thyristors if multiple PFUs operate together with discharge time intervals.

However, the reverse recovery currents in the thyristors in both type I and II topologies are relatively high when sustaining instantaneous overvoltages caused by load mutations. Therefore, the reliability of conventional PPS is relatively low in the worst cases and it is imperative to enhance its robustness.

References [6,19] indicate that adopting a resistance-capacitance (RC) snubber circuit in parallel with the thyristors can protect them from being damaged. It can reduce the reverse recovery currents but consumes more useful energy because it is more sensitive to the voltage variation of the thyristors and less sensitive to the voltage variation of the load side. If multiple PFUs operate together with discharge time intervals, the voltage variation of the thyristors is sharper than that of the load side, which causes more useful energy to be consumed by the snubber resistor due to the snubber capacitor presenting a low resistance at a high-frequency range. Reference [20] proposed that an impedance matching method can be used to protect the semiconductor devices from being damaged, which requires that the ratio of the internal resistance of the load to its inductance is far greater than that of the inductance of the pulse-shaping inductor to its internal resistance. In practical application, it is difficult to meet this requirement because the load always changes under different working conditions but the pulse-shaping inductor stays the same.

An improved PPS topology is developed in this paper. The improved PPS topology applies coupling inductor and RC snubber circuit techniques, which consumes less useful energy compared with conventional approaches because it is less sensitive to the voltage variation of the thyristors and more sensitive to the voltage variation of the load side. Besides, it also reduces the reverse recovery currents in the thyristors by absorbing the surge energy from the load side when load mutation occurs. Furthermore, the size of the pulse-shaping inductor can be reduced due to the coupling technique being applied. Finally, the simulation technique has been applied to validate theoretical analysis and the proposed model.

2. Reverse Recovery Current Models of the Thyristors in PPS

2.1. Description of PPS

The PPS is composed of multiple PFUs in parallel. Type I PFU topology and its photograph are shown in Figure 1. In Figure 1, the capacitance of the energy-storage capacitor Ci is C_i ($i = 1, 2, \ldots, n$), the inductance of the pulse-shaping inductor Li is L_i ($i = 1, 2, \ldots, n$), and the resistance of the load R_{load} is R_{load}. Generally, $C_1 = C_2 = \ldots = C_n$, $L_1 = L_2 = \ldots = L_n$, R_{load} is at mΩ level, the thyristor SCRi ($i = 1, 2, \ldots, n$) is used as a switch, and the fast recovery diode Di ($i = 1, 2, \ldots, n$) provides an after-flow path for the pulse-shaping inductor Li. If a PFU is discharging and the fast recovery diode is in a cut-off state, the current equation can be described as [21]:

$$L_i \frac{di(t)}{dt} + \frac{1}{C_i} \int i(t)dt + R_{load} i(t) = 0, \tag{1}$$

where $i(t)$ represents the current in the circuit.

The initial conditions are as follows:

$$\begin{cases} i(t)\big|_{t=0} = 0 \\ \frac{di(t)}{dt}\Big|_{t=0} = \frac{U_0}{L_i} \end{cases}, \tag{2}$$

where U_0 is the initial voltage of the energy-storage capacitor.

Typically, R_{load} is in the range of 5~50 mΩ, L_i is in the range of 5~40 µH, and C_i is in the range of 1~6 mF, indicating that $\frac{R_{load}{}^2}{L_i{}^2} - \frac{4}{L_i C_i} < 0$. The angular frequency is $\omega_0 = 1/\sqrt{L_i C_i}$ and the attenuation factor is $\alpha = R_{load}/(2L_i)$. Based on Equations (1) and (2), it can be deduced:

$$i(t) = \frac{U_0}{L_i \sqrt{\omega_0^2 - \alpha^2}} e^{-\alpha t} \sin(\sqrt{\omega_0^2 - \alpha^2} t). \tag{3}$$

In practical application, the discharge time of each PFU is not synchronous [6,20]. The discharge time of PFU1 is t_1, the discharge time of PFU2 is t_2, and the discharge time of PFUn is t_n. Therefore, there are $n - 1$ discharge time intervals: $t_{s1} = t_2 - t_1$; $t_{s2} = t_3 - t_2$; ... ; $t_{sn-1} = t_n - t_{n-1}$. The current in the load side is a superposition of the currents in the discharging PFUs.

2.2. The Reverse Recovery Current Models of the Thyristors in PPS

The equivalent type I PFU topology and the reverse recovery current models of the thyristors are shown in Figure 2. Figure 2a shows that if the PPS is operating, a PFU sustains backward voltages, equivalent to a reverse voltage source $V_P(t)$ and R_{load} is regarded as its internal resistance, from other discharging PFUs. L_P is an equivalent parasitic inductance in the energy-storage capacitor and lead wires, and its typical value is in the range of 0.05~0.2 µH. Figure 2b,c show two models of the reverse recovery currents in the thyristors in the type I PFU topology.

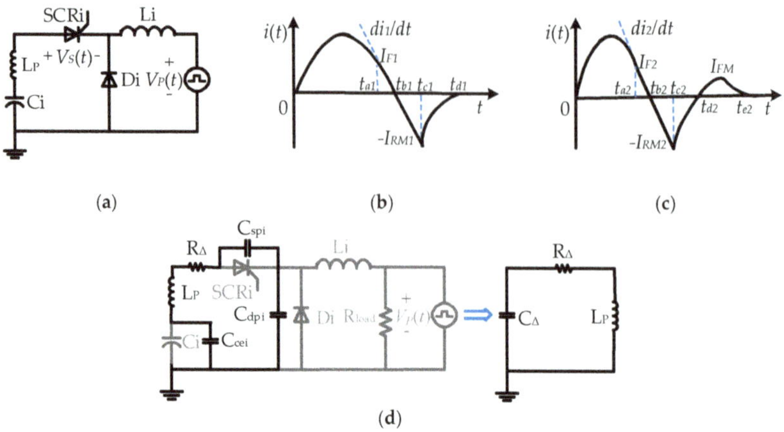

Figure 2. Equivalent type I PFU topology and reverse recovery current models of the thyristors. (**a**) Equivalent type I PFU topology; (**b**) reverse recovery current model No. 1 of the thyristor; (**c**) reverse recovery current model No. 2 of the thyristor; (**d**) equivalent circuit of the reverse recovery current model No. 2 when forward oscillation current occurs.

The current model No. 1 is suitable for the situation where the fast recovery diodes are not turned on in the reverse recovery phases of the thyristors. The reverse recovery current loop is mainly composed of: thyristor, energy-storage capacitor, load, and pulse-shaping inductor. When $0 < t < t_{a1}$, the energy-storage capacitor is in a discharge phase, and the current $i_1(t)$ in the thyristor is positive; when $t = t_{a1}$, $i_1(t)$ begins to decrease with a slope of $di_1/dt = -V_P(t)/L_i$ (in practical application, $L_i >> 10L_P$, thus L_p is relatively small compared with L_i and can be ignored), and $i_1(t_{a1}) = I_{F1}$; when $t = t_{b1}$, $i_1(t_{b1}) = 0$; when $t_{b1} < t \le t_{c1}$, $i_1(t)$ increases inversely with the same slope of di_1/dt until it reaches the maximum reverse value $-I_{RM1}$; when $t_{c1} < t \le t_{d1}$, $i_1(t)$ decreases in the form of exponential function until it returns to zero at $t = t_{d1}$. The reverse recovery time can be written as: $t_{rr1} = t_{d1} - t_{b1}$.

The current model No. 2 is suitable for the situation that the fast recovery diodes have been turned on in the reverse recovery phases of the thyristors. There are two reverse recovery current loops: thyristor, energy-storage capacitor, load, and pulse-shaping inductor; thyristor, energy-storage capacitor, and fast recovery diode. When $t = t_{a2}$, the fast recovery diode has been turned on, and $i_2(t)$ begins to decrease with a slope of $di_2/dt = -[V_P(t)/L_i + V_S(t)/L_P]$; when $t = t_{b2}$, $i_2(t)$ decreases to zero; when $t = t_{c2}$, $i_2(t)$ reaches the maximum reverse value $-I_{RM2}$; when $t_{b2} < t \le t_{d2}$, the current change law is similar to that in Figure 2b; when $t_{d2} < t \le t_{e2}$, a forward oscillation current occurs due to RLC series resonance caused by parasitic inductance, capacitance, and resistance. The equivalent circuit of the reverse recovery current model No. 2 when forward oscillation current occurs is shown in Figure 2d. Due to higher reverse recovery current change rate in the current model No. 2, effects of parasitics are more obvious and the load side presents an open circuit at this moment. C_{spi} represents parasitic capacitance in the thyristor SCRi, C_{dpi} represents parasitic capacitance in the fast recovery diode Di, C_{cei} represents equivalent capacitance of the energy-storage capacitor Ci, R_Δ represents the total parasitic resistance, and C_Δ represents the total parasitic capacitance. R_Δ is far less than R_{load}, and $C_\Delta = \frac{C_{spi}C_{cei}C_{dpi}}{C_{spi}C_{cei}+C_{cei}C_{dpi}+C_{dpi}C_{spi}}$.

The reverse recovery current in the thyristor shown in Figure 2b can be written as [22]:

$$i_1(t) = \begin{cases} I_{F1} - (t - t_{a1})V_P(t)/L_i & t_{a1} < t \le t_{c1} \\ -I_{RM1}\exp(-\frac{t-t_{c1}}{\tau_1}) & t_{c1} < t \end{cases}, \tag{4}$$

where τ_1 is the minority carrier lifetime in the base region of the thyristor.

The relationship between the current $i_1(t)$ and the charge Q(t) stored in the N^- base region of a thyristor can be written as:

$$i_1(t) = \frac{dQ(t)}{dt} + \frac{Q(t)}{\tau_1}. \quad (5)$$

Based on Equations (3) and (5), it can be deduced:

$$Q(t) = \frac{GU_0}{L_i\omega^2(1+b^2)}(e^{-t/\tau_1} + be^{-\alpha t}\sin\omega t + e^{-\alpha t}\cos\omega t), \quad (6)$$

where G is the equivalent current gain of the thyristor and $b = \frac{1-\alpha\tau_1}{\omega\tau_1}$.

$Q(t)$ equals zero at t_{c1} and $t_{s1} = t_{c1} - t_{b1}$, then $t_{c1} = t_{s1} + \pi/\omega$. Since $\omega t_{s1} << \pi/2$ and $t_{c1} >> \tau_1$, Equation (7) is deduced:

$$\tau_1 = \frac{t_{s1}}{1+\alpha t_{s1}}. \quad (7)$$

The current model of the thyristor shown in Figure 2c is described as:

$$i_2(t) = \begin{cases} I_{F2} - (t-t_{a2})[V_P(t)/L_i + V_S(t)/L_P] & t_{a2} < t \le t_{c2} \\ -I_{RM2}\exp(-\frac{t-t_{c2}}{\tau_2}) & t_{c2} < t \le t_{d2}, \\ 2E\sqrt{C_\Delta/(4L_P - R_\Delta{}^2C_\Delta)}\exp[-R_\Delta(t-t_{d2})/(2L_P)]\sin\omega_\Delta(t-t_{d2}) & t_{d2} < t \le t_{e2} \end{cases} \quad (8)$$

where E is the initial voltage of the total parasitic capacitance C_Δ; τ_2 is the minority carrier lifetime of the base region of the thyristor; angular frequency is $\omega_\Delta = \frac{1}{\sqrt{L_PC_\Delta}}\sqrt{1-\frac{R_\Delta{}^2C_\Delta}{4L_P}}$.

Generally, the breakdown risk increases with the reverse recovery current in a thyristor, so it is reasonable to investigate the reverse recovery current characteristic of the thyristor. The relationship between the breakdown risk and the reverse recovery current characteristic of the thyristor will be discussed in Section 2.3. Which case is suitable for the reverse recovery current model No. 1 or No. 2 of the thyristor will be identified in Section 3.

2.3. Breakdown Risks of the Thyristors in PPS

If initial conditions and devices' parameters keep the same, assuming that: in current model No. 1, the instantaneous voltage of a thyristor is V_{S1} and its instantaneous power is P_{SCR1} when $i_1(t)$ reaches its maximum reverse value $-I_{RM1}$; in current model No. 2, the instantaneous voltage of a thyristor is V_{S2} and its instantaneous power is P_{SCR2} when $i_2(t)$ reaches its maximum reverse value $-I_{RM2}$.

Generally, the instantaneous voltage of a thyristor approximately equals to its turn-on voltage when the current reaches its maximum reverse value, therefore, $V_{S1} = V_{S2}$. If $I_{RM1} << I_{RM2}$, $P_{SCR2} = V_{S2}I_{RM2} >> P_{SCR1} = V_{S1}I_{RM1}$, indicating that the instantaneous power of a thyristor with current model No. 2 is higher than that with current model No. 1. Therefore, the breakdown risk of the thyristor with current model No. 2 is greater than that with current model No. 1 under the same working conditions.

Documents [17,20,23] indicate that the thyristors are prone to be breakdown by sustaining overvoltages in the reverse recovery phases. Therefore, it is meaningful to investigate the factors affecting the reverse recovery currents in the thyristors and find approaches to reduce their peak values, which will be explored in Section 3.

3. Factors Affecting the Reverse Recovery Currents in Thyristors and Improvement of PPS Topology

If the breakdown of a thyristor occurs, the general way is to remove and replace it with a new one, which spends massive manpower and financial resources in a long term. Due to the randomness of load mutation, a typical phenomenon in a multiple uninterruptible discharge process is that the thyristors in conventional PPS topologies are not always but randomly damaged especially in the case of numerous PFUs operating together with discharge time intervals, although all initial conditions and devices' parameters keep the same. Load mutation can lead to an open circuit or sharp increase in resistance of the load side, which results in instantaneous overvoltage and high current change rate *di*/*dt* in pulse-shaping inductors. The instantaneous overvoltages increase reverse recovery currents in the thyristors and cause breakdowns of them. Breakdowns of the thyristors often occur in their reverse recovery phases, which are also related to the factors such as the initial voltage of the energy-storage capacitor, the discharge time intervals, and the load resistance.

Figure 3 shows two simplified PFU topologies. Figure 3a shows the type I PFU topology and Figure 3b shows the type II PFU topology. It is displayed that the difference in the two types of PFU topologies is the locations of the fast recovery diodes.

Figure 3. Two simplified PFU topologies. (**a**) Simplified type I PFU topology; (**b**) simplified type II PFU topology.

3.1. Validation of the Two Current Models of Thyristors

To validate the circuit models No. 1 and No. 2 of thyristors, the simulation circuit of type I PPS is established in software Simplorer as shown in Figure 4. ESL_i represents the series parasitic inductance of the energy-storage capacitor Ci, ESR_i represents the series parasitic resistance of the energy-storage capacitor Ci, R_{Li} represents the series parasitic resistance of the pulse-shaping inductor Li. The devices' parameters of PPS are shown in Table 1, and n is the number of PFUs.

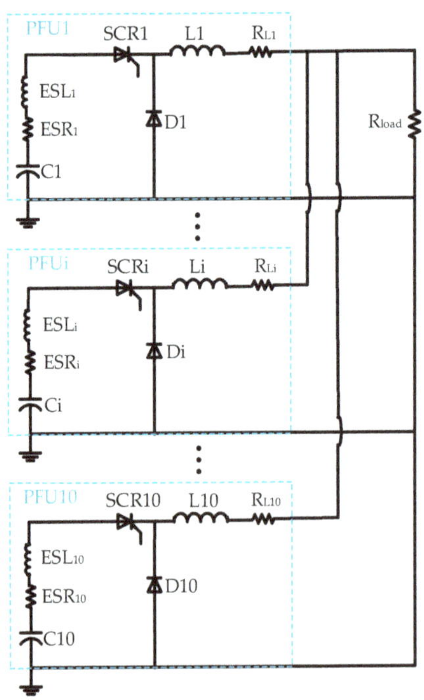

Figure 4. Simulation circuit of type I PPS.

Table 1. Devices' parameters of PPS.

C_i (mF)	ESL_i (µH)	ESR_i (mΩ)	L_i (µH)	R_{Li} (mΩ)	n
1	0.2	2	5	0.5	10

Figure 5 shows the simulation current curves of the thyristors and fast recovery diodes in type I PPS topology. The initial voltage of the energy-storage capacitor U_0 = 5 kV, the discharge time intervals $t_{s1} = t_{s2} = \ldots = t_{sn-1} = t_{sn} = t_s$ = 100 µs, the load resistance R_{load} = 50 mΩ, and other devices' parameters keep the same as those in Table 1. SCR1 represents the current curves of the thyristor SCR1, ... , SCR10 represents the current curves of the thyristor SCR10, and D1 represents the current curves of the fast recovery diode D1, ... , D10 represents the current curves of the fast recovery diode D10.

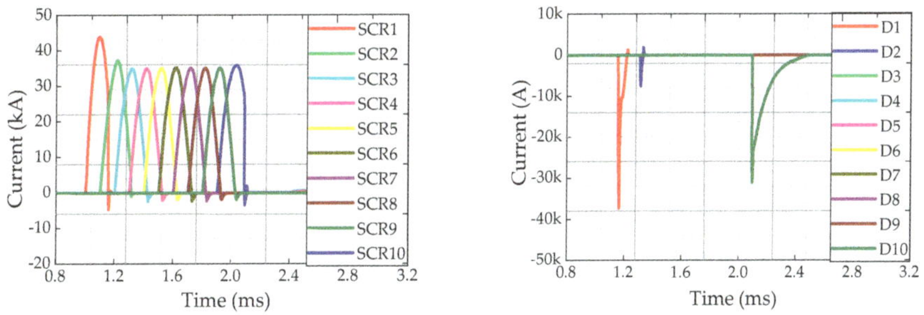

(**a**) (**b**)

Figure 5. Simulation current curves of the thyristors and fast recovery diodes in type I PPS topology. (**a**) Current curves of the thyristors; (**b**) current curves of the fast recovery diodes.

The two current models of thyristors mentioned in Section 2.2 can be validated in Figure 5a. It is displayed in Figure 5b that the currents of the fast recovery diodes D3–D9 are zero, indicating that they are not turned on; the current of the fast recovery diode D2 is relatively small, indicating it is slightly turned on; the currents of the fast recovery diodes D1 and D10 are relatively large, indicating that they are fully turned on. Based on Figure 5a,b, it is found that if the peak values of the reverse recovery currents in the thyristors SCR1 and SCR10 are greater and their reverse recovery time values are shorter than those of the others due to the fast recovery diodes D1 and D10 being fully turned on. It is revealed that there are forward oscillation currents in the thyristors SCR1 and SCR10, thus the current models No. 1 and No. 2 of the thyristors in PPS are validated. It is also exhibited that if the fast recovery diodes, especially D1 and D10, are turned on in type I PPS topology, the peak values of the reverse recovery currents in the thyristors are increased, which results in appearance of the forward oscillation currents and increases the breakdown risks of the thyristors.

3.2. Effects of the Initial Voltage of the Energy-Storage Capacitor on the Reverse Recovery Currents in Thyristors

Figure 6 shows the simulation current curves of the thyristors in the type I PPS topology with different initial voltages of the energy-storage capacitor based on the simulation circuit shown in Figure 4. The initial conditions and devices' parameters keep the same as those in Section 3.1, except that the initial voltage of the energy-storage capacitor varies. Since there are two thyristor current models in the type I PPS topology, the thyristors SCR1 and SCR6 are selected as the research objects to observe the effects of the initial voltage of the energy-storage capacitor on the reverse recovery currents. The thyristor SCR1 and the fast recovery diode D1 are in PFU1. The fast recovery diode D1 is turned on under this condition. The thyristor SCR6 and the fast recovery diode D6 are in PFU6. The fast recovery diode D6 is not turned on under this condition.

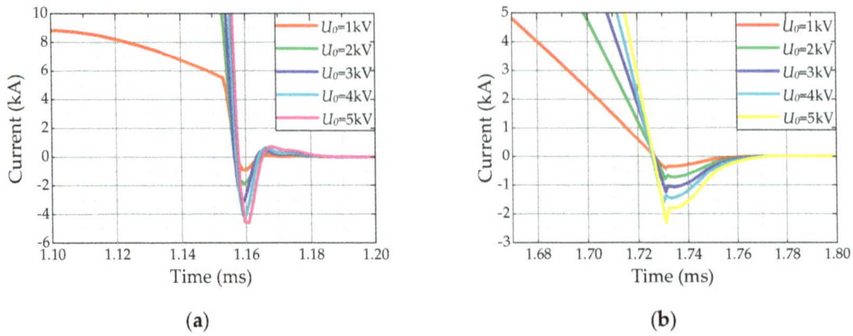

Figure 6. Simulation current curves of the thyristors in the type I PPS topology with different initial voltages of the energy-storage capacitor. (**a**) Current curves of the thyristor SCR1; (**b**) current curves of the thyristor SCR6.

Figure 6a illustrates that the peak value of the forward oscillation current in the thyristor SCR1 increases with the increase of the initial voltage of the energy-storage capacitor. It is also revealed that if the initial voltage of the energy-storage capacitor increases, the peak value of the reverse recovery current in the thyristor SCR1 and its reverse recovery time increase. Besides, the peak value of the reverse recovery current in the thyristor SCR1 is larger and its reverse recovery time is shorter than those of the thyristor SCR6. It can be inferred that the smaller the initial voltage of the energy-storage capacitor is, the less likely the thyristor will be damaged.

3.3. *Effects of the Discharge Time Intervals on the Reverse Recovery Currents in Thyristors*

Figure 7 shows the simulation current curves of the thyristors and voltage curves of the load side in the type I PPS topology with different discharge time intervals. The initial conditions and devices' parameters keep the same as those in Section 3.1 except that the discharge time intervals vary. The thyristors SCR1 and SCR6 are also selected as the research objects.

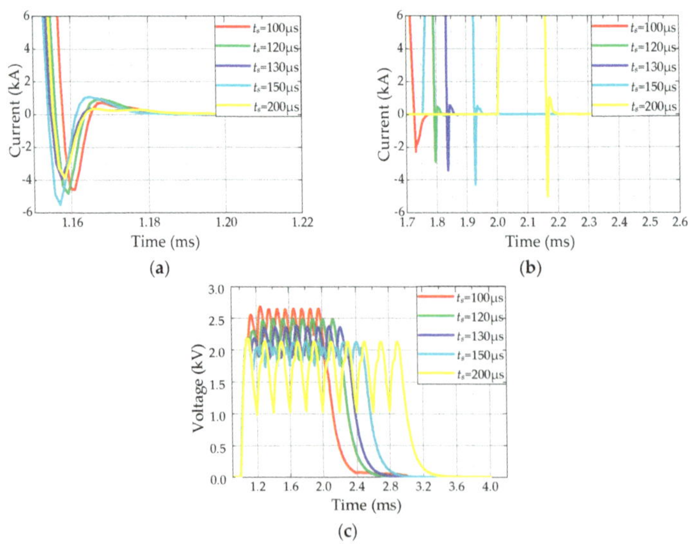

Figure 7. Simulation current curves of the thyristors and voltage curves of the load side in the type I PPS topology with different discharge time intervals. (**a**) Current curves of the thyristor SCR1; (**b**) current curves of the thyristor SCR6; (**c**) voltage curves of the load side.

Figure 7a illustrates that the peak values of the reverse recovery and forward oscillation currents, and its reverse recovery time of the thyristor SCR1 do not increase but vibrate with the increase of the discharge time intervals. Figure 7b illustrates that the peak values of the reverse recovery and forward oscillation currents in the thyristor SCR6 increase with the increase of the discharge time intervals. It is also displayed that the forward oscillation current gradually occurs with the increase of the discharge time intervals, indicating that the fast recovery diode D6 is gradually turned on. Figure 7c illustrates that the voltage amplitude of the load side decreases and its ripple increases with the increase of the discharge time intervals, which indicates that the reverse voltage source $V_P(t)$ sustained by the thyristor is not stable, resulting in the phenomenon shown in Figure 7a.

It can be inferred that with the increase of the discharge time intervals, the number of the turned-on fast recovery diodes increases, and the number of the thyristors with forward oscillation currents increases, but the amplitude of $V_P(t)$ reduces and tends to a stable amplitude finally, which indicate that the breakdown risks of the thyristors increase, and a similar result is given in [20]. It should be noted that with the increase of the discharge time intervals to a certain extent, $V_P(t)$ is no longer a flat-topped wave, which will fail to meet the working requirements of the load side in practical application.

3.4. Effects of the Load Resistance on the Reverse Recovery Currents in Thyristors

Figure 8 shows the simulation current curves of the thyristors in the type I PPS topology with different load resistances. The initial conditions and devices' parameters keep the same as those in Section 3.1 except that the load resistance varies. The thyristors SCR1 and SCR6 are also selected as the research object.

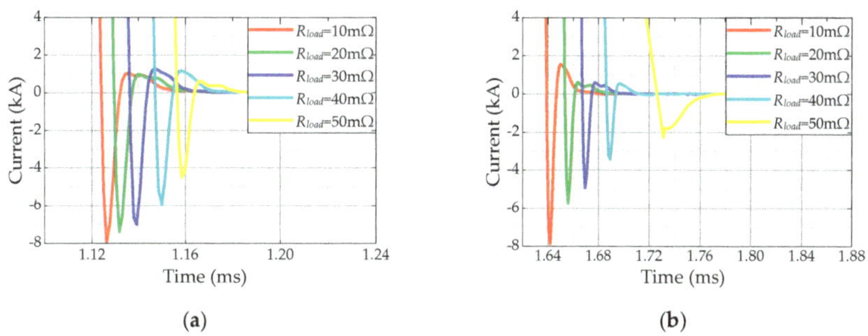

Figure 8. Simulation current curves of the thyristors in the type I PPS topology with different load resistances. (**a**) Current curves of the thyristor SCR1; (**b**) current curves of the thyristor SCR6.

Figure 8a,b illustrates that with the increase of load resistance, there is a decrease in the peak values of the reverse recovery currents in the thyristors SCR1 and SCR6. Similarly, their forward oscillation currents exhibit reduction in peak values, and the moments of entering the reverse recovery phases are delayed. In Figure 8b, the forward oscillation current in the thyristor SCR6 even disappears when R_{load} = 50 mΩ. It can be inferred that with the increase of the load resistance, the number of the thyristors with forward oscillation currents decreases. Therefore, the breakdown risks of the thyristors in PPS can be reduced by increasing the load resistance.

3.5. The Reverse Recovery Currents in the Thyristors in Type II PPS Topology

The simulation circuit of type II PPS is shown in Figure 9 and the simulation results are shown in Figure 10. Figure 10a shows the simulation current curves of the thyristors in the type II PPS topology, and its initial conditions and devices' parameters keep the same as those in Section 3.1. Figure 10a illustrates that the peak values of the reverse recovery currents in all thyristors are almost consistent in the type II PPS topology. Similarly, their reverse recovery time values are almost consistent, and there are no forward oscillation currents. Comparing Figure 10a,b, it is found that no matter whether the fast recovery diodes are turned on or not, there is no increase in the peak values of the reverse recovery currents and appearance of the forward oscillation currents in the thyristors in the type II PPS topology. Comparing Figures 5a and 10a, it is found that the peak values of the reverse recovery currents in the thyristors, especially SCR1 and SCR 10 in the type II PPS topology, are lower than those in the type I PPS topology.

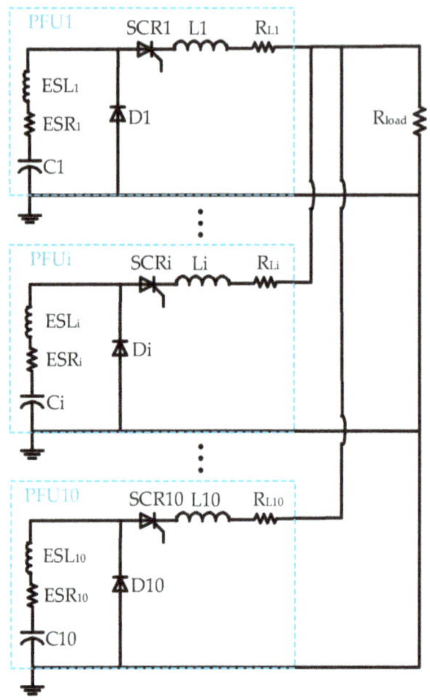

Figure 9. Simulation circuit of type II PPS.

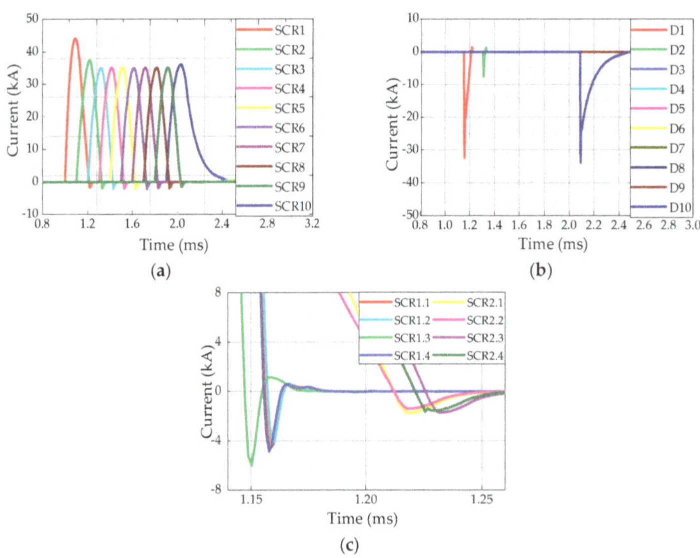

Figure 10. Simulation current curves of the thyristors and fast recovery diodes in conventional PPS. (**a**) Current curves of the thyristors in the type II PPS topology; (**b**) current curves of the fast recovery diodes in the type II PPS topology; (**c**) current curves of the thyristor SCR1 in type I and II PPS topologies.

To analyze the reverse recovery currents in the thyristors in type I and II PPS topologies, the thyristor SCR1 in PFU1 is selected as a research object, and the simulation results are shown in Figure 10c. SCR 1.1 represents the simulation current curve of the thyristor SCR1 in the type I PPS topology when the initial voltage of the energy-storage capacitor U_0 = 5 kV, the discharge time intervals t_r = 100 μs, and the load resistance R_{load} = 50 mΩ. SCR 1.2 represents the simulation current curve of the thyristor SCR1 in the type I PPS topology when the initial voltage of the energy-storage capacitor U_0 = 4 kV, the discharge time intervals t_r = 100 μs, and the load resistance R_{load} = 50 mΩ. SCR 1.3 represents the simulation current curve of the thyristor SCR1 in the type I PPS topology when the initial voltage of the energy-storage capacitor U_0 = 5 kV, the discharge time intervals t_r = 100 μs, and the load resistance R_{load} = 40 mΩ. SCR 1.4 represents the simulation current curve of the thyristor SCR1 in the type I PPS topology when the initial voltage of the energy-storage capacitor U_0 = 5 kV, the discharge time intervals t_r = 120 μs, and the load resistance R_{load} = 50 mΩ. SCR 2.1 represents the simulation current curve of the thyristor SCR1 in the type II PPS topology when the initial voltage of the energy-storage capacitor U_0 = 5 kV, the discharge time intervals t_r = 100 μs, and the load resistance R_{load} = 50 mΩ. SCR 2.2 represents the simulation current curve of the thyristor SCR1 in the type II PPS topology when the initial voltage of the energy-storage capacitor U_0 = 4 kV, the discharge time intervals t_r = 100 μs, and the load resistance R_{load} = 50 mΩ. SCR 2.3 represents the simulation current curve of the thyristor SCR1 in the type II PPS topology when the initial voltage of the energy-storage capacitor U_0 = 5 kV, the discharge time intervals t_r = 100 μs, and the load resistance R_{load} = 40 mΩ. SCR 2.4 represents the simulation current curve of the thyristor SCR1 in the type II PPS topology when the initial voltage of the energy-storage capacitor U_0 = 5 kV, the discharge time intervals t_r = 120 μs, and the load resistance R_{load} = 50 mΩ.

It can be revealed that the peak values of the reverse recovery currents in the thyristor SCR1 in the type II PPS topology are smaller than those in the type I PPS topology under the same working conditions, and no forward oscillation currents in the thyristor SCR1 in the type II PPS topology are found. Comparing Figures 5 and 10, the peak values of the reverse recovery and forward oscillation currents in the thyristors in the type I PPS topology are also influenced by the fast recovery diodes. However, the influence of the fast recovery diodes on the peak values of the reverse recovery currents in the thyristors in the type II PPS topology can be ignored, and the forward oscillation currents are even not influenced by the fast recovery diodes.

It can be concluded that the influence of the initial voltage of the energy-storage capacitor, the discharge time intervals, and the load resistance on the reverse recovery currents in the thyristors in the type II PPS topology is smaller than that in the type I PPS topology, thus the breakdown risks of the thyristors can be reduced.

3.6. *Improvement of PPS Topology*

The breakdown risks of the thyristors in both type I and II PPS topologies, which are caused by the instantaneous overvoltages from the pulse-shaping inductors of other discharging PFUs due to load mutations, are relatively high. The load mutation occurs in the EML experiment due to the armature is out of or broken in the railgun barrel, which leads to an open circuit or sharp increase in resistance of the load side. The load mutation can cause high current change rates, which lead to instantaneous overvoltages of the pulse-shaping inductors. The instantaneous overvoltages can result in high reverse recovery currents in the thyristors, and cause damages of them in conventional topologies.

Based on the simulation circuits of PPS shown in Figures 4 and 9, it is found that the thyristors in both type I and II PPS topologies easily suffer from the instantaneous overvoltages from the pulse-shaping inductors. Therefore, it is meaningful to explore approaches to protect the thyristors from being damaged.

The common method is to increase the number of thyristors in series. It is a practicable way, but requires higher output power and increases the complexity of trigger circuits. Furthermore, due to the amplitudes of the instantaneous overvoltages from the pulse-shaping inductors being different in different working conditions, which results in different amplitudes of the reverse recovery currents in the thyristors, it is not easy to determine the number of the thyristors in series, especially when the cost is considered.

To overcome the shortcomings of the conventional PPS topologies, a new PPS topology is developed. By applying the coupling technique, the pulse-shaping inductor is designed as a coupling inductor. Then, an RC snubber circuit is added between the common connection point of the two windings of the coupling inductor and the ground, which absorbs the surge energy from the load side. Based on the decoupling theory, the simplified model of the coupling inductor in PFUi is shown in Figure 11. If the inductance of one winding is L_{ai}, the other is L_{bi}, and their mutual inductance is M_i, then a negative inductance with value $-M_i$ is formed between the common connection point of the two windings and the RC snubber circuit. The negative inductance can be equivalent to a capacitance with value C_{eqi}, which can be deduced as:

$$-j\omega M_i = \frac{1}{j\omega C_{eqi}}, \tag{9}$$

where ω represents angular frequency.

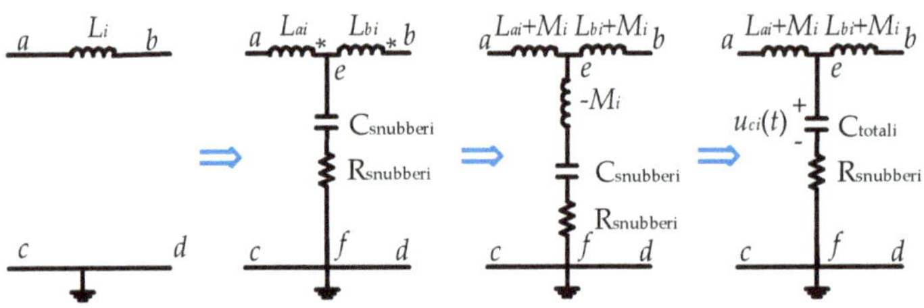

Figure 11. Simplified model of the coupling inductor in PFUi.

Based on (9), the equivalent capacitance is written as: $C_{eqi} = \frac{1}{\omega^2 M_i}$, thus the total capacitance C_{totali} of the ef branch is:

$$C_{totali} = \frac{C_{eqi} C_{snubberi}}{C_{eqi} + C_{snubberi}}. \tag{10}$$

If the voltage of the equivalent capacitance C_{totali} is $U_{ci}(t)$, then the energy absorbed by the ef branch from t_0 to t is:

$$W(t) = \frac{1}{2} C_{totali} [u_{ci}{}^2(t) - u_{ci}{}^2(t_0)] + C_{totali}(R_{snubberi} + R_{ci}) \int_{t_0}^{t} [\frac{du_{ci}(\tau)}{d\tau}]^2 d\tau, \tag{11}$$

where R_{ci} is the parasitic resistance of the equivalent capacitance C_{totali}, and $R_{snubberi}$ is the resistance of the snubber resistor $\mathrm{R_{snubberi}}$.

Due to the ef branch consuming a relatively small part of the useful energy, it can be ignored when PFUi is operating and there is:

$$L_i = L_{ai} + L_{bi} + 2M_i. \tag{12}$$

Figure 12 shows the simulation circuit of the improved PPS. The initial conditions and devices' parameters keep the same as those in Section 3.1, $L_{ai} = L_{bi} = 2$ μH, coupling factor of the inductor $k = 0.25$, $C_{snubberi}$ =10 μF, and $R_{snubberi}$ = 30 Ω. The switch S is open at t = 2 ms, resulting in an open circuit at the load side.

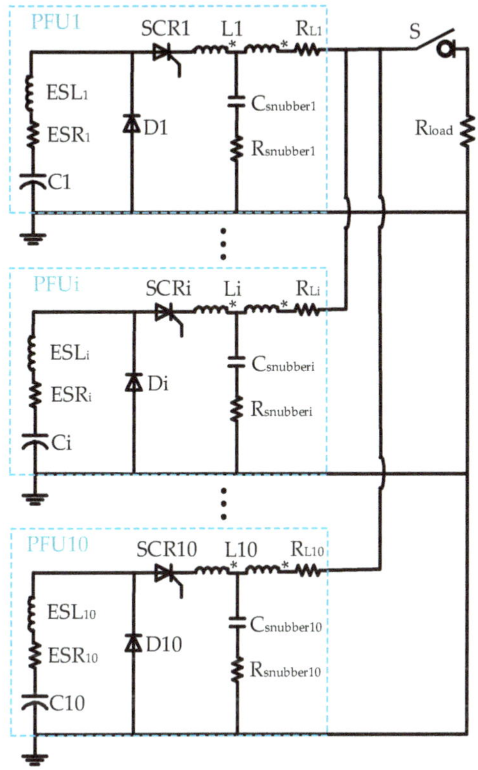

Figure 12. Simulation circuit of improved PPS.

The simulation voltage curves of pulse-shaping inductors with high voltage spikes in the type II PPS topology are shown in Figure 13a. It is displayed that the high voltage spike of the pulse-shaping inductor L9 is formed in the after-flow phase when load mutation occurs, while the high voltage spike of the pulse-shaping inductor L10 is formed before its after-flow phase. It is also exhibited that the value of the high voltage spike of the pulse-shaping inductor L10 is greater than that of the pulse-shaping inductor L9, which will inevitably cause the thyristor SCR9 to suffer from the high voltage spike of the pulse-shaping inductor L10.

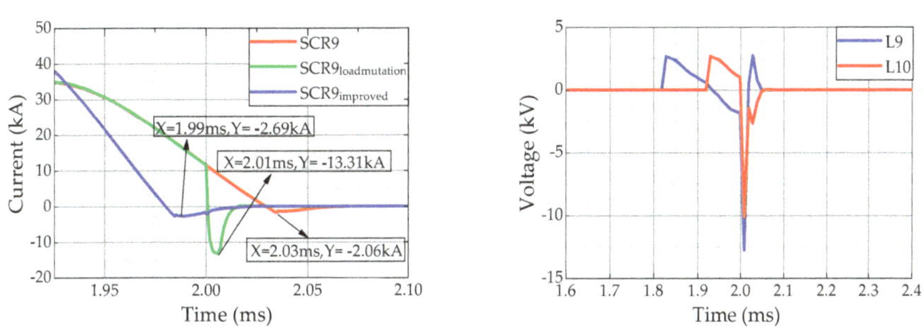

(a) (b)

Figure 13. Simulation current curves of the thyristor SCR9 and voltage curves of the pulse-shaping inductors L9 and L10. (**a**) Voltage curves of the pulse-shaping inductors L9 and L10 in type II PPS topology; (**b**) current curves of the thyristor SCR9 in type II and improved PPS topologies.

The simulation current curves of the thyristor SCR9 is shown in Figure 13b. In Figure 13b, SCR9 represents the current curve of the thyristor SCR9 in the type II PPS topology without load mutation, $SCR_{loadmutation}$ represents the current curve of the thyristor SCR9 in the type II PPS topology with load mutation, and $SCR_{improved}$ represents the current curve of the thyristor SCR9 in the improved PPS topology with load mutation.

Figure 13b illustrates that the peak value of the reverse recovery current in the thyristor SCR9 in the type II PPS topology with load mutation at about 5 times larger than that without load mutation, and the peak value of the reverse recovery current in the thyristor SCR9 in improved topology with load mutation is almost equal to that in the type II PPS topology without load mutation. Therefore, it is reasonable to apply the improved PPS topology to reduce the breakdown risks of the thyristors and improve the reliability of conventional PPS.

To sum up, the thyristors in conventional PPS topologies have larger peak values of the reverse recovery currents if sustaining instantaneous overvoltages from the pulse-shaping inductors of other discharging PFUs caused by load mutation in the reverse recovery phases, while the thyristors in the improved PPS topology have lower peak values of the reverse recovery currents due to the RC snubber circuits can absorb the surge energy from the load side when load mutation occurs.

3.7. Test Results of EML Experiments

The test layout of the EML experiment is shown in Figure 14. In the experiment, the load is an electromagnetic railgun, which is mainly composed of a copper barrel and an aluminum armature. The resistance of the electromagnetic railgun is very small, which is at the mΩ level. The control system sends charging signals to the chargers initially; then, the chargers provide energy for the energy-storage capacitors; next, when the preset voltage of the energy-storage capacitors is reached, the control system sends charging-stop signals to the chargers, and the charging process is over; ultimately, when the PPS receives a firing signal from the control system, the PFUs discharge according to the preset discharge time intervals. The currents in the thyristors can be measured by Rogowski current probes which can send their collected results to the data analysis system. The data analysis system can automatically analyze the measured current data and present it for the testers.

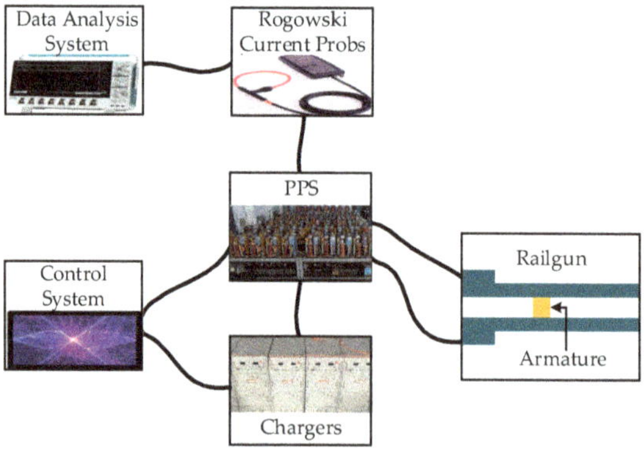

Figure 14. Test layout of the EML experiments.

The measured current curves of the thyristors in type I PPS topology are shown in Figure 15. Figure 15a shows the measured current curve of a thyristor in the type I PPS topology when the initial voltage of the energy-storage capacitor U_0 = 5 kV and the discharge time intervals t_r = 200 μs. It is observed that the thyristor in Figure 15a has a reverse recovery current with the peak value −3.93 kA. The thyristor has not been damaged in this case. Figure 15b shows the measured current curve of a thyristor in the type I PPS topology when the initial voltage of the energy-storage capacitor U_0 = 8 kV and the discharge time intervals t_r = 200 μs. It found that the armature is broken in the barrel and the thyristor is damaged. Figure 15b illustrates that the damaged thyristor is in a low resistance state and loses its reverse blocking capability.

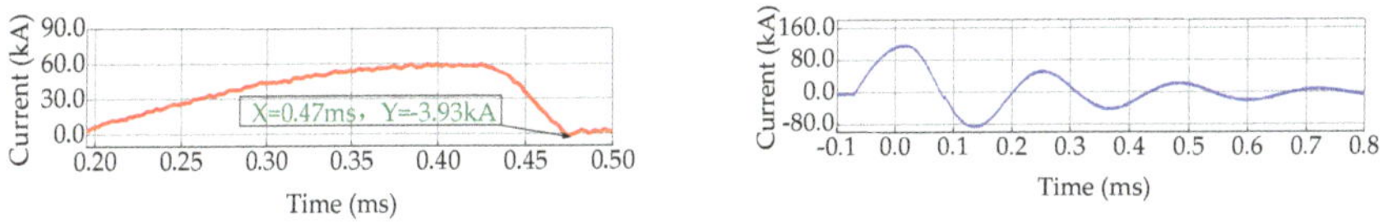

(a) (b)

Figure 15. Measured current curves of the thyristors in the type I PPS topology. (**a**) Current curve of the thyristor without being damaged; (**b**) current curve of the damaged thyristor.

The photograph of the damaged thyristor is shown in Figure 16. The maximum forward and reverse breakdown voltages of the thyristor are far greater than its working voltage. It can be inferred that the breakdown of the thyristor occurs in its reverse recovery phase. The breakdown of the thyristor is caused by the instantaneous overvoltages from the pulse-shaping inductors of other discharging PFUs. Once the breakdown of the thyristor occurs, it will suffer from a large current in a long time due to the disappearance of the reverse recovery blocking capability, which also inevitably causes damage to it. Therefore, it is meaningful for the designers of PPS to consider the load mutation and develop new topologies.

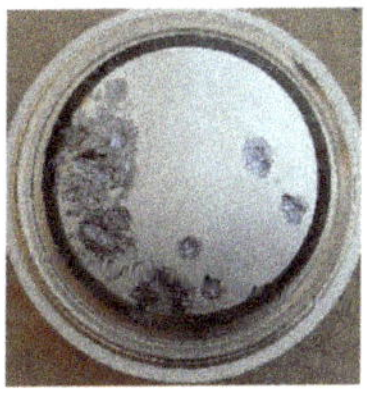

Figure 16. Photograph of the damaged thyristor.

4. Conclusions

In this paper, the reverse recovery currents in the thyristors in PPS are investigated, which is helpful to reduce the breakdown risks of the thyristors. Two cases of the reverse recovery currents are analyzed initially. Then, the effects of the initial voltage of the energy-storage capacitor, the discharge time intervals, and the load resistance on the reverse recovery currents are explored.

The research results are as follows: large initial voltage of the energy-storage capacitor leads to large reverse recovery currents, which increases breakdown risks of the thyristors; large discharge time intervals increase the number of the thyristors with forward oscillation currents, which increases the breakdown risks of the thyristors; large load resistance leads to less number of the thyristors with the forward oscillation currents, which reduces the breakdown risks of the thyristors. It is also found that the type II PPS topology is more helpful to reduce the breakdown risks of the thyristors under the same working conditions. However, the reverse recovery currents in the thyristors in both type I and II topologies are relatively high when sustaining instantaneous overvoltages caused by load mutations.

To overcome the shortcomings of the conventional PPS topologies, an improved PPS topology is proposed, which consumes less useful energy compared with conventional approaches and absorbs the surge energy from the load side more quickly when load mutation occurs. Therefore, the improved PPS topology can reduce the reverse recovery currents and breakdown risks of the thyristors.

Author Contributions: Conceptualization, J.W. and B.L.; methodology, J.W. and Z.L.; formal analysis, J.W., Z.L. and B.L.; investigation, J.W. and B.L.; data curation, J.W.; simulation, J.W.; validation, J.W. and Z.L.; writing—original draft preparation, J.W.; writing—review and editing, J.W., Z.L. and B.L.; supervision, Z.L. and B.L. All authors have read and agreed to the published version of the manuscript.

Funding: This research received no external funding.

Conflicts of Interest: The authors declare no conflict of interest.

References

1. Cho, C.H.; Park, S.H.; Choi, Y.; Kim, B.G. Production of nanopowders by wire explosion in liquid media. *Surf. Coat. Technol.* **2007**, *201*, 4847–4849. [CrossRef]
2. Inoue, H.; Lisitsyn, I.V.; Akiyama, H.; Nishizawa, I. Drilling of hard rocks by pulsed power. *IEEE Electr. Insul. Mag.* **2000**, *16*, 19–25. [CrossRef]
3. Jin, Y.; Lee, H.; Kim, J.; Rim, G.; Kim, J.; Lee, Y.; Moon, H. Performance of 2.4-MJ pulsed power system for electrothermal-chemical gun application. *IEEE Trans. Magn.* **2003**, *39*, 235–238.
4. Jack, S.B.; Michael, F.S.; Thomas, E.J. Analysis of a Capacitor-Based Pulsed-Power System for Driving Long-Range Electromagnetic Guns. *IEEE Trans. Magn.* **2003**, *39*, 486–490.
5. Li, J.; Li, S.Z.; Liu, P.Z.; Gui, Y.C. Design and Testing of a 10-MJ Electromagnetic Launch Facility. *IEEE Trans. Plasma Sci.* **2011**, *4*, 1187–1191.
6. Dai, L.; Wang, Y.; Zhang, Q.; Li, W.; Lu, W. Effect of Sequence Discharge on Components in a 600-kJ PPS Used for Electromagnetic Launch System. *IEEE Trans. Plasma Sci.* **2013**, *41*, 1300–1306. [CrossRef]
7. Guo, X.; Dai, L.; Zhang, Q.; Lin, F.; Huang, Q.; Zhao, T. Influence of Electric Parameters of Pulsed Power Supply on Electromagnetic Railgun System. *IEEE Trans. Plasma Sci.* **2015**, *43*, 3260–3267. [CrossRef]
8. Liu, J.B.; Bu, L.P.; Qin, X.X. Study on Reverse Charging of Stored Energy Capacitor in Pulsed Power Supply System. *Appl. Mech. Mater.* **2014**, *556*, 2008–2012. [CrossRef]
9. Liu, Y.; Lin, F.; Dai, L.; Zhang, Q.; Lee, L.; Han, Y.; Lu, W. Development of a Compact 450-kJ Pulsed-Power-Supply System for Electromagnetic Launcher. *IEEE Trans. Plasma Sci.* **2011**, *39*, 304–309. [CrossRef]
10. Jin, Y.S.; Kim, Y.B.; Kim, J.; Cho, C.; Lim, S.W.; Lee, B.; Koo, I.S. Fabrication and Testing of a 600-kJ Pulsed Power System. *IEEE Trans. Plasma Sci.* **2013**, *41*, 2671–2674. [CrossRef]
11. Kim, J.; Choi, Y.; Chu, J.; Sung, G. Analysis on high surge voltages generated in paralleled capacitor banks. *IEEE Trans. Magn.* **2003**, *39*, 422–426.
12. Han, Y.; Lin, F.; Dai, L.; Li, H.; Wang, L.; Bo, L.; Peng, B. Research on applications of TVS and thyristor in a pulsed power supply system used for EMG. *IEEE Trans. Dielectr. Electr. Insul.* **2009**, *16*, 979–984.
13. Li, W.; Tang, W.; He, H.; Rao, J.; Zhang, Y. Pulse power switch centered on SCR for the launch system. In Proceedings of the International Conference on Electrical Machines and Systems, Beijing, China, 20–23 August 2011.
14. Li, Z.X.; Yang, C.X.; Li, B.M. Research on the Application of Thyristor in the Pulsed Power Supply. *Power Electron.* **2010**, *44*, 87–90.
15. Spahn, E.; Buderer, G.; Brommer, V.; Sterzelmeier, K.; Welleman, A. Novel 13.5 kV Multichip Thyristor with an Enhanced DI/DT for Various Pulsed Power Applications. In Proceedings of the IEEE International Pulsed Power Conference, Monterey, CA, USA, 13–17 June 2005.
16. Li, Z.X.; Zhang, Y.Z.; Ni, Y.J.; Li, B.M. Analysis on the Influence of Turn-off Characteristics of Thyristor on Augmented Railgun. *Acta Armamentarii* **2016**, *37*, 1599–1605. [CrossRef]
17. Dai, L.; Tian, S.Y.; Jin, C.L.; Yang, Y.; Lei, Y.Q.; Lin, F.C. Reverse recovery characteristics of pulse power thyristor. *High Power Laser Part. Beams* **2016**, *28*, 1–5.
18. Zhang, Y.Z.; Li, Z.X.; Tian, H.; Li, H.Y.; Li, B.M. Application of Pulse Power Supply with RSD Switch in Electromagnetic Launch. *Acta Armamentarii* **2017**, *38*, 658–663.
19. Li, Z.X.; Li, B.M.; Lin, Q.H. Research of Protection for High Voltage Diode-stacks in Pulse Power Supply. *Power Electron.* **2009**, *43*, 47–49.
20. Li, Z.C.; Jin, C.L.; Dai, L.; Chen, C.; Ju, L.; Lin, F.C. Breakdown and Protection of Semiconductor Device in a Sequentially Fired Pulse Forming Network. *Acta Armamentarii* **2017**, *38*, 2349–2353.
21. Wen, Y.L.; Dai, L.; Zhu, Q.; Wang, S.J.; Lin, F.C. Efficiency of Distributed Energy Sstorage Electromagnetic Railgun. *High Power Laser Part. Beams* **2020**, *32*, 1–5.

22. Zhang, J.; Tang, G.F.; Wen, J.L.; Zha, K.P. Dynamic Modelling of the Turn off Performance of High Voltage Power Thyristor. *Power Electron.* **2015**, *49*, 50–53.
23. Li, Z.X.; Zhang, Y.Z.; Gao, L.; Jin, Y.; Li, B.M. Test and Analysis of Silicon Stack Failure in Electrothermal-chemical Gun. *Acta Armamentarii* **2015**, *36*, 577–581.

electronics

MDPI

Article

A Fusion Algorithm for Online Reliability Evaluation of Microgrid Inverter IGBT

Chuankun Wang, Yigang He *, Chenyuan Wang, Xiaoxin Wu and Lie Li

School of Electrical Engineering and Automation, Wuhan University, Wuhan 430072, China; wck980108@whu.edu.cn (C.W.); wcy960810@whu.edu.cn (C.W.); xin638943@163.com (X.W.); lilie2017282070166@whu.edu.cn (L.L.)

* Correspondence: yghe1221@whu.edu.cn

Received: 26 July 2020; Accepted: 10 August 2020; Published: 12 August 2020

Abstract: Due to the diversity of distributed generation sources, microgrid inverters work under complex and changeable conditions. The core device of inverters, an insulated gate bipolar transistor (IGBT), bears a large amount of thermal stress impact, so its reliability is related to the stable operation of the microgrid. The effect of the IGBT aging process cannot be considered adequately with the existing reliability evaluation methods, which have not yet reached the requirements of online evaluation. This paper proposes a fusion algorithm for online reliability evaluation of microgrid inverter IGBT, which combines condition monitoring and reliability evaluation. Firstly, based on the microgrid inverter topology and IGBT characteristics, an electrothermal coupling model is established to obtain junction temperature data. Secondly, the segmented long short-term memory (LSTM) algorithm is studied, which can accurately predict the aging process of the IGBT and judge the aging state via the limited monitoring data. Then, the parameters of the electrothermal coupling model are corrected according to the aging process. Besides, the fusion algorithm is applied to the practical case. Finally, the data comparison verifies the feasibility of the fusion algorithm, whose cumulative damage degree and estimated life error are 5.10% and 5.83%, respectively.

Keywords: segmented LSTM; microgrid inverter; IGBT reliability; online evaluation; fusion algorithm

1. Introduction

The microgrid can take full advantage of the high efficiency and flexibility of the distributed generation, which can maintain the balance of load power and achieve a certain degree of optimal management. Inverter in microgrid plays a key role in power conversion, transmission, and storage, showing its reliability particularly crucial in practical application [1,2]. Insulated gate bipolar transistors (IGBTs), with fast switching speed, simple driving circuits, and large current capacity, have been widely used in microgrid inverters [3]. Due to the diversity of distributed generation sources and the complexity of operation mode, IGBT, the core device of a microgrid inverter, often bears a lot of thermal stress cycles. Under complex working conditions, the performance of IGBT will gradually degrade, which is a critical factor of inverter fault [4–6].

Generally, the IGBT reliability analysis is carried out from Physics-of-Failure (PoF). Studies have shown that the fluctuation of junction temperature is the main reason for IGBT failure. Owing to the different coefficients of thermal expansion (CTE), the thermal stress inside the IGBT structure is uneven, resulting in damage to the bond wires, the solder layer, and the interior of the chip [7,8]. There are two ways to obtain junction temperature online: direct measurement and indirect measurement [9,10]. Direct measurement is to obtain junction temperature data by embedding integrated sensors inside the IGBT module. In the process of designing and producing IGBTs, manufacturers need to consider the electromagnetic compatibility of integrated sensors. Hence, this method has the disadvantages of data transmission delay and high cost in actual projects. Indirect measurement means to estimate the junction temperature of the IGBT in real-time by establishing an electrothermal coupling model, which has the advantages of low delay and strong online monitoring capabilities. In addition, the infrared thermometer can also be used to measure the IGBT 3-D temperature distributions, but it is often disturbed by the package structure [11]. The electrothermal coupling model estimates the junction temperature in real-time through power loss without intruding into the package. However, the electrothermal coupling model of the IGBT is generally established based on the IGBT's technical manual. The IGBT is constantly aging owing to fatigue damage during operation, which makes the pre-established electrothermal coupling model no longer adapt to the current IGBT state. In [12], the author proposed a mathematical analysis method for fundamental frequency junction temperature fluctuation based on equivalent sine half-wave loss. However, the converter IGBT junction temperature was estimated without updating the parameters of the electrothermal coupling model in time. In [13], when predicting the life of the IGBT in a static synchronous compensator (STATCOM) via the electrothermal coupling model and rainflow counting algorithm, the influence of the IGBT aging process was not considered. In [14], an adaptive thermal equivalent circuit model for estimating the junction temperature of the IGBT is proposed, which can correct the parameter deviation of the electrothermal coupling model caused by the aging of the solder layer. However, multiple temperature sensors between the substrate and heat sink must be installed, which has a weak anti-interference ability to the external environment. In general, the electrothermal coupling model is very suitable for online monitoring of the IGBT junction temperature, but there is currently no effective means to correct the effect of the aging process in the online monitoring process.

Since IGBT aging has a non-negligible influence on reliability evaluation, IGBT condition monitoring can provide new ideas for the correction of electrothermal coupling parameters. IGBT state parameters include gate threshold voltage, module thermal resistance, collector current, collector–emitter voltage, short-circuit current, etc., which can reflect the aging state of IGBT [15,16]. In [17], the on-state collector–emitter voltage at the inflection point was used to detect the degradation of the bonding wire, and experiments had shown that the method is not disturbed by the external environment temperature. In [18], the aging process of the solder layer was monitored through the thermal resistance of the IGBT module, and the equivalent thermal network model parameters were updated in real-time accordingly. In [19], the author detected the IGBT chips in the multi-chip IGBT power module through the gate turn-on threshold voltage and accurately judged the number of faulty chips for early warning. In [20], the aging state of the IGBT was monitored by monitoring the difference of the short-circuit current, and the experiment proved that this parameter was little affected by the junction temperature. The above studies show that the IGBT state parameters can accurately reflect the state of health and have strong anti-interference, but these studies are carried out under the conditions of sufficient monitoring data.

The aging cycle of IGBT is very long, which means that a large amount of condition monitoring data is needed to identify the aging stage. In the process of online evaluation of IGBT reliability, it is impossible to obtain a large amount of monitoring data in a short time. In order to evaluate the IGBT state more efficiently, the application of data-driven (DD) can extract more health information from the historical data of the state parameters. DD is to predict the time series of observation parameters through traditional numerical techniques (Kalman filters [21], particle filters (PF) [22], regression [23], and statistical methods [24]), machine learning (neural networks [25], decision trees [26], and support vector machines [27]), and other approaches. In [22], a prognostic method based on Mahalanobis distance (MD) and PF methods were used to predict the remaining useful life (RUL) of IGBT, with an error of 20%. In [25], two machine learning methods, neural network (NN) and adaptive neuro-fuzzy inference system (ANFIS), were adopted to predict the RUL of IGBT via information beyond half-life. The errors calculated using NN and ANFIS are 19.04% and 30.91%, respectively. At present, few studies are adopting NN to predict and analyze the aging process of IGBTs, and the accuracy of prediction needs to be improved [28].

Motivated by the analysis described above, this paper improves the long short-term memory (LSTM) algorithm according to IGBT aging characteristics to obtain the segmented LSTM prediction network. Combined condition monitoring and reliability evaluation, a PoF and DD fusion algorithm is proposed for online reliability evaluation of microgrid inverter IGBT. First, the segmented LSTM accurately predicts the aging state curve of the IGBT based on the limited monitoring data, and the IGBT aging state is judged in real-time. Next, parameters of the electrothermal coupling model are updated to correct the influence of the aging process, which guarantees the junction temperature data is consistent with the actual working conditions. Further, the rainflow counting algorithm makes statistics on the thermal stress distribution via the junction temperature data. Finally, combined with fatigue damage theory and the life prediction model, a reliability evaluation is carried out.

This paper is organized as follows. Section 2 establishes an electrothermal coupling model for the topology of the microgrid inverter and verifies the model by the power loss and junction temperature data of the manufacturer. Section 3 illustrates IGBT fatigue damage theory and accelerated aging experiments, and studies the segmented LSTM prediction network suitable for the IGBT aging process. Section 4 introduces the proposed fusion algorithm flow and analyzes an actual case via the fusion algorithm. Verification and comparison are presented in Section 5. Section 6 draws the conclusion.

2. IGBT Reliability Modeling

2.1. Electrothermal Coupling Model

Bonding wire peeling and solder layer cracking are two main failure modes of IGBTs, which mainly caused the junction temperature inside the device [29]. Therefore, the junction temperature is the crucial data to study the performance degradation and reliability analysis of microgrid inverter IGBTs. Figure 1 illustrates the flow chart of establishing the electrothermal coupling model, which can output real-time junction temperature data. The IGBT power loss model is derived by combining the microgrid inverter topology and IGBT operating characteristics. The equivalent model of the IGBT thermal network is derived based on the physical structure of the IGBT module and the internal heat conduction process. Eventually, the real-time junction temperature fluctuation data of the IGBT under the current operating conditions can be output.

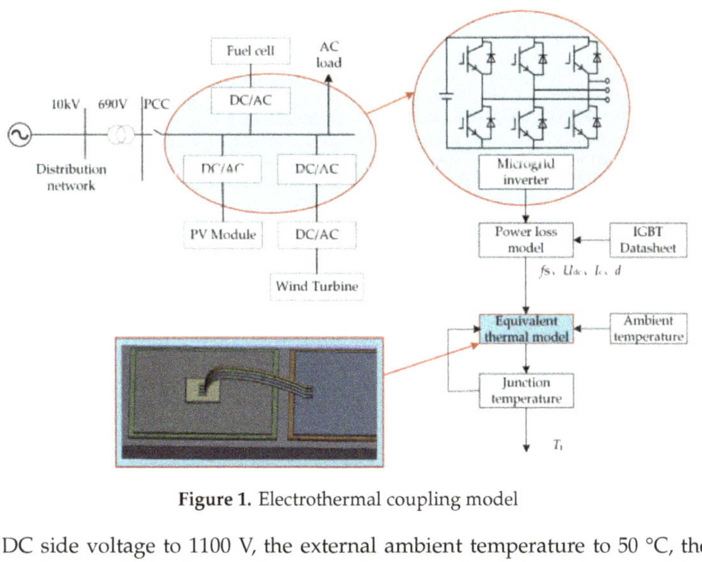

Figure 1. Electrothermal coupling model

Set the DC side voltage to 1100 V, the external ambient temperature to 50 °C, the switching frequency to 10 kHz, and the duty cycle to 0.4. The electrothermal coupling model refers to the IRG4BC30K IGBT datasheet, which is produced by Infineon. In Table 1, compared with the data output by the Infineon-IGBT simulation tool under the same conditions, the model is verified by power loss and junction temperature.

Table 1. Electrothermal coupling model verification.

Verify Content	IGBT Switching Loss (W)	IGBT Conduction Loss (W)	Junction Temperature (°C)
Infineon-IGBT simulation tool	8.010	7.682	68.83
Electrothermal coupling model	8.276	7.799	69.81
Error	3.32%	1.52%	1.42%

2.2. *Life Prediction Model*

In order to evaluate the reliability and life of IGBT modules, manufacturers and researchers generally carry out accelerated aging tests and put forward a series of IGBT module life models based on test data and failure mechanisms, such as the Lesit model, Bayerer model, and Coffin–Manson model.

- The Lesit model considers both the junction temperature fluctuation and the average value, and its mathematical expression is:

$$N_f = A\cdot\left(\Delta T_j\right)^{-\alpha}\cdot \exp\left(\frac{E_\alpha}{k\cdot T_m}\right) \tag{1}$$

  where A is the model correction coefficient, α is the junction temperature fluctuation index, k is the Boltzmann constant, and E_α is the excitation energy of the IGBT module chip [30].
- The Coffin–Manson model involves three factors: maximum temperature, junction temperature fluctuation, and cycle frequency [31]. Its mathematical expression is:

$$N_f = A\cdot f^{-a}\cdot\left(\Delta T_j\right)^{-b}\cdot G(T_m) \tag{2}$$

  where A is the fitting constant, a is the cycle frequency index (typical value is about 1/3), and b is the junction temperature fluctuation index (standard value is about 2).
- The Bayerer model takes many other variables into account, in addition to the maximum junction temperature and junction temperature fluctuations considered by the Coffin–Manson model. Its mathematical expression is:

$$N_f = K\cdot\left(\Delta T_j\right)^{-\beta_1}\cdot e^{-\frac{\beta_2}{T_{jmax}+273}}\cdot t_{on}^{\beta_3}\cdot I^{\beta_4}\cdot V^{\beta_5}\cdot D^{\beta_6} \tag{3}$$

  where K is the model correction coefficient, β_1–β_2 is the junction temperature fluctuation index and the maximum junction temperature index, and β_3–β_6 is the power cycle heating time, the device withstand voltage rating, the bonding wire current, and the index of the bond wire diameter [32].

Currently, the Lesit model is mostly applied in the life prediction of IGBT life. Its expression is simple and consistent with the results of the aging experiment. Moreover, it has an excellent online monitoring capability in conjunction with rainflow counting. Although the Bayerer model is more accurate than other models, it is impossible to apply to online monitoring due to many parameters being difficult to measure.

3. IGBT Aging Monitoring

3.1. *Fatigue Damage Theory*

IGBT needs to withstand a large number of thermal stress cycles during operation. Assume that N_f is the number of failure cycles of the IGBT under a stress cycle with constant amplitude. When the number of cycles that it bears the stress cycle is N (N is less than N_f), the fatigue damage of the IGBT can be expressed by the cumulative damage degree as follows [33]:

$$D = \frac{N}{N_f} \tag{4}$$

If the device endures multiple constant amplitude stresses, and the number of impacts generated by each constant amplitude stress is N_i, the cumulative damage degree can be expressed as follows:

$$D = \sum_{i=1}^{k} D_i = \sum_{i=1}^{k} \frac{N_i}{N_{f,\,i}} \tag{5}$$

when the cumulative damage D reaches 1, it indicates that the device fails due to fatigue damage. In the actual working process of the microgrid inverter, the solder layer of the IGBT is prone to fatigue damage, and the thermal resistance increases with the aging process of the device material. Due to the influence of the IGBT aging process on the life prediction, it is necessary to update the thermal network parameters of the electrothermal coupling model in time. According to the aging law of the IGBT module, increasing the thermal model parameters by 10%–50% can simulate different aging stages [34].

3.2. Accelerated Aging Test

Since the aging process of IGBT is very long, in order to obtain relatively accurate aging data in a short time, it is necessary to conduct an accelerated aging test on IGBT. During the experiment, the device is in a state of high-speed switching, so the device can withstand a large number of thermal stress cycles in a short time, which accelerates the aging failure. The aging process of IGBT will change the electrical parameters, so parameters measured easily can be selected to predict the aging process of IGBT. Among these parameters, collector–emitter voltage is the most suitable precursor for aging prognostic, considering online measurement, calibration, accuracy, linearity, and sensitivity [26]. When IGBT is turned off, parasitic transistors produce a transient voltage, which interacts with the IGBT collector voltage to produce the transient peak voltage in Figure 2a. The analysis of aging data in the National Aeronautics and Space Administration (NASA) laboratory shows that the collector–emitter peak voltage (V_{ce_peak}) is closely related to the degradation, so V_{ce_peak} can be employed to monitor the aging process of IGBT.

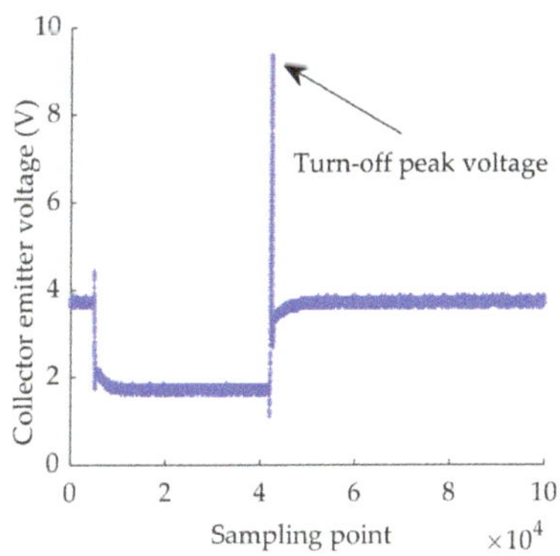

(a)

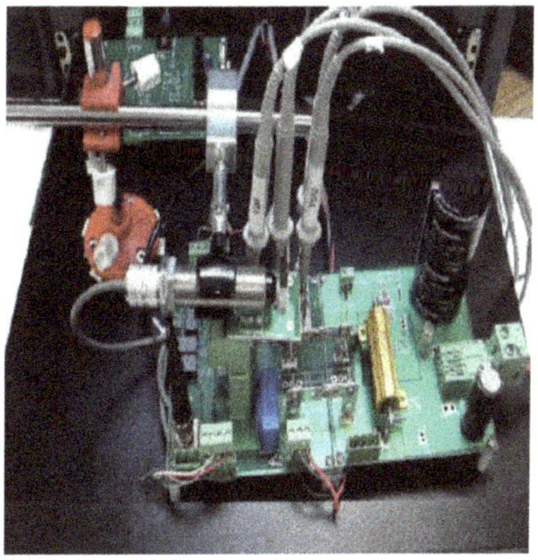

(b)

Figure 2. (**a**) Collector–emitter turn-off voltage waveform. (**b**) Insulated gate bipolar transistor (IGBT) accelerated aging experiment platform.

This paper uses the accelerated aging experimental data set published by NASA Prognostics Center of Excellence (PCoE) to study the aging prediction network model and apply the best model in the actual example [35]. The experimental platform of the NASA accelerated aging test for IRG4BC30K is shown in Figure 2b, and the test parameters are listed in Table 2. Therefore, the setting of the accelerated aging test is very close to the actual working condition of IGBT. The parameters measured by the accelerated aging experiment contain collector current, collector–emitter voltage, gate voltage, and packaging temperature. Vce_peak was sampled and extracted to obtain the aging monitoring data set until the device failed.

Table 2. Accelerated aging test parameters.

Types	Setting
IGBT type	IRG4BC30K
PWM duty cycle	0.4
Switching frequency (kHz)	10
Package temperature (°C)	260–270
Gate voltage (V)	10

3.3. *Segmented LSTM Algorithm*

In recent years, with the continuous development of deep learning (DL), relevant models are gradually applied to fault time series data. DL is a kind of deep neural network model with multiple nonlinear mapping levels, which can abstract and extract features of input signals layer by layer and dig out more profound potential laws [36]. As one of the DL networks, having the inherent potential of fully mining the data time-series information, a recurrent neural network (RNN) has been widely used in time series prediction [37]. A convolution neural network (CNN) can also achieve the purpose of time series prediction by constructing samples, but vast amounts of data are needed, which cannot match the aging characteristics of IGBTs.

RNN introduces the concept of a time sequence into the design of a network structure, making it more adaptable in the analysis of time-series data. However, RNN has many disadvantages, such as gradient disappearance, gradient explosion, and reduced long-term memory. As an improved model of RNN, LSTM can make up for the shortcomings of RNN. IGBTs can be affected by factors such as high temperature, high pressure, and harsh environment during the working process. Besides, its continuous stress impact will cause fatigue damage to the device. Therefore, the IGBT aging state gradually changes with time, which indicates that the current device aging state will be affected by the previous one. So, IGBTs aging prediction can be abstracted as a time series prediction problem.

In real projects, the monitoring device cannot collect a large amount of aging data in a short time, which is not conducive to real-time evaluation of the aging state of IGBTs. The LSTM algorithm can predict the aging process of IGBT according to the monitoring data. The prediction framework is shown in Figure 3, and the parameter settings are listed in Table 3. The first 25% of the data set was used as training data to train the LSTM time series prediction network, and the remaining 75% data verified the prediction results in Figure 4. It can be found that the forecast data in the early stage coincided with the observed value, but the forecast data in the later stage deviated far from the observed data. Finally, the root mean squared error (RMSE) of the LSTM prediction was 0.27824.

Table 3. Hyperparameter settings of long short-term memory (LSTM) networks.

Types	Setting
Initial Learn Rate	0.001
Learn Rate Schedule	piecewise
Max Epochs	50
Gradient Threshold	1
Learn Rate Drop Period	25
Learn Rate Drop Factor	0.1
Execution environment	GPU
Optimizer	Adam
Verbose	0

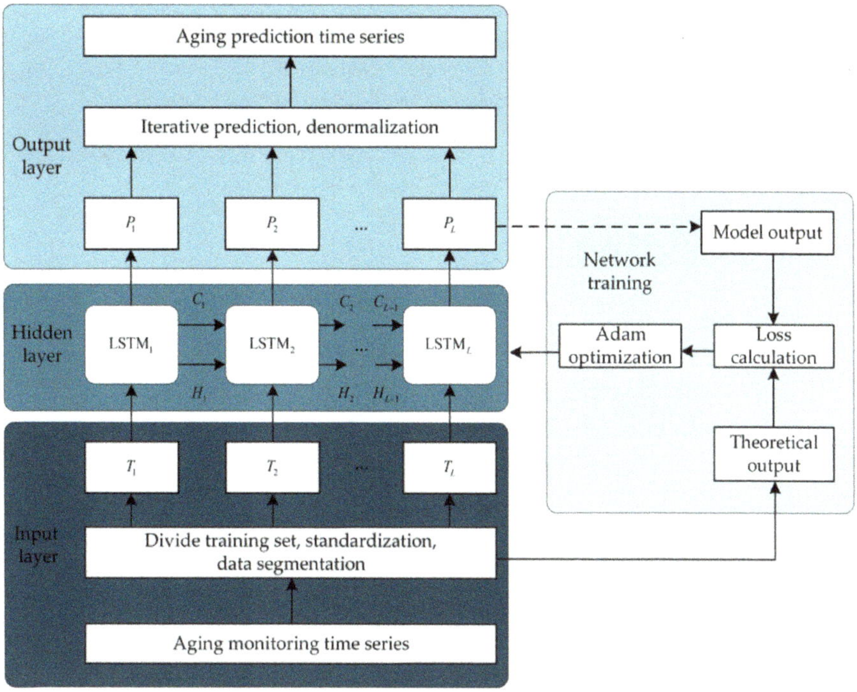

Figure 3. Aging time series prediction framework based on LSTM.

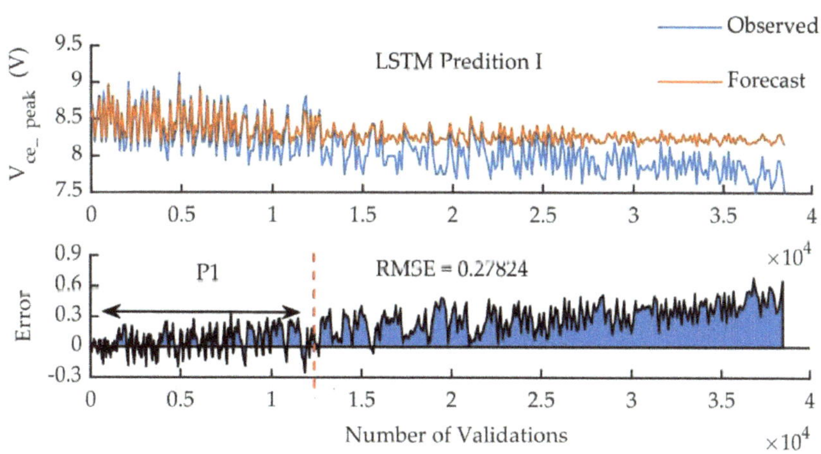

Figure 4. The first LSTM prediction.

In response to this problem, this paper proposed a segmented LSTM prediction algorithm, and the flowchart is illustrated in Figure 5. During the first LSTM prediction, the first 1/3 data of the prediction data (P1) was retained and combined with the original training data (T1) to form the training data of the second LSTM prediction (T2). During the second LSTM prediction in Figure 6, the first 1/2 of the prediction data (P2) and T2 were combined into the training data set for the third LSTM prediction (T3). During the third LSTM prediction shown in Figure 7, all aging prediction data (P3) was consistent with the measured data. After three-segmented LSTM predictions, the final comparison of prediction data and monitoring data is presented in Figure 8, and the RMSE was only 0.1153. The segmented LSTM algorithm had higher prediction accuracy under the condition of limited training data, which meets the needs of microgrid inverter IGBT condition monitoring and reliability analysis.

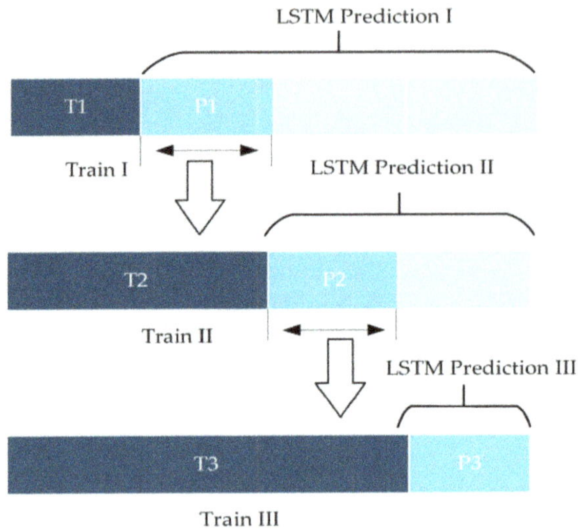

Figure 5. The segmented LSTM prediction flowchart.

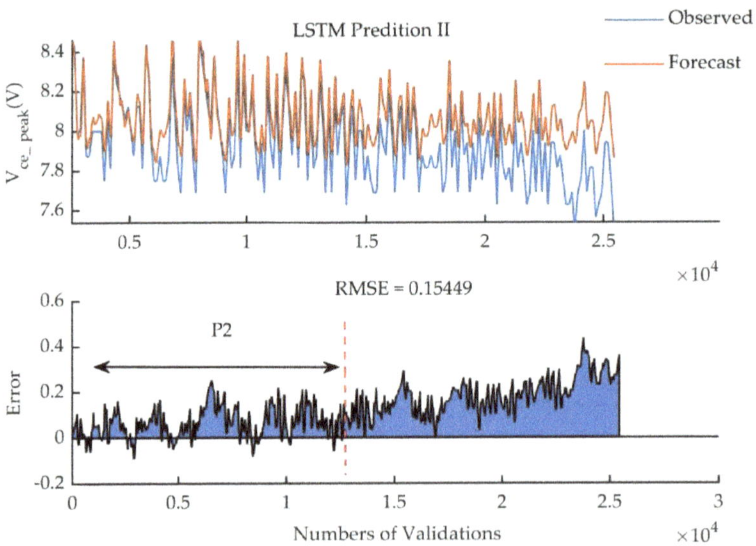

Figure 6. The second LSTM prediction.

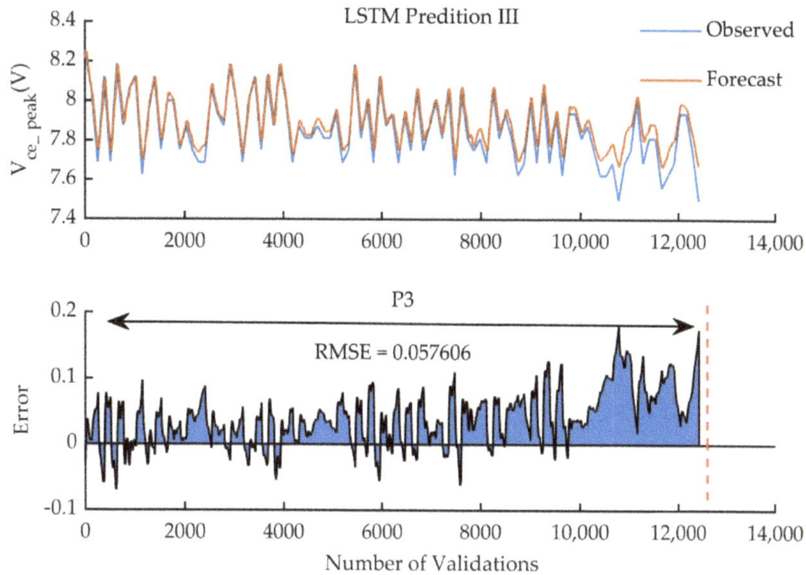

Figure 7. The third LSTM prediction.

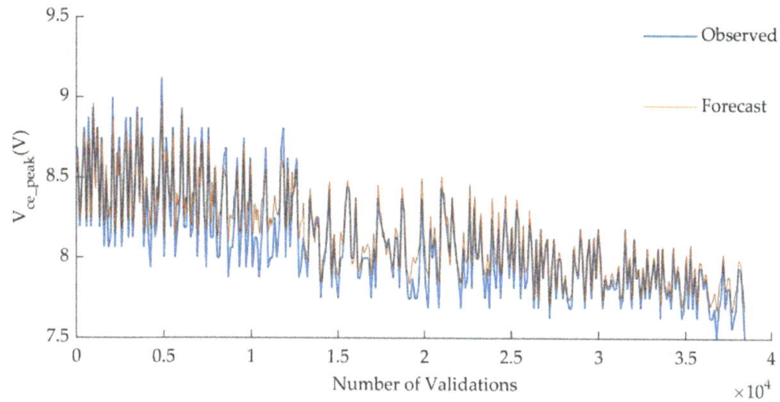

Figure 8. The result of the segmented LSTM prediction.

4. IGBT Reliability Online Evaluation Fusion Algorithm

4.1. Algorithm Flow

As shown in Figure 9, the framework of the online reliability evaluation fusion algorithm proposed in this paper consisted of three parts. The first part is the reliability analysis based on PoF. The influence of wind speed and ambient temperature is considered through SCADA data, and then the wind turbine model and electrothermal coupling model are used to obtain real-time junction temperature data. The second part is the DD condition monitoring. Vce_peak monitoring data is used for segmented LSTM prediction to estimate the aging process of IGBT. According to the obtained aging curve, the thresholds of different aging stages are divided. Compare the threshold and monitoring data to determine the aging stage in real-time, and select the corresponding electrothermal coupling model, which considers the impact of the aging process and improves the accuracy of microgrid IGBT reliability assessment. The third part is the life prediction. The junction temperature fluctuation data is processed by the rainflow counting algorithm to obtain the distribution of the fluctuation amplitude and mean value. In the end, the Lesit life prediction model could efficiently use the output data of the rainflow counting algorithm and combine fatigue damage theory for life prediction.

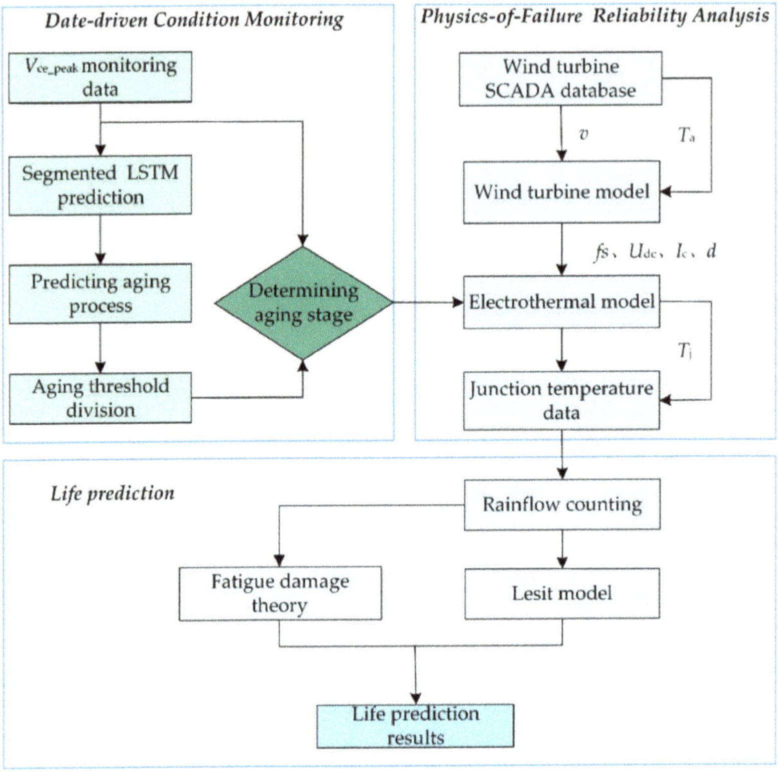

Figure 9. Flowchart of the reliability evolution fusion algorithm.

4.2. Case Analysis

In order to show the working principle of the fusion algorithm more clearly, this paper took the wind power generation system in the microgrid as an example (specific parameters are listed in Table 4) and analyzed the reliability of its inverter IGBT. The wind speed and ambient temperature recorded in the SCADA database within one year were imported into the wind turbine model and the electrothermal coupling model to derive the real-time junction temperature curve of the inverter IGBT. Meanwhile, the segmented LSTM algorithm predicted the aging process through the monitoring data of Vce_peak. Perform zero-order retention and averaging of aging data to extract more obvious aging trends. The estimated aging process is presented in Figure 10, and the thresholds for different aging stages are enlisted in Table 5.

Compare the threshold value and the monitoring data to determine the aging stage of the IGBT in real-time, and then select the corresponding aging correction factor in Table 5. Substitute the aging correction coefficient into the thermal resistance update equation to update the thermal network parameters of the electrothermal coupling model, which ensures the accuracy of the junction temperature data. The thermal resistance update equation is as follows:

$$R = R_{initial}(1 + a \cdot r^{m}) \tag{6}$$

where $R_{initial}$ is the initial value of thermal resistance, a is the aging factor (typical value is about 0.5), r is the aging process coefficient (regular value is 0–1), and m is temperature stress factor (standard value is 1) [38].

Table 4. Wind power system parameters.

Types	Parameters
Rated power (kW)	20
Cut-in wind speed (m/s)	3
Rated wind speed (m/s)	11
Cut out wind speed (m/s)	25
Grid-side voltage (V)	690
DC side voltage (V)	1100
Grid-side frequency (Hz)	50
IGBT switching frequency (kHz)	10
IGBT Type	IRG4BC30K

Table 5. The threshold voltage of IGBT aging stages.

Threshold Voltage	Forecast (V)	Observed (V)	Aging Correction Factor
Health status	10.323	10.323	-
Aging stage 1	9.939	9.927	0.2
Aging stage 2	9.554	9.532	0.4
Aging stage 3	9.170	9.134	0.6
Aging stage 4	8.786	8.741	0.8
Aging stage 5	8.402	8.346	1.0

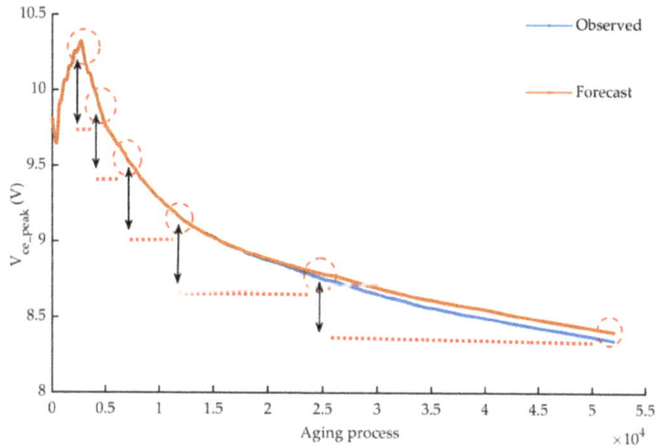

Figure 10. IGBT aging process.

After the above steps, the junction temperature data after aging correction can be obtained. As shown in Figure 11, the rainflow counting method was used to extract the mathematical distribution of junction temperature fluctuation ΔT_j and the average value of junction temperature Tm, which are the critical data for the next reliability evaluation and life prediction. Compared with the uncorrected junction temperature data, the corrected junction temperature data was more in line with the actual working conditions, and the thermal stress load distribution obtained accordingly was more accurate.

Finally, the Lesit life prediction model was used to predict the maximum number of thermal stress cycles Nf, i that the aging IGBT can withstand under various operating conditions, and then the cumulative damage and estimated life of IGBT were calculated with Equation (5).

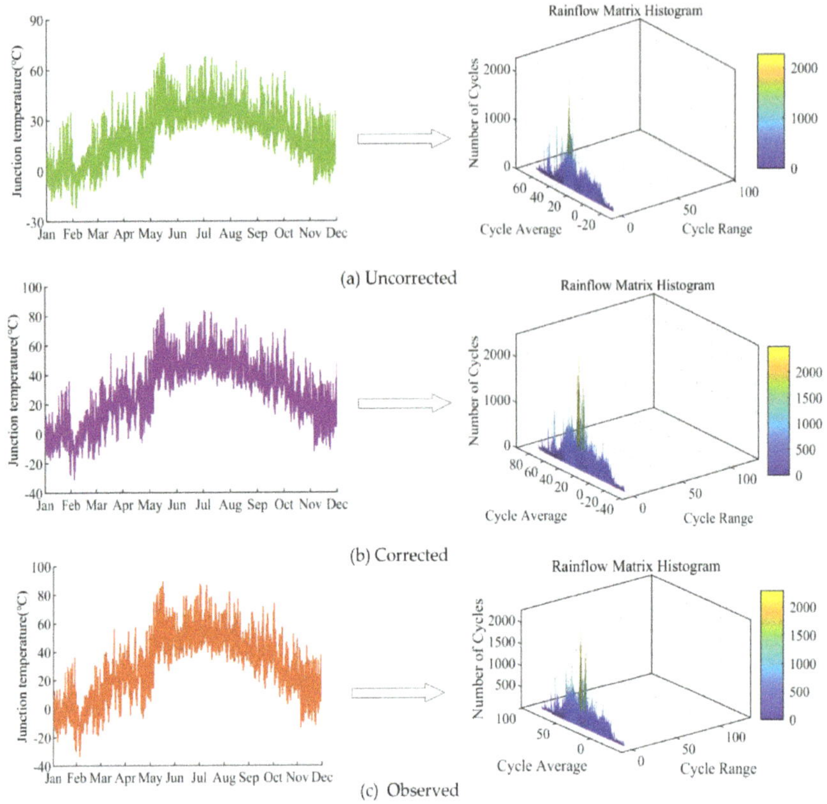

Figure 11. (**a**) Uncorrected thermal stress load distribution. (**b**) Corrected thermal stress load distribution. (**c**) Observed thermal stress load distribution.

5. Comparison and Verification

Previously, the proposed fusion algorithm has been used to evaluate the reliability of the microgrid inverter IGBT. In order to fully explain the advantages of the proposed fusion algorithm for online evaluation of IGBT reliability, which combines PoF reliability analysis with DD condition monitoring, the results of a health assessment were first compared with historical statistical data. Then, the life prediction results were compared with that of other correction algorithms.

In order to study the reliability of IGBTs, scientific research institutions and scholars have made statistics on the failure and damage causes of a large number of IGBTs. Table 6 lists the average cumulative damage degree statistics for the same type of IGBT in one year. View the weighted average cumulative damage and life prediction results as the mathematical expectation of the evaluation results.

Table 6. Historical statistics of IGBT cumulative damage.

Data Sources	Year	Number of Samples	Cumulative Damage Degree
WMEP [39]	1998–2000	209	0.189
	1989–2006	1028	0.085
LWK [39]	1993–2006	5719	0.042
CARROLL [40]	2005–2010	9110	0.020
FISCHER [41]	2003–2017	2316	0.150

5.1. Comparison and Verification of a Health Assessment

This paper used DL algorithms (LSTM), traditional time series prediction algorithms (ARIMA), and the proposed algorithm (the segmented LSTM) to predict the aging of the monitoring data, and make aging corrections in the process of reliability evaluation. Table 7 shows a comparison of the results of aging correction using different prediction algorithms. Without aging correction, the cumulative damage error reached 45.51%, which overestimated the IGBT health status. After aging correction based on the actual aging parameter observation data, the cumulative damage error was only 3.88%, which indicates the effectiveness of the fusion algorithm. It can be found that monitoring the aging state and updating the parameters of the electrothermal coupling model in time could significantly improve the accuracy of the reliability evaluation.

Compared with the LSTM algorithm and ARIMA algorithm, the segmented LSTM algorithm could predict the aging process of IGBT more accurately, and the cumulative damage error after correction was only 5.10%. Thus, the fusion algorithm based on the segmented LSTM still had excellent adaptability in the case of insufficient monitoring data. The fusion algorithm could effectively correct the influence of the aging process on the reliability evaluation of the IGBT, and the evaluation result could genuinely reflect the health status of the IGBT.

Table 7. Comparison of health assessment.

Algorithm Type	Cumulative Damage	Error
Mathematical Expectation	0.0490	-
No correction	0.0267	45.51%
Observation data correction	0.0471	3.88%
Segmented LSTM	0.0465	5.10%
LSTM	0.0398	18.78%
ARIMA	0.0341	30.41%

5.2. Comparison and Verification of Life Prediction

In the existing literature, some correction methods have been tried to predict the life of the same type of inverter IGBT. The expected life span is between 17.71 and 28.50 years. The predicted lifetime obtained by the fusion algorithm proposed in this paper is 22.22 years, which is consistent with the prediction results of the existing literature. So as to compare the accuracy of each correction algorithm, the errors between the life prediction results and the mathematical expectation are calculated in Table 8. The prediction error of the fusion algorithm is only 5.83%, which is far lower than other correction algorithms. Therefore, the fusion algorithm based on the segmented LSTM can correctly predict the life of the IGBT based on its health status.

Table 8. Comparison of life prediction.

Algorithm Type	Life Perdition (Year)	Error
Mathematical Expectation	20.408	-
Proposed fusion Algorithm	21.598	5.83%
Multiple PoF algorithm [42]	22.000	7.80%
Segmented network update algorithm [43]	28.500	39.65%
Single PoF algorithm [34]	17.710	22.63%

6. Conclusions

This paper proposed an online IGBT reliability evaluation fusion algorithm for microgrid inverters, which combined the PoF reliability analysis with DD condition monitoring to eliminate the influence of aging on a reliability analysis. For solving the contradiction between the IGBT aging cycle and observation scale, the segmented LSTM algorithm was studied on the framework of the original LSTM algorithm. Based on limited aging monitoring data, it could accurately predict the aging process of the device and judge the aging stage, which is the basis for real-time updating of the electrothermal coupling model parameters. The case analysis shows that the combination of condition monitoring and reliability analysis dramatically improved the accuracy of the assessment. In the case of limited monitoring data in actual projects, segmented LSTM can more accurately correct the impact of the IGBT aging process than other traditional algorithms. Statistical data and algorithm comparison verify the feasibility and superiority of the fusion algorithm. The proposed fusion algorithm reduces the dependence on the length of the monitoring data time series and improves the accuracy of reliability evaluation, which meets the requirements of the online reliability evaluation.

Author Contributions: Conceptualization, methodology, software, validation, formal analysis, investigation, resources, C.W. (Chuankun Wang) and Y.H.; Data curation writing—original draft preparation, C.W. (Chenyuan Wang); Writing—review and editing, X.W.; Visualization, L.L. All authors have read and agreed to the published version of the manuscript.

Funding: This research was supported by the National Natural Science Foundation of China (Grant No. 51977153, 51977161, and 51577046), the State Key Program of National Natural Science Foundation of China (Grant No. 51637004), the national key research and development plan "important scientific instruments and equipment development" of China (Grant No. 2016YFF0102200), the Equipment research project in advance of China (Grant No. 41402040301).

Conflicts of Interest: The authors declare no conflict of interest.

References

1. Bani-Ahmed, A.; Rashidi, M.; Nasiri, A.; Hosseini, H. Reliability analysis of a decentralized microgrid control architecture. *IEEE Trans. Smart Grid.* **2019**, *10*, 3910–3918. [CrossRef]
2. Bahramirad, S.; Reder, W.; Khodaei, A. Reliability-Constrained optimal sizing of energy storage system in a microgrid. *IEEE Trans. Smart Grid.* **2012**, *3*, 2056–2062. [CrossRef]
3. Choi, U.; Blaabjerg, F.; Jørgensen, S.; Munk-Nielsen, S.; Rannestad, B. Reliability improvement of power converters by means of condition monitoring of IGBT modules. *IEEE Trans. Power Electron.* **2017**, *32*, 7990–7997. [CrossRef]
4. Sangwongwanich, A.; Yang, Y.; Sera, D.; Blaabjerg, F. Mission profile-oriented control for reliability and lifetime of photovoltaic inverters. *IEEE Trans. Ind. Appl.* **2019**, *56*, 601–610. [CrossRef]
5. Choi, U.; Blaabjerg, F.; Lee, K. Study and handling methods of power IGBT module failures in power electronic converter systems. *IEEE Trans. Power Electron.* **2015**, *30*, 2517–2533. [CrossRef]
6. Choi, J.; Khalsa, A.; Klapp, D.A.; Baktiono, S.; Illindala, M.S. Survivability of prime-mover powered inverter-based distributed energy resources during microgrid islanding. *IEEE Trans. Ind. Appl.* **2019**, *55*, 1214–1224. [CrossRef]

7. Oh, H.; Han, B.; McCluskey, P.; Han, C.; Youn, B.D. Physics-of-failure, condition monitoring, and prognostics of insulated gate bipolar transistor modules: A Review. *IEEE Trans. Power Electron.* **2015**, *30*, 2413–2426. [CrossRef]
8. Sabbah, W.; Arabi, F.; Avino-Salvado, O.; Buttay, C.; Théolier, L.; Morel, H. Lifetime of power electronics interconnections in accelerated test conditions: High temperature storage and thermal cycling. *Microelectron. Reliab.* **2017**, *76*, 444–449. [CrossRef]
9. Eleffendi, M.A.; Johnson, C.M. Application of Kalman filter to estimate junction temperature in IGBT power modules. *IEEE Trans. Power Electron.* **2016**, *31*, 1576–1587. [CrossRef]
10. Musallam, M.; Johnson, C.M. Real-Time compact thermal models for health management of power electronics. *IEEE Trans. Power Electron.* **2010**, *25*, 1416–1425. [CrossRef]
11. Van der Broeck, C.H.; Lorenz, R.D.; De Doncker, R.W. Monitoring 3-D Temperature Distributions and Device Losses in Power Electronic Modules. *IEEE Trans. Power Electron.* **2019**, *34*, 7983–7995. [CrossRef]
12. Wang, X.; Li, Z.; Yao, F.; Tang, S. Simplified estimation of junction temperature fluctuation at the fundamental frequency for IGBT modules considering mission profile. *IEEE Access* **2019**, *7*, 149308–149317. [CrossRef]
13. GopiReddy, L.R.; Tolbert, L.M.; Ozpineci, B.; Pinto, J.O.P. Rainflow algorithm-based lifetime estimation of power semiconductors in utility applications. *IEEE Trans. Power Electron.* **2019**, *51*, 3368–3375. [CrossRef]
14. Hu, Z.; Du, M.; Wei, K.; Hurley, W.G. An adaptive thermal equivalent circuit model for estimating the junction temperature of IGBTs. *IEEE J. Emerg. Sel. Top. Power Electron.* **2019**, *7*, 392–403. [CrossRef]
15. Astigarraga, D.; Ibanez, F.M.; Galarza, A.; Echeverria, J.M.; Unanue, I.; Baraldi, P.; Zio, E. Analysis of the results of accelerated aging tests in insulated gate bipolar transistors. *IEEE Trans. Power Electron.* **2016**, *31*, 7953–7962. [CrossRef]
16. Mohamed Sathik, M.H.; Sundararajan, P.; Sasongko, F.; Pou, J.; Natarajan, S. Comparative analysis of IGBT parameters variation under different accelerated aging tests. *IEEE Trans. Electron. Devices* **2020**, *67*, 1098–1105. [CrossRef]
17. Singh, A.; Anurag, A.; Anand, S. Evaluation of Vce at inflection point for monitoring bond wire degradation in discrete packaged IGBTs. *IEEE Trans. Power Electron.* **2017**, *32*, 2481–2484. [CrossRef]
18. Hu, Z.; Zhang, W.F.; Wu, J. An improved electro-thermal model to estimate the junction temperature of IGBT module. *Electronics* **2019**, *8*, 1066. [CrossRef]
19. Mandeya, R.; Chen, C.; Pickert, V.; Naayagi, R.T.; Ji, B. Gate-Emitter pre-threshold voltage as a health-sensitive parameter for IGBT chip failure monitoring in high-voltage multichip IGBT power modules. *IEEE Trans. Power Electron.* **2018**, *34*, 9158–9169. [CrossRef]
20. Sun, P.; Gong, C.; Du, X.; Peng, Y.; Wang, B.; Zhou, L. Condition monitoring IGBT module bond wires fatigue using short-circuit current identification. *IEEE Trans. Power Electron.* **2017**, *32*, 3777–3786. [CrossRef]
21. Zhu, P.; Chen, B.; Príncipe, J.C. Learning nonlinear generative models of time series with a Kalman filter in RKHS. *IEEE Trans. Signal Process.* **2014**, *62*, 141–155. [CrossRef]
22. Patil, N.; Das, D.; Pecht, M. A prognostic approach for non-punch through and field stop IGBTs. *Microelectron. Reliab.* **2012**, *52*, 482–488. [CrossRef]
23. Brown, D.; Kalgren, P.; Roemer, M. Electronic Prognostics—A Case Study Using Switched-Mode Power Supplies (SMPS). *IEEE Instrum. Meas. Mag.* **2007**, *10*, 20–26. [CrossRef]
24. Li, R.; Verhagen, W.J.C.; Curran, R. A comparative study of Data-driven Prognostic Approaches: Stochastic and Statistical Models. In Proceedings of the IEEE International Conference on Prognostics and Health Management (ICPHM), Seattle, WA, USA, 11–13 June 2018; pp. 1–8. [CrossRef]
25. Ahsan, M.; Stoyanov, S.; Bailey, C. Data driven prognostics for predicting remaining useful life of IGBT. In Proceedings of the 39th International Spring Seminar on Electronics Technology (ISSE), Pilsen, Czech Republic, 18–22 May 2016; pp. 273–278. [CrossRef]
26. Zheng, Z.; Peng, J.; Deng, K.; Gao, K.; Li, H.; Chen, B.; Yang, Y.; Huang, Z. A Novel Method for Lithium-Ion Battery Remaining Useful Life Prediction Using Time Window and Gradient Boosting Decision Trees. In Proceedings of the 10th International Conference on Power Electronics and ECCE Asia (ICPE 2019-ECCE Asia), Busan, Korea, 27–30 May 2019; pp. 3297–3302.
27. Wei, J.; Dong, G.; Chen, Z. Remaining Useful Life Prediction and State of Health Diagnosis for Lithium-Ion Batteries Using Particle Filter and Support Vector Regression. *IEEE Trans. Ind. Inform.* **2018**, *65*, 5634–5643. [CrossRef]

28. Fang, X.; Lin, S.; Huang, X.; Lin, F.; Yang, Z.; Igarashi, S. A review of data-driven prognostic for IGBT remaining useful life. *Chin. J. Electron.* **2018**, *4*, 73–79. [CrossRef]
29. .Zhang, J.; Du, X.; Wu, Y.; Luo, Q.; Sun, P.; Tai, H. Thermal parameter monitoring of IGBT module using case temperature. *IEEE Trans. Power Electron.* **2019**, *34*, 7942–7956. [CrossRef]
30. Held, M.; Jacob, P.; Nicoletti, G.; Scacco, P.; Poech, M.H. Fast power cycling test for insulated gate bipolar transistor modules in traction application. *Int. J. Electron.* **1999**, *86*, 1193–1204. [CrossRef]
31. Cui, H. Accelerated temperature cycle test and Coffin-Manson model for electronic packaging. In Proceedings of the Annual Reliability and Maintainability Symposium, Alexandria, VA, USA, 24–27 January 2005; pp. 556–560. [CrossRef]
32. Bayerer, R.; Herrmann, T.; Licht, T.; Lutz, J.; Feller, M. Model for power cycling lifetime of IGBT Modules—Various factors influencing lifetime. In Proceedings of the 5th International Conference on Integrated Power Electronics Systems, Nuremberg, Germany, 11–13 March 2008; pp. 1–6.
33. Choi, U.; Ma, K.; Blaabjerg, F. Validation of lifetime prediction of IGBT modules based on linear damage accumulation by means of superimposed power cycling tests. *IEEE Trans. Power Electron.* **2018**, *65*, 3520–3529. [CrossRef]
34. Chen, M.; Chen, Y.; Gao, B.; Lai, W.; Xu, S. Lifetime evaluation of IGBT module considering fatigue accumulation of solder layers. *Proc. CSEE* **2018**, *38*, 6053–6061. [CrossRef]
35. NASA Ames Research Center. Available online: https://ti.arc.nasa.gov/tech/dash/groups/pcoe/prognostic-data-repository/#igbt (accessed on 2 May 2020).
36. Chen, B.; Liu, Y.; Zhang, C.; Wang, Z. Time series data for equipment reliability analysis with deep learning. *IEEE Access* **2020**, *8*, 105484–105493. [CrossRef]
37. Hajiabotorabi, Z.; Kazemi, A.; Samavati, F.F.; Ghaini, F.M.M. Improving DWT-RNN model via B-spline wavelet multiresolution to forecast a high-frequency time series. *Expert Syst. Appl.* **2019**, *138*, 112842. [CrossRef]
38. Lai, W.; Chen, M.; Ran, L.; Alatise, O.; Xu, S.; Mawby, P. Low ΔTj stress cycle effects in IGBT power module die-attach lifetime modeling. *IEEE Trans. Power Electron.* **2016**, *31*, 6575–6585. [CrossRef]
39. Chung, H.; Wang, H.; Blaabjerg, F.; Pecht, M. *Reliability of Power Electronic Converter Systems*; The Institution of Engineering and Technology: London, UK, 2015; pp. 174–187, ISBN 978-1-84919-901-8.
40. Carroll, J.; Mcdonald, A.; Mcmillan, D. Reliability comparison of Wind Turbines with DFIG and PMG drive trains. *IEEE Trans. Energy Convers.* **2015**, *30*, 663–670. [CrossRef]
41. Fischer, K.; Pelka, K.; Bartschat, A.; Tegtmeier, B.; Coronado, D.; Broer, C. Reliability of power converters in wind turbines: Exploratory analysis of failure and operating data from a worldwide turbine fleet. *IEEE Trans. Power Electron.* **2019**, *34*, 6332–6344. [CrossRef]
42. Wang, H.; Liserre, M.; Blaabjerg, F.; de Place Rimmen, P.; Jacobsen, J.B.; Kvisgaard, T.; Landkildehus, J. Transitioning to Physics-of-Failure as a Reliability Driver in Power Electronics. *IEEE J. Emerg. Sel. Top. Power Electron.* **2014**, *2*, 97–114. [CrossRef]
43. Huang, H.; Mawby, P.A. A lifetime estimation technique for voltage source converters. *IEEE Trans. Power Electron.* **2013**, *28*, 4113–4119. [CrossRef]

MDPI
St. Alban-Anlage 66
4052 Basel
Switzerland
Tel. +41 61 683 77 34
Fax +41 61 302 89 18
www.mdpi.com

Electronics Editorial Office
E-mail: electronics@mdpi.com
www.mdpi.com/journal/electronics

www.ingramcontent.com/pod-product-compliance
Ingram Content Group UK Ltd.
Pitfield, Milton Keynes, MK11 3LW, UK
UKHW060124010826
14089UKWH00006B/140

9 783036 511771